# The Structural Engineer's Professional Training Manual

## ABOUT THE AUTHOR

**David K. Adams, S.E.,** is a registered civil and structural engineer in California who graduated from the University of California at San Diego with a degree in Structural Engineering and has practiced with Lane Engineers, Inc. since 1990. A typical workday for Mr. Adams includes completion of structural calculations, drawings, and reports on buildings and other structures for commercial, residential, educational, institutional, and industrial uses. In addition to providing plan-checking services for local municipalities to determine building compliance with life safety, accessibility, and structural requirements of current codes, the author also provides expert review services for California's engineer licensing board, has participated in the development of national and California-specific structural engineering licensing examinations, and is currently an instructor for the American Society of Civil Engineers.

# The Structural Engineer's Professional Training Manual

David K. Adams, S.E.

New York  Chicago  San Francisco  Lisbon  London  Madrid  Mexico City
Milan  New Delhi  San Juan  Seoul  Singapore  Sydney  Toronto

The **McGraw·Hill** Companies

Cataloging-in-Publication Data is on file with the Library of Congress.

McGraw-Hill books are available at special quantity discounts to use as premiums and sales promotions, or for use in corporate training programs. For more information, please write to the Director of Special Sales, McGraw-Hill Professional, Two Penn Plaza, New York, NY 10121-2298. Or contact your local bookstore.

Copyright ©2008 by The McGraw-Hill Companies, Inc. All rights reserved. Printed in the United States of America. Except as permitted under the United States Copyright Act of 1976, no part of this publication may be reproduced or distributed in any form or by any means, or stored in a data base or retrieval system, without the prior written permission of the publisher.

1 2 3 4 5 6 7 8 9 0 DOC/DOC 0 1 2 1 0 9 8 7

ISBN 978- 0-07-148107-6
MHID 0-07-148107-9

This book is printed on acid-free paper.

**Sponsoring Editor**
Cary Sullivan

**Acquisitions Coordinator**
Alexis Richard

**Editorial Supervisor**
David E. Fogarty

**Project Manager**
Gita Raman

**Copy Editor**
Anju Panthari

**Proofreader**
Upendra Prasad

**Indexer**
Valerie Perry

**Production Supervisor**
Pamela A. Pelton

**Composition**
International Typesetting and Composition

**Art Director, Cover**
Jeff Weeks

---

Information contained in this work has been obtained by The McGraw-Hill Companies, Inc. ("McGraw-Hill") from sources believed to be reliable. However, neither McGraw-Hill nor its authors guarantee the accuracy or completeness of any information published herein, and neither McGraw-Hill nor its authors shall be responsible for any errors, omissions, or damages arising out of use of this information. This work is published with the understanding that McGraw-Hill and its authors are supplying information but are not attempting to render engineering or other professional services. If such services are required, the assistance of an appropriate professional should be sought.

For Jim and Miriam Luster

# Contents

*Preface xv*
*Acknowledgments xix*

## 1 The Dynamics of Training 1

**1.1** Making the Transition from Academics to Practice 1
 1.1.1 Making Sense of It All 2
 1.1.2 The Engineer's Toolbox 6
**1.2** Training and Being Trained 10
 1.2.1 A Philosophy of Training 11
 1.2.2 Mentoring 14
 1.2.3 How to Teach Others 16
**1.3** What is Structural Engineering? 19
 1.3.1 Analysis and Design 20
 1.3.2 Uncertainty and Error 20
 1.3.3 The Experience 24

## 2 The World of Professional Engineering 25

**2.1** The Road to Licensure 25
**2.2** Affiliations and Societies 27
**2.3** Ethics 28
**2.4** Civil Liability 30
 2.4.1 Standard of Care 30
 2.4.2 Managing Risk 32
 2.4.3 Who's Responsible? 33
**2.5** Design Regulations 33
 2.5.1 The Role of Government 35
 2.5.2 Codes 40
 2.5.3 Design and Construction Standards 42
 2.5.4 Other Regulations and Considerations 44
**2.6** Responsibility to Society 46
 2.6.1 A Healthy Workforce 46
 2.6.2 The Ring Ceremony 47

**2.7** International Issues 48
**2.8** Advocacy 49

# 3 The Business of Structural Engineering 51

**3.1** Places of Employment 51
   3.1.1 *Typical Hierarchy* 52
   3.1.2 *Government* 54
   3.1.3 *Private Consulting Firms* 54
   3.1.4 *Industry* 55
   3.1.5 *Colleges and Universities* 56
**3.2** How Does an Engineering Business Survive? 57
   3.2.1 *Management* 58
   3.2.2 *Employees* 58
   3.2.3 *Projects* 58
   3.2.4 *Procedures* 59
   3.2.5 *Communication* 59
   3.2.6 *Qualifications* 60
**3.3** Clients and Consultants 61
**3.4** Engineering Services 62
   3.4.1 *Contracts* 63
   3.4.2 *Scope of Services* 64
   3.4.3 *Estimating Your Work: Time and Cost* 68
   3.4.4 *Estimating Your Worth: Fair Compensation* 68
**3.5** Crisis Management 69
   3.5.1 *Philosophy of Conflict Resolution* 69
   3.5.2 *Working with Difficult People* 71
   3.5.3 *Legal Means of Resolution* 73
   3.5.4 *Litigation* 74

# 4 Building Projects 75

**4.1** Building Systems 75
   4.1.1 *Structural* 77
   4.1.2 *Plumbing* 79
   4.1.3 *Mechanical* 81
   4.1.4 *Electrical* 81
   4.1.5 *Fire Protection* 82
   4.1.6 *Egress and Circulation* 84
   4.1.7 *Weatherproofing* 85
**4.2** The Building Team 89
   4.2.1 *Owners* 90
   4.2.2 *Architects* 91

  4.2.3 Engineers 93
  4.2.4 Contractors and Subcontractors 94
**4.3** Land Development 94
  4.3.1 Ownership and Legal Interests 95
  4.3.2 Surveying 95
  4.3.3 Civil Engineering Work 96
**4.4** Project Phases 97
  4.4.1 Design Phase 97
  4.4.2 Approval Phase 98
  4.4.3 Bidding Phase 100
  4.4.4 Construction Phase 101
  4.4.5 Occupancy and Continued Use 105

## 5 Bridge Projects 107

**5.1** Types of Bridges 107
  5.1.1 Highway 107
  5.1.2 Railway 110
  5.1.3 Others 111
**5.2** Size and Function of Bridges 112
  5.2.1 Reasons to Span 112
  5.2.2 Scour 114
  5.2.3 Elements of Bridges 114
**5.3** Bridge Systems 121
  5.3.1 Slab Spans 121
  5.3.2 Steel 121
  5.3.3 Prestressed Concrete 122
  5.3.4 Timber 123
  5.3.5 Movable 123
**5.4** Other Issues 125
  5.4.1 Drainage 126
  5.4.2 Joints 127
**5.5** Project Phases 128
  5.5.1 Approval Phase 128
  5.5.2 Design Phase 130
  5.5.3 Construction Phase 132

## 6 Building Your Own Competence 139

**6.1** Technical Growth 139
  6.1.1 Continuing Education Regulations for Licensure 140
  6.1.2 Advanced Educational Degrees 141
  6.1.3 Active Professional Involvement 142

6.1.4　Seminars, Conferences and Personal Research　142
6.1.5　Making Proper Use of Technical Research　143
**6.2**　The Art of Problem Solving　146
6.2.1　Critical Thinking　147
6.2.2　Reaching a Conclusion　149
**6.3**　Improving Your Productivity　149
6.3.1　How Quickly Can (or Should) You Design?　151
6.3.2　Time Management　152
6.3.3　Developing Consistency and Clarity　155
**6.4**　Building Your Confidence　156
6.4.1　Working Within Your Means　157
6.4.2　Computer Usage　158
6.4.3　Defending Your Results　159
**6.5**　Communication Skills　159
6.5.1　Philosophy of Good Communication　160
6.5.2　Verbal　161
6.5.3　Writing　161

# 7　Communicating Your Designs　165

**7.1**　Structural Calculations　166
7.1.1　Analysis and Design　167
7.1.2　Presentation　169
**7.2**　Project Specifications　171
7.2.1　General Organization　173
7.2.2　Bidding Documents and General Project Conditions　173
7.2.3　Technical Section　174
7.2.4　Special Sections or Conditions　175
**7.3**　Project Drawings　175
7.3.1　Goals and Methods　177
7.3.2　Presentation　179
7.3.3　Reviewing the Work of Other Consultants or Clients　180
7.3.4　Responsibility　181
**7.4**　Engineering Reports　182

# 8　Engineering Mechanics　185

**8.1**　Static Loads　185
8.1.1　Dead　186
8.1.2　Live　186
8.1.3　Snow　187
8.1.4　Soil Pressure　188
8.1.5　Others　189

**8.2** Dynamic-Type Loads   190
    *8.2.1 Understanding Structural Dynamics   190*
    *8.2.2 Wind   191*
    *8.2.3 Seismic   194*
    *8.2.4 Blast, Impact, and Extreme Loads   201*

**8.3** Combining Loads and Forces   202
    *8.3.1 Design Methods   202*

**8.4** Introduction to Building Materials   203
    *8.4.1 Common Construction Materials   203*
    *8.4.2 Environmentally Sensitive Materials   204*

**8.5** General Behavior of Structural Elements   204
    *8.5.1 Solid Body Mechanics   205*
    *8.5.2 Serviceability   208*

**8.6** General Behavior of Structural Systems   209
    *8.6.1 Horizontal Systems   209*
    *8.6.2 Vertical Systems   210*
    *8.6.3 Redundancy and Reliability   214*

**8.7** General Behavior of Completed Structures   215
    *8.7.1 Buildings   215*
    *8.7.2 Bridges   216*
    *8.7.3 Progressive Collapse   217*

# 9 Soil Mechanics   219

**9.1** Character of Different Soil Types   220
    *9.1.1 Rock/Granite   221*
    *9.1.2 Gravel   221*
    *9.1.3 Sand   222*
    *9.1.4 Silt and Clay   222*
    *9.1.5 Other Soil Types   223*

**9.2** Preparing a Site for Construction   224
    *9.2.1 Geotechnical Reports   224*
    *9.2.2 Clearing and Excavation   226*
    *9.2.3 Grading   227*
    *9.2.4 Compaction   228*

**9.3** Behavior of Foundation Types   230
    *9.3.1 Spread Footings   231*
    *9.3.2 Continuous (Strip) Footings   232*
    *9.3.3 Combined or Mat-Type Footings   233*
    *9.3.4 Deep Foundations   235*
    *9.3.5 Other Types or Systems   238*

**9.4** Buried or Retaining Structures   238

**9.5**  Factors to Consider in Foundation Design  241
  9.5.1  Consequences of Poor Soils  241
  9.5.2  Settlement  242
  9.5.3  Risk  242
**9.6**  Codes and Standards  243

## 10  Understanding the Behavior of Concrete  245

**10.1**  Common Terms & Definitions  246
**10.2**  Elements of Concrete  247
  10.2.1  Aggregate  247
  10.2.2  Hydraulic Cement  249
  10.2.3  Water  249
  10.2.4  Admixtures  250
**10.3**  Characteristics of a Final Mix  251
**10.4**  Behavior of Concrete Elements  252
  10.4.1  Plain Concrete  252
  10.4.2  Reinforced Concrete  254
  10.4.3  Precast and Prestressed Concrete  261
**10.5**  Behavior of Concrete Systems  266
  10.5.1  Rigid Frames or Cantilevered Columns  266
  10.5.2  Shear Walls  268
  10.5.3  Horizontal Diaphragms  270
  10.5.4  Shell-Type Structures  271
**10.6**  Construction  272
  10.6.1  Risks in Design and During Service  275
**10.7**  Quality Control  276
  10.7.1  Crack Control  277
**10.8**  Codes and Standards  277

## 11  Understanding the Behavior of Masonry Construction  279

**11.1**  Common Terms and Definitions  279
**11.2**  Elements of Masonry Assemblies  281
  11.2.1  Masonry Units  281
  11.2.2  Mortar  288
  11.2.3  Grout  289
  11.2.4  Reinforcement  291
**11.3**  Behavior of Masonry Assemblies  293
  11.3.1  Beams and Columns  294
  11.3.2  Walls  297

|       | 11.3.3 Frames  298 |
|-------|---|
|       | 11.3.4 Prestressed Assemblies  299 |
| **11.4** | Construction  300 |
|       | 11.4.1 Constructability  301 |
|       | 11.4.2 Risk in Design and During Service  303 |
| **11.5** | Quality Control  304 |
| **11.6** | Codes and Standards  305 |

# 12 Understanding the Behavior of Structural Steel  307

| **12.1** | Common Terms and Definitions  307 |
|---|---|
| **12.2** | Where Does Steel Come From?  308 |
|  | 12.2.1 Mining and Refining  310 |
|  | 12.2.2 Mills and Suppliers  312 |
|  | 12.2.3 Regulations  313 |
| **12.3** | Behavior and Characteristics of Steel Shapes  315 |
|  | 12.3.1 Hot-Rolled Shapes  315 |
|  | 12.3.2 Plate Girders  316 |
|  | 12.3.3 Tubular and Pipe Sections  317 |
|  | 12.3.4 Composite Members  318 |
| **12.4** | Behavior and Characteristics of Steel Connections  318 |
|  | 12.4.1 Bolts  319 |
|  | 12.4.2 Welds  320 |
|  | 12.4.3 High Strength Bolted Connections  324 |
| **12.5** | Behavior of Steel-Framed Systems  325 |
|  | 12.5.1 Stability of Beams  325 |
|  | 12.5.2 Stability of Columns and Plates  326 |
|  | 12.5.3 Frames  328 |
|  | 12.5.4 Steel-Panel Shear Walls  330 |
| **12.6** | Fabrication and Erection  331 |
|  | 12.6.1 Risks in Design and During Service  334 |
| **12.7** | Quality Control  335 |
| **12.8** | Codes and Standards  336 |

# 13 Understanding the Behavior of Wood Framing  337

| **13.1** | Common Terms and Definitions  337 |
|---|---|
| **13.2** | Where Does Sawn Lumber Come From?  338 |
|  | 13.2.1 Lumber Supply and Harvest  338 |
|  | 13.2.2 Milling and Finishing  339 |

    *13.2.3 Species 340*
    *13.2.4 Grading Rules and Practices 341*
**13.3** General Characteristics of Wood 342
    *13.3.1 Structure 342*
    *13.3.2 Mechanics 343*
    *13.3.3 Moisture Content, Temperature, and Chemical Treatment 344*
    *13.3.4 Engineered Lumber 345*
**13.4** Behavior of Wood Elements 346
    *13.4.1 Panels or Sheathing 347*
    *13.4.2 Connections 349*
    *13.4.3 Influence of Defects 352*
**13.5** Behavior of Wood-frame Systems 353
    *13.5.1 Horizontal Diaphragms 354*
    *13.5.2 Laminated Decks 357*
    *13.5.3 Frames 357*
    *13.5.4 Trusses 358*
    *13.5.5 Structural Wood Panel Shearwalls 360*
    *13.5.6 Nonwood Panel Shearwalls 362*
    *13.5.7 Wood Systems Combined with Other Materials 363*
**13.6** Construction 363
    *13.6.1 Risk in Design and During Service 364*
**13.7** Quality Control 366
**13.8** Codes and Standards 366

    *References 369*
    *Index 383*

# Preface

One of the most difficult things about writing this book was coming up with a title, believe it or not. It needed to be something descriptive, yet succinct; imaginative, yet practical; and inspiring, yet memorable. The book's title, simple as it may be, gives an indication that a variety of topics will be covered; being a manual for such a broad-based subject as training engineering graduates. The process of training is a human one, where individual personalities can either hamper or enhance effort of both mentor and protégé, and each party must be certain of his or her role in the experience. It is for graduate and experienced professional alike. It is for licensed engineers from all ranges of society to refresh their knowledge of business practice, material behavior, and personal improvement in communicating with others. Most importantly of all, I trust that this book will benefit the profession of structural engineering, as we all work together to advance a solid reputation for service to others from all walks of life, all races and creeds, and all economic backgrounds.

Over the years, I have gathered articles, clippings, books, videos, and other information to help in the process of mentoring new graduates who are eager to begin their careers on the right foot. This information was also helpful to my own professional growth, as teaching others caused me to examine my own life, resulting in a benefit for everyone involved in that process. I will be the first to admit that I have not always consistently remembered or applied principles found in this book, nor have I had 100% success with bringing out the best in others that I have trained over the years. However, the information is timely and relevant to all skill levels within the profession.

I'm quite certain that each chapter will have its own share of critics who ask, "Why didn't you cover this?" or "You didn't spend enough time on that," and I am in agreement with many objections that could be raised. There is so much to talk about, but a choice had to be made as to what seems to be most important and what might be easily found in another resource. Within these pages, the reader will not find detailed instructions for designing a wood-frame shear wall, nor a reinforced concrete drilled pier, nor even for monitoring the financial health of a sole proprietorship, as all of these duties are exhaustively presented in other references. Rather, this book has three main objectives: (1) introduce the reader to subject matter that is important to know in order to discover the best solutions to real engineering problems; (2) provide a logical, comprehensive collection of recommendations, facts, and figures to help a mentor guide protégés along an accelerated path of career growth in confidence; and (3) bring together information that is usually scattered around in different references, yet important to consider in a unified manner to develop an intimacy with the practice of structural engineering.

Chapter 1 begins with the essence of training, including a look inside an engineer's toolbox that must be opened for solving real-world problems. These tools are defined in such a way that a mentor learns how to recognize and pass them along in the context of building upon a basic engineering education. The practice of structural engineering is also described as not just a career, but an experience that can only be lived out through reasonable knowledge of what makes everything around us work and why care is necessary in all dealings.

Chapters 2 and 3 introduce the big world and business of professional engineering in a manner that encourages active participation in the betterment of society. The subject of ethical practice is not learned to any great degree in college, but it is a key point of behavior in this profession, as engineers learn how to responsibly apply design regulations to the affairs of business. Different players in the business world affect the health of this profession and wise associations, as described in this book, will help to relieve stressful situations. Chapter 3 closes with an important presentation on managing business-related crises, though the material certainly has application in one's own personal life as well.

Chapters 4 and 5 paint a picture of the major trades involved in completing the design and construction of a building and a bridge, respectively. Different systems are described that may at first seem to have little relevance to the work of a structural engineer, but the more intimately we are connected with our structures, the better able we will be to design them or to solve problems when something goes wrong. Each chapter reviews phases that exist in the creation of these types of structures, and the reader will discover that some terms and conditions certainly are relevant to both buildings and bridges.

Chapters 6 and 7 present material useful for personal advancement in confidence and competence as related to engineering practice. Engineers must have a strong ability to solve problems strategically, to be as productive as humanly possible, and to communicate well enough to leave little room for misinterpretation. This is also where Chapter 7 comes in: we communicate to others through project documents such as structural calculations, specifications, drawings, and reports, all of which should be clear and presentable. An ability to translate three-dimensional thought into two dimensions is challenging, but nonetheless critical for success, and some of the principles described in these chapters will help for training on this matter.

Chapters 8 through 13 describe the background, composition, and behavior characteristics of common construction materials when subjected to load, which make up the technical backbone of everything that a structural engineer does, and is therefore relevant to the whole process of training. The most obvious starting point is a remembrance of engineering principles learned at college, plus an introduction to further patterns of material behavior that may not have been covered to a great extent elsewhere. Each of the material chapters begins with a list of some common definitions used within that particular industry (by no means exhaustive lists), continues with a discussion on the origin of those materials, and is laced with different references to model codes or design standards for description.

I decided to reference the most current editions of model structural codes and material standards so the reader can look up the stated sections for further review, as well as to establish a mindset toward specifics rather than randomness. I felt that generic code titles with specific sections referenced didn't make a lot of sense.

In closing, this book will be beneficial as a tool in a variety of different situations: (1) to show a mentor the material necessary to pass along and how to teach it; (2) to further the education of a new graduate, or a student nearing graduation, in personal study on relevant subjects they will have to deal with during the course of their career; (3) as a course book, where a mentor can assign a number of pages or sections to a trainee, then get together at a later time to discuss or work on practical applications of the material; (4) as further reading during undergraduate or graduate work in classes relevant to each topic covered; or (5) as a reference for practical knowledge of subjects certain to broaden an engineer's problem solving capabilities. Any way you choose to use it, I trust that it will make a difference in your understanding of this dynamic career.

*Happy mentoring,*
*Dave K. Adams, S.E.*
Tulare, California

# Acknowledgments

I would like to take this opportunity to express my sincere gratitude to Nestor Agbayani, Stan Caldwell, Kevin Dong, Jonathan Mallard, and Barry Welliver for providing technical review of a number of these chapters and suggesting content bits. I would also like to thank Derek Damko for his work on many of the figures, my brother Don Adams for some organizational assistance, and my kids, Megan and Josh, for their work on printing photos and writing captions. There are many other wonderful individuals and companies that contributed artwork or suggestions for this book and credit is given where it is due.

*Dave K. Adams, S.E.*
Tulare, California

# The Structural Engineer's Professional Training Manual

# 1 The Dynamics of Training

Structural engineers usually begin training long before they've even dreamed of joining the profession. As kids, they were the ones who studied roller coasters, such as that shown in Fig. 1-1, a bit more closely than their pals. They tended to have a certain creativity in everything they did, whether it was flying to Mars in a tall bush or solving mysteries that always seemed to plague the neighborhood. They were curious about how things worked: The world held certain constants (ice cream always tended to splatter onto the sidewalk if you didn't eat it fast enough), yet there were also things that seemed unpredictable (a baseball could either dribble across a field or fly over a fence depending on how it was hit).

## 1.1 MAKING THE TRANSITION FROM ACADEMICS TO PRACTICE

To put an engineering education in the right perspective, a student must ask, "Why do I want to become a professional engineer?" The answer may begin on a personal level (money, power, prestige) or from a more service-oriented mindset (create a better world for future generations), but a new graduate will quickly learn that both perspectives work together to drive a professional engineer forward. Society recognizes the contribution an engineer makes to improve quality of life and rewards service with job security and satisfaction.

If a child's imagination creates the skeleton of a career in engineering, it is university study that adds muscle to the bones. The dictates of business and legal restrictions form the protective surface of skin and an engineer's growth within

**Figure 1-1** Possibly the most enjoyable application of structural engineering! *(Photo by Gustavo Vanderput)*

the industry fires the neurons to animate this career, causing it to mature and develop a personality of its own. The process of designing a structure cannot be truly understood within textbooks or example problems, but should be experienced in a workplace setting where skeleton, muscle, skin, and neurons work together to discover the best solution to a problem. Skills that define an engineer's personality and contribution within the profession are acquired through a change in thought (taking exams versus taking responsibility) and attitude (benefiting self versus benefiting society) from what was learned during the days of college life.

### 1.1.1 Making Sense of It All

The most obvious way for a graduate to ease the mental transition between theory and application is to be trained to think like an engineer. The mind processes data in different ways for different purposes and being aware of how these things work will give a person an edge in career development. It is critical that sound thinking skills be developed and enhanced in order to move any career forward with the ever-changing times.

1. *Knowledge retention and recall*: Over the course of their lives, engineers learn a great deal of information covering a wide variety of subjects and it can be difficult to recall an important piece of data at the right time if there is no organization of thought. An engineer must have the ability to retain knowledge through an association with some sort of practical purpose. Facts and figures

are useless to an engineer if they do not benefit his or her work in some way, therefore it is far easier for an engineer to recall an important bit of knowledge when it has an application.

For example, one of the earliest occurrences of a concrete beam reinforced with some type of tensile element (in this case, a bronze rod), in recognition of the familiar tension-compression couple which resists an applied bending moment, was discovered over the door of a Roman tomb, dating to about 100 BC. Is this some useless bit of outdated information? Absolutely not! Recalling this bit of historical/archaeological knowledge gives an engineer some sense of comfort in a world that changes so rapidly, that even though technology continues to advance by leaps and bounds, solid principles of structural engineering have remained constant, predictable, and memorable.

2. *Comprehension*: Before an application can be linked to factual information, useable data must be properly understood. Sometimes a dictionary or thesaurus might need to be consulted if there is a question as to meaning. Mathematical formulas and scientific principles often contain coefficients, relationships, or sequences that need to be researched and simplified in order to understand the purpose of the information given.

For example, the Ideal Gas Law is remembered as follows:

$$PV = nRT \tag{1-1}$$

To understand the equation, however, four important characteristics should be realized: The equation is also a relationship; each term has a specific value; one term is constant; and other terms are variable depending on what is known and what is desired to be known. Such intimate understanding causes the equation to be useful, which is why *understanding* is so much more crucial than simply storing and plugging in data.

3. *Applying knowledge to a variety of different situations*: One of the exciting things about being an engineer is that there are many different types of problems to solve, whether it involves safely carrying a farmer from one side of a creek to the other or designing a skyscraper to stand steady under 140 mph (225 km/h) wind speeds. Some real-world problems follow the pattern of examples found in a college textbook, but the great majority of them do not, which means an engineer must be adept at translating known information into a different situation than might have originally fit a completely unrelated application.

4. *Defining multiple solutions to a given problem*: An advanced level of understanding and applying knowledge is the process of considering multiple applications and choosing one solution over another. This involves careful observation of technical data in order to weigh the strengths and weaknesses of different procedures and decisions, and also to consider the effects of one choice over another. For example, a very old proverb claims that there are more than "one hundred ways to skin a cat" and there may certainly be more than one hundred ways to connect a beam to a column. Each method defines a solution and each solution

must be carefully chosen in response to probing questions: What are the implications of a bolted connection over one that is welded? What about the influence of surrounding members? Is there a cost or time benefit to one method over another? An engineer must analyze solutions and decide which one best meets the needs of the client, but one that also does not violate the engineer's duty to the public.

5. *Elaborating on an idea or solution*: The ability to take a simple problem and apply a wide base of knowledge, even in subject areas that are not dominant to the eventual solution, is one which separates engineers from technicians. Elaboration is the application of an ability to think flexibly, which is necessary in many instances when a solution to a particular problem is not immediately apparent. Engineers must not only be able to elaborate on initial, somewhat automatic, thoughts, but to also consider alternative solutions and elaborate on their respective merits. This is difficult to do, as all human beings have their own preconceived notions regarding life and how the world operates, but fixation on single-minded solutions doesn't encourage growth in this career.

6. *Originality of thought*: Solutions do not come neatly wrapped in a box. Sometimes a problem or issue should be seen in a new and refreshing light, opening the doors to better, more creative solutions. Structural engineering is truly a profession of science married to art, where creative expression of antitypical, original designs instills confidence in the practitioner as well as those who must build the system. Textbook examples can only bring us so far and many theories only leave us at the doorstep, but it takes originality to work outside the box of what is known. Solutions must not violate the rules of science, but there are many ways to meet those rules and use them to our advantage.

It is easy for an engineering student to be completely in awe of surrounding structures. From the tall but gracefully simple Empire State Building in New York, New York to the somewhat complex Royal Canadian Mint in Winnipeg, Manitoba (see Figs. 1-2 and 1-3), the constructed environment has much to offer curious eyes. Theoretical principles and merciless all-night study sessions leave an impression of complexity in the mind of graduates, but that need not continue throughout a budding career: Young engineers must remember the basics of sound engineering doctrine in a vital effort to simplify the world of complexity. Introductory science classes remind the student of the foundational gospel of engineering, namely that gravity will always move objects toward the earth. A complementary teaching quickly and logically follows, which is, for every action there is an equal and opposite reaction. The weight of a structure drives it into the ground, but the size of a properly designed foundation system reacts in an equal and opposite way to maintain stability. We are able to understand this behavior through a proper use of math and the physical sciences, but sometimes it takes a bit of philosophy to bring young graduate engineers into the light of simplicity, which is a vital move to understanding how this profession, and the structures that are produced, properly function.

The Dynamics of Training | 5

| **Figure 1-2**  Empire State Building.

| **Figure 1-3**  Royal Mint Building in Winnipeg, Manitoba, Canada.

Simplicity can be found in an engineer's first realization that he is, in fact, competent to do something useful beyond graduation. First-time employment has its charm, especially after months of fretting about it, but receiving a job is not nearly as satisfying as completing the very first task on a real project and doing it well. One of the first projects I ever worked on was to complete the structural design of a single story, wood-framed office building. It was similar to another office building produced by the same architect, in the same plaza of structures, for which my boss had recently completed the structural calculations. My boss was meticulous in his work, therefore I couldn't have had a better set of calculations to use as a guide. All of the site specific loading conditions were predetermined, the basic lateral force resisting system had a preestablished pattern (for the most part), and there were only a few different beam loading conditions to design members for. I'm quite certain it took me much longer to complete the design than it should have, but after I completed making corrections to the calculations after my boss' first review, I was thoroughly impressed at the skill of our drafting department in creating working drawings from my own calculations and even more satisfied to see the finished product standing tall after construction. That was confidence! That was real achievement for a young graduate who still wasn't completely certain what had just taken place...and all was good in the world.

Sometimes during the course of an engineering career, it is important to bring to mind those days of first achievement, a time when things were fresh but uncertain. There was a strong desire to understand every nail, every stud, every hanger— a true love affair as a theoretical foundation was lovingly courted by practical application, resulting in the wondrous birth of a living, breathing structure. It is the basis of simplicity that helps an engineer understand what a structure intends to do with the loads that it must carry. The ability to see the simple within the highly complex can save an engineer from sleepless nights because that ability brings about understanding of the behavior of structures. The majesty of the supporting towers of a suspension bridge need not seem so unapproachable when an engineer chooses to see them as glorified moment-resisting frames with special loads applied by the suspended cables and additional environmental effects. The frightening image of a heavy cantilevered floor system can be identified in terms of simpler pieces, discovered during a semester of college instruction, that function together to form a unique work of art.

## 1.1.2  The Engineer's Toolbox

One of the most important aspects of having a successful career in structural engineering is to recognize the tools that are available for completing projects, confirming the adequacy of a particular design, or defending the conclusions of an investigative report. A tool is most often pictured as a type of device used to accomplish a task, but a tool can also be thought of as a particular field of study and knowledge that engineers must make use of in order to accomplish their objectives. For example, during a career day presentation to high school students, a professional engineer often explains how math and science (tools) work to create wonderful designs that benefit society. The tools that an engineer uses can be divided into two categories: Those that are somewhat psychological in nature and those that are classified as academic.

**Psychological Tools**   The first two psychological tools have already been introduced in this chapter, *curiosity* and *creativity*. Blood brothers to the very end, these instruments have unlimited potential and have proven over thousands of years to solve even the most plaguing of problems that society has faced. Call them keys to unlocking the secrets of the practical application of engineering theory. Think of them as a bridge to islands yet uncharted, only dreamed about, both as a child and as a graduating student. They cannot be separated, yet each serves an individual function: Curiosity recognizes the truth of everything that is seen, but does not understand it completely; creativity explores this truth in a tangible way in order to discover how it can be repeated and passed along for all ages to benefit. Unfortunately, children learn that the poor old cat died because of curiosity and somehow that knowledge pushes this important tool into a locked trunk, rarely retrieved for fear of what happened to the cat.

When curiosity is rarely exercised, creativity also suffers. They depend on one another for survival. Structural engineers must remember that we are a curious species and this curiosity unlocks vast storehouses of creativity, eventually leading to incredible discoveries, including the finding of our own confidence. Creative effort also depends a great deal on the freedom of the dreamer—freedom to explore different avenues of thought and to verify new discoveries with all available resources, though it is to be tempered with proper understanding of geometrical and analytical limitations of the profession and structures in general. Certain things can be done by young engineers to enhance creative thought, including the continual creation of hand-drawn structural details using appropriate degrees of hardness of pencil leads for different line types, purposely determining several different solutions to an engineering problem, and perusal of different technical magazines.

The third tool is *confidence*, often thought of as a personality trait. An engineer's confidence begins with a simple identification of something obvious that will give value, purpose, and importance to any project regardless of its size or complexity. When there is an intrinsic value to the work that someone does, a value that will affect thousands of lives for many years, a successful engineer immediately recognizes the need for quality which can only be attained through a confident exercise of duties. For example, the goal of a successful bridge design is to safely support the load of all persons and things that use it to get across a particular obstacle, while remaining stable under the effects of the environment. This goal is easily recognized, understood, and comprehended. Confidence, therefore, encourages excellence and develops an ability to think "outside the box."

Confidence in the societal benefit of a project encourages designers to work diligently with building or bridge owners, the public, and governmental agencies to assure that needs are met, financial as well as operational, and that all relevant codes are complied with. A desire for confidence will lead a professional engineer to carefully exercise creativity in selecting one of the best structural systems and that same confidence will be used in the design of every individual component and connection until construction documents are ready to be delivered to the owner. A well-known axiom of engineering ethics recognizes the use of confidence as a tool in all work that a professional chooses to undertake, stating that a licensed engineer must only work within his or her area of expertise and practical confidence. In the earlier example of the bridge project, the engineer in responsible charge may be confident in all areas except the design of a prestressed concrete element or system incorporated into the overall structure. He may have learned about it at the university, but has not made enough practical use of that knowledge to undertake a real project where the lives of thousands will be affected. If he does not feel that he can adequately research and study the subject to boost the needed confidence, then he will hire a qualified consultant to complete that portion of the design.

The final psychological tool to discuss is *courage*, which brings all others to the front of a situation. It takes courage to stand on convictions. Facts, gut feelings, and a desire to draw out the truth of a situation enhances any project through courage. It is crucial to recognize that the responsible use of psychological tools is absolutely dependant on an effective use of academic tools. Scientific fact, that which has been proven over and over again under most conceivable conditions, provides an unambiguous foundation upon which to make use of these sometimes subjective personal tools. It is this foundation that allows practicing engineers to exercise courage in their convictions and proposed solutions to real-world problems.

**Academic Tools** *Mathematics* always seems to be one of the most frightening subjects to students. Perhaps because it is an exact science: there is a right answer and there is a wrong answer, though there are different ways of finding a solution to an exercise or problem. An engineer depends on the repeatability of that solution, whether in formulas that define the behavior of a simple beam or from a more complicated finite element analysis that is performed to understand the distribution of load applied to a shell-type structure. If a solution can be repeated through a variety of different methods, it becomes more credible to a suspicious mind and promotes confidence.

Engineers also depend on the reliability of solutions discovered through the use of all mathematical levels learned during the course of education, from simple counting to exciting realms of complexity. A student understands this reliability through proper application of the more basic levels by memorization and constant drilling, under the premise that the less a student has to commit thought to procedures that are basic to the overall solution of a problem, the more that student will be able to understand the heart and soul of the actual solution. There are important similarities between a young mathematician and a young engineer in this regard. A mathematics student should have a solid command of multiplication facts without having to expend effort on a calculator because it distracts attention away from the real solution to a problem—that which is not found in the math itself, but is discovered through application of the math. Professor Ethan Akin of the City College in New York speaks of different practices; such as cooking, carpentry, playing a musical instrument and horseback riding; that demand a certain foundation of physical skills which can be performed automatically (Akin 2001, p. 1), arguing that the same idea applies to the solution of mathematical problems.

This same principle applies to the practice of engineering, that the behavior of a simply supported beam can be understood immediately only after many of them have been designed without the aid of a computer program. It becomes common knowledge because of the reliability of work done so many times before and the engineer may direct effort away from these simple pieces into the more complex regions or assemblies of a structure. The entire spectrum of mathematics may be used strategically, from simple counting of load combinations to advanced forms of trigonometry for defining load patterns to partial differential equations used for defining the behavior and distributive effects of internal forces.

*Physics* is another tool that begins an interesting path into practical application of mathematical principles and provides an initial glimpse into the world of structural engineering, where the behavior of bodies and fluids comes alive through mathematical relationships. The motion of a falling body or a launched projectile introduces the concept of three-dimensional (3-D) behavior, not only in terms of length but also position, speed, and acceleration. Newton's Third Law, as has already been briefly introduced, is often remembered as the foundation of structural behavior: "To every action there is always an equal and opposite reaction." A column exerts force onto a pad footing, which in turn is resisted by the soil below pushing against that footing (an equal and opposing reaction), causing it to bend upwards, sometimes requiring steel reinforcing bars to resist tension imposed on the concrete section by this reactive force.

Through the use of *chemistry*, an engineer quickly learns of the limitations and strengths of materials used in design. It is here that one learns about corrosion and the effects of impurities within elements. All useful things are held together by chemical bonds and reactions are understood by the heat produced. Principles of conservation of mass and energy help the engineer understand how thermodynamics, chemistry, and physics work together in the behavior of materials that go into construction, aiding in a better knowledge of the fundamentals of design. Chemistry is what ultimately dictates the behavior of building materials, as the interaction between atomic particles ultimately dictates mechanical properties such as strength and ductility, therefore an engineer will not abandon principles and strategies that were learned before truly appreciating their relevance to the world of engineering.

Sound *logic* is necessary for proving the adequacy of a design, whether that proof comes in the form of calculations or a well-researched report, and is based on a simple progression of "premise-inference-conclusion." To logically prove a given statement or condition, *premises* are offered in the form of known, factual data that are easy for most people to accept as being true. For a structural engineer, these facts would include clear directives taken from a model building code, laboratory test results, and independently-verified observations. *Inferences*, on the other hand, are statements of supposed fact that can be reasonably extrapolated from the factual premises originally presented and are necessary for filling gaps of understanding, perhaps involving a thought of the intention behind a complicated code provision. These may also be gleaned from the conclusions of other researchers or structural calculations that may not be directly related to the issue being analyzed or proven. *Conclusions* are to logically follow the string of premises and inferences, provided the engineer can assemble all necessary information coherently. Much engineering knowledge is founded on this logical progression, which becomes more critical in addressing problems that do not have an immediate and obvious solution.

*Linguistics* involves the written, verbal, or artistic expression of an idea that must be explained in a way that people from a variety of backgrounds will understand. A design or engineering conclusion is useless if it cannot be accurately transmitted to the people who must make use of it, such as laborers or government officials. It is not simply a matter of a particular language, but the appropriate use of that language. An engineer who is communicating a principle to peers will use more technical jargon, whereas a report given to a homeowner after inspecting a pesky crack in the concrete uses far simpler terminology, or may include some type of glossary (though that is not the best approach). Professionals learn to respect and honor the power of communication, as it can quickly raise the spirits or crush the opposition. Control of tongue and pen leads to the responsible use of linguistics, as patience, tolerance, and a genuine desire to bring out the truth of any given situation can only be pursued through discipline.

## 1.2　TRAINING AND BEING TRAINED

Training can be a frustrating experience, both for the trainer and the trainee. New graduates enter the workplace with uncertain expectations: they know the direction in which to proceed, but aren't sure how to do it effectively. They know that what is learned during this uncertain time will have an effect on their entire career. In fact, one of the goals of training also happens to be a concern of many employers, that young engineers must be trained toward self-management to the extent that if they were to begin their own engineering company, they would be successful at it. Employers certainly don't want to lose an engineer that they've spent time and money to train, but that employee will only be valuable to the company if properly trained to become independent: Able to secure and keep clients, to handle problems, and to manage staff.

In order to be successful, a graduate engineer must be trained in the following areas:

1. *Ethics and liability*: Because an engineer is expected to create a product that safeguards the life and welfare of the public, this profession can have some painful legal penalties when negligence is proven. Engineers need to understand their responsibility to the public, employers, clients, and their families, keeping ethical practice firmly embedded within the process of earning a living. As technical ability projects the course of a professional career, adherence to ethical practice will give it an air of satisfaction and of peace.
2. *Business knowledge*: All engineers need clients, whether the government funding research or a local homeowner with a dream to fulfill. Managing clients and business aspects, including scheduling, deadlines, and resource management, is not only a business owner's concern, but that of every employee who is instrumental in delivering a product. Those who begin a professional career will eventually come to the point where they can seriously consider becoming their own boss through climbing the company ladder or building one of their own, therefore a solid grounding in business skills opens the door to great possibilities.
3. *Communicating and delivering a product*: An engineer's work will be reviewed by an agency having the right to give or refuse a building permit. A building, for example, requires a set of structural calculations to prove that a particular design works and complies with adopted codes, notes, or specifications to indicate a desired product to use in the construction, and a set of drawings to show the complete assembly of the building from foundation and roof framing to means of weather-protection. These documents must be organized, straightforward, and easy to follow through.
4. *Technical knowledge*: There will always be room to learn new things and to expand on existing knowledge related to the technical aspects of structural engineering including new technologies and discoveries, building- or bridge-code changes, new design standards, or design methods. Another important aspect to understand is material and assembly behavior, which can be understood through mechanical principles and research.

## 1.2.1  A Philosophy of Training

A program of training must also be one of mentoring, where a life investment is made for the purpose of advancing another person's growth. It is not a single instance, but an ongoing experience that can be both difficult and rewarding, consistently followed through for each person who is being trained. The most effective philosophy of training deals in the short and long term with two specific subjects: Attitude and method. This philosophy also recognizes the process itself, which is usually handled at the time questions or issues come up during normal operation according to a new graduate's ability and inclination.

**Attitude** Mentoring involves a willingness to consider the needs of others as more important than your own, which means that a project being worked on should be temporarily set aside when a protégé requires assistance. It is difficult to stop making progress on a project that has a looming deadline, so patience and strategic coordination of duties are certainly called for. Sometimes it is helpful for mentors to remember the days when concepts and solutions to problems did not come as easily or quickly, especially when a trainee repeatedly seeks help on the same issues. Engineering is not a "microwave" profession, where numbers and relationships are blindly plugged into some machine that spits out a solution. The human mind doesn't operate that way. Engineering concepts are partially abstract, partially concrete, and it takes time to manage these differences—time and careful direction. As time is taken with a new engineer, mentors ask questions to stimulate thought rather than offer quick, stale solutions in an effort to speed up encounters. A protégé needs to experience a genuinely caring attitude toward advancement in the profession.

Good mentors are "confidently humble," which means they are willing to set aside a preformed notion or idea in order to honor the value of another's input, yet they will only do so if there is technical merit to an alternative view. Trainees will gain confidence in their own solutions to a problem only as their mentor expresses confidence that they will work. Different solutions to an engineering problem that may come to the forefront need to be analyzed logically, with each step being carefully explained (or discovered). Those who offer a solution, including mentors themselves, should be able to explain the steps taken in arriving at that conclusion in such a way as to convince someone else that it is perhaps one of the best solutions. A mentor must be willing to accept previously unknown, but sound conclusions to a problem and to offer appropriate praise of encouragement.

**Method** The right attitude leads to a more productive method of training, involving not merely a series of conferences and in-house instruction strung together, but a global approach that focuses on several key areas:

1. *Teaching*: Because a mentor has so much to teach, it's often difficult to know where to start. One of the first things that a trainee needs to learn is how to apply engineering theory to real world problems. Using calculations and drawings of existing projects as practical examples work best for this instruction, where a mentor takes time to explain similarities in what was done before as

applicable to a current project (such as finding code requirements, setting up a problem from a variety of angles, laying out framing, and the like.). It is not necessary to "reinvent the wheel" and for the practice of structural engineering, there is likely an existing example for anything that can be dreamed up: The key is to recognize common elements of any design that can be brought forward, even from the most complicated of geometries. This is not a skill easily achieved in college simply because it takes time to develop, but being shown how engineering applications remain consistent over time goes a long way toward driving away fear and inhibition.

Teaching is done both by direct instruction and by living example. To effectively learn how to interact with clients, a trainee needs to participate in face-to-face meetings and observe how an engineer directs the conversation. Business skills are sometimes learned through failure, where budgets are blown or employee hours are ridiculously extended for the sake of meeting a deadline that came from out of nowhere because it was not properly scheduled. Technical skills are best learned by first filling in the gaps of knowledge leftover from a university engineering curriculum, then by enhancing areas of competence that an individual already possesses, and finally by practicing how to move theories into practical engineering situations. These strategies will overlap at different times, and some individuals may be better suited to a different arrangement, but these stages will form the basis of technical training in all cases.

2. *Rebuking*: A strong word, but sometimes required for a person's positive growth. To rebuke simply means to reprimand sharply and it is sometimes called for in getting a point across. In an effort to please clients and make money for the company, it would be disastrous for an engineer to lose sight of the importance of ethics and of the obligation held to society. Blatant dishonor of authority or disinterest in company procedures falls under the purview of rebuke, as disharmony and lack of respect for persons and operational philosophy can tear a company apart. Repetitive mistakes either show a deficient technical ability or a lack of attention, both of which need to be addressed with a tone that brings the seriousness of the issue to the forefront. A lighter form of rebuke also means to simply point out that something is wrong, delivered with an explanation of why.

3. *Correcting*: Young and well-seasoned engineers alike make mistakes. Part of the process of correcting a mistaken conclusion or method is to encourage others to take calculated risks, to pick themselves up after failing, to take skills to the next level, and to find the courage to keep moving forward. History shows innumerable examples of how engineers have learned from mistakes, both their own and those of others, and that progression is based on how setbacks are dealt with. Correction implies that the reason something went wrong will be adequately explained by a mentor and the solution will be properly linked to that underlying cause. A mistake that isn't turned into a teaching opportunity through correction festers in the memory and exists as a roadblock toward considering unusual or innovative solutions.

4. *Evaluating*: Evaluations serve as opportunities for employer and employee to take a closer look at what has taken place during the time spent together and to make an informed decision on whether the relationship should be continued. They do not have to be restricted to semiannual or annual occurrences, but can be offered anytime a new project is completed. At reasonable intervals, mistakes can be corrected in a more timely manner and any philosophical differences can be addressed quickly rather than become embedded in a shroud of discontent. Progress evaluations can be done verbally or in writing, as long as an employee understands what improvements are being called for and how to incorporate change.

## 1.2.2 Mentoring

Mentoring is encouraged in many areas of life as an effective way to preserve a message or teaching for future generations. Spiritual leaders of all faiths are instructed to teach followers the truths found within their scriptures. Government leaders bring along interns to closely observe how they handle affairs in order that their own methods and philosophies will be remembered and repeated. Likewise, the experience and wisdom of engineers must be passed along to sustain the health of their own companies, the engineering profession, and society as a whole.

But what exactly is a mentor? The first thing to understand is that a mentor does more than simply pass the torch. A teacher instructs others in how to use a particular body of knowledge, typically for the purpose of passing some type of exam, whether it is a class test or a placement exam. However, teachers become mentors when they act not on a purely academic level, but out of genuine concern for their students' futures. They take extra time to carefully explain a difficult concept until they are certain that a pupil is truly ready to move on to the next level. They show a loving concern for the success of their students and always find ways to encourage and build them up. They are also not afraid to correct or discipline, always with patience and a clear description of what action is being addressed, because such correction will have a positive effect on that student's progress.

Mentors are concerned about growth in all areas of a person's life—physical, intellectual, spiritual, or emotional—and though there are obvious limitations to intimacy in a business relationship, all of these areas are certainly relevant to the practice of engineering. The *physical* well-being of an employee helps to maintain alertness, leading to fewer mistakes in judgment. *Intellectual* health moves a person forward in any profession as new information adds to a solid base of knowledge and experience. An engineer-of-record maintains a close, *spiritual* connection to a building's design and behavior because of a need to understand what to do if something goes wrong during construction. The engineer will care for that building like a close friend—an *emotional* attachment that can only be learned through experience.

Mentors lead by example, and those who do so successfully display the following qualities, concerns, and abilities:

1. Recognizes a person's potential for growth
2. Turns a mistaken notion into a teaching opportunity
3. Flexible with employees, supervisors, clients, and circumstances
4. Clearly envisions goals and defines how to achieve them
5. Maintains a proper perspective when things go wrong
6. Raises curiosity in a particular project or subject through active involvement or encouragement
7. Accepts criticism with grace, humility, and eagerness to correct errors
8. Sorts through available resources to select that which will offer the greatest benefit to professional growth
9. Strategically manages and directs the course of projects
10. Coordinates the efforts of a design team, including correction and admonition

Listening is one of the most important skills that a mentor should strive to enhance, especially in engineering. There are often many valid solutions to an engineering problem and different answers are reached because of different ways of thinking. Different types and sizes of beams, for example, can be designed to support a particular load and there may be reasonable choices other than the default "most-economical solution." If a trainee discovers a different solution than a mentor might have arrived at, there is an inherent responsibility to listen carefully and evaluate reasons and methods used in arriving at that particular answer. Good listening skills are usually coupled with good question-forming skills. Questions are important tools for getting another person to think about something. For example, if a trainee has detailed a beam and post, but does not indicate how the two framing members are to be connected, a mentor can respond to the inadequate detail in different ways: "This is wrong. Go back and do it again," "The beam and column aren't connected together. Go back and do it again," or "How does the load from the beam get transferred to the column?" By offering a question, the student can be steered in the right direction without giving away the answer, thereby empowering that person to discover independence and confidence, which will be remembered with far greater intensity than if the solution was simply handed over.

Many times, issues that require further learning can be discovered in the way a graduate approaches an engineering problem if a mentor instills a sense of quality and consistency in work habits. Because of human variety, we do not always have the same idea as to level of quality of work produced, but product consistency is demanded by clients who use engineering services and cannot be easily dismissed. If everyone is on a different page regarding the goods a company produces, it will not be successful, therefore definitions are of great importance.

A graduate should also remember that, although information gathered in the course of university study serves as an important foundation for a budding and prosperous career, much more remains to be learned. Coursework required to obtain a bachelor's or a master's degree is comprehensive, but some have argued that additional classes should be added to the curriculum in order to produce an engineering graduate who is better prepared to apply what has been learned in a practical way. On the other hand, there is also some benefit to plugging a graduate into the profession in as timely a manner as possible. Practical information that a young graduate should know—personal skills, understanding the economics of a growing company, learning how to synthesize the benefits of different disciplines of study to achieve successful designs—can truly be learned only through experience. Personal skills not only involve casual conversation, which leads to a level of comfort between two parties, but also skillful negotiation and tactical salesmanship. When things go wrong, strong alliances with clients can help to save a company from legal ruin and personal skills, not technical prowess, will often be the best defense.

### 1.2.3 How to Teach Others

Teaching is a skill that comes naturally to a fortunate few, but is difficult to grasp effectively for many others. Most people have to work at not only teaching a particular subject, but making the presentation interesting (even exciting) and critically relevant. The challenge facing an engineer in terms of teaching others is to combine important technical knowledge with confident human interaction and patient persistence: Engineers simply want to get things done in the most efficient manner possible, but teaching doesn't always fit a comfortable, predictable mold. A teacher must not be driven by frustration if things do not go according to plan. Students are human and they will behave as such, with different speeds of learning, a variety of temperaments, and a wide range of skill levels.

An engineer becomes a teacher when it is understood that knowledge and experience are important to pass along to others for the sake of the profession, the company, and society. Teaching prepares future generations for great things and is a truly enjoyable experience when it is done with the proper perspective and in the right frame of mind. No one will ever grow and mature without someone to teach them how to do so. This skill for teaching becomes easier with increasing personal familiarity of the subject being taught. If an engineer decides to teach others how to design bridges, then that person better know the ins and outs of such a structure, in addition to the latest technologies, theories, and methods. Continual growth in knowledge of a particular subject helps an instructor to stay ahead of students and be regarded as an expert, even though there is still plenty to learn.

Students of all ages and backgrounds approach new material with a variety of different learning styles. The American Association of School Administrators identifies four types of learners: Imaginative, analytic, common sense, and dynamic (AASA 1991, pp. 10–16). *Imaginative learners* grow more through the process of sensory perception and visual observation of events, such as may occur during a jobsite visit. *Analysts* absorb information better through visual identification and stimulation of deep thought. Students of this type thrive on considering different solutions to an engineering problem that are offered during a brain storming session. Those who learn better through a *common sense* approach can integrate theory into practice with pragmatism, coupling careful thought with actually performing a task, which translates into an excellent ability to read and interpret construction documents. *Dynamic learners* learn by trial and error as they integrate experience with application (sensing and doing) and seem to arrive at the right conclusion even in the absence of logic. These engineers are particularly adept at thinking "outside the box." The challenge that lies before an engineering mentor is to teach all of these principles and skills, though they will not come natural to some.

To effectively teach new things, a mentor must be familiar with the extent to which a trainee has put a university education to practical use in order to define new stages of advancement in that new career. If successive stages can be defined in a precise manner, with identifiable goals clearly laid out, a teacher's job will be much easier. For example, suppose an engineering graduate brings knowledge of basic design methods and characteristics of different building materials forward into employment. During the course of study, however, a concept of load path may only have been understood as related to specific arrangements, such as the connection of elements in the Brooklyn Bridge (see Fig. 1-4). This preliminary experience that is brought into the company would be identified by the teacher as *Stage 1*, the foundation upon which to build. *Stage 2* will inevitably introduce business constraints such as client interaction, an engineering budget, and a multifaceted time schedule into the process that the student likely has not yet experienced—now a simple building becomes more complicated even though the original engineering problem remains the same. This business stage begins at employment and will involve every future project regardless of size, complexity, or role.

*Stage 3* must continue to develop the student's knowledge and experience with building industry regulations, including codes and standards that establish a visible path an engineer must follow with every project. It is difficult to understand, but regulations actually offer freedom to a practicing engineer by setting an example of how things work in the real world of material behavior, economics, and human interaction (for the most part). *Stage 4* may be described as professional technical development that is either dictated by licensing agencies, company policy, or personal desire to learn new things or expand on existing knowledge. Technical skills such as combining the effects of different building materials in the design of structures must continually be added to a student's repertoire. *Stage 5* of an engineer's development is thought by some to be the most frightening of all: Passing along knowledge and experience to others

**Figure 1-4** Brooklyn Bridge tower, cables, deck, and suspenders.

through mentoring, involvement in committees related to the profession, and teaching or leading classes for the benefit of others.

It is immediately obvious that all of these developmental stages not only build upon one another, but they will often be experienced at the same time, even on a single project. However, if a teacher is able to effectively monitor a student's growth within each one of these areas, being able to distinguish between the stages, it will be easier to identify the immediate and future needs of a student, thereby simplifying the process of teaching all together.

Suppose a particular graduate is most comfortable with steel structures. A teacher may then provide a simple one-story steel building project and have the student design the gravity members and their connections. Stage 2 is introduced by review of the architect's drawings to determine member layouts and materials that contribute to the carried weight. The work of other consultants, such as mechanical or plumbing plans, is also reviewed to completely determine loading requirements and the space with which structural members must fit into. The graduate will interact with other members of the design team to be certain that design restrictions are met, including a schedule of delivery. A teacher assists by explaining (or showing) proper communication methods, demonstrating how a set of plans can be properly understood, and providing resources needed to accomplish the goals of this stage as they relate to this particular project. Taking things one step at a time, though not always practical when a project is being rushed by the client, will keep a teacher's job manageable.

Stage 3 for this sample project typically requires much supervision. Building codes do not always follow a logical engineering progression and students must be guided through relevant portions in order to stay on track and to help alleviate frustration. Contacting the local building official to ask about site specific engineering requirements, such as wind exposure category and ground snow load, will serve as a proper introduction to the local design community as names, voices, and faces are assigned to positions that would otherwise be impersonal. As progress is made through actual design of elements, Stage 4 will immediately be realized as example problems learned at college must be broadened to include a different scope or assembly of members. Textbook problems become springboards — a means to an end — that the student is taught to work through.

Working with draftspersons or junior technicians introduces the student to Stage 5, as communication of design intent is carried out to others, teaching them how to interpret results for the purposes at hand. The student will also learn such things as company procedures and project development by working with administrative staff in a mixed teaching/learning environment. The supervising structural engineer must oversee the entire project and make sure the student understands each stage of progression as best as possible.

## 1.3  WHAT IS STRUCTURAL ENGINEERING?

Structural engineering, being considered a field of specialty within the realm of civil engineering, is the application of math and science to the design of structures, including buildings, bridges, storage tanks, transmission towers, roller coasters, aircraft, space vehicles, and much more, in such a way that the resulting product will safely resist all loads imposed upon it. The design of structures has always involved theory, buttressed by testing and direct observation, and a professional engineer is able to make wise use of intuition and experience to bring theoretical truths into reality. In order to develop an adequate understanding of structures that are designed, an engineer must make justifiable approximations and assumptions in regards to materials used and loading imposed and must also simplify the problem in order to develop a workable mathematical model.

## 1.3.1 Analysis and Design

In order to solve a problem related to their field of expertise, an engineer must use tools that have been discussed earlier in this chapter to analyze the situation at hand and design an appropriate solution to meet as many of the stipulated parameters as possible. These parameters include predefined pathways, such as the restrictions imposed by regulatory agencies and documents; or boundaries, including economics and the limitations of science and theoretical knowledge. An engineer analyzes materials through the use of research and testing to determine their suitability in creating the final structure. Materials will be formed into shapes, or combined with other materials to improve their own properties, and these new geometries function together with the material's physical qualities to define structural behavior, which an engineer will analyze in an effort to model the complete interaction of all pieces and systems. Once the anticipated reaction of a building or bridge has been analyzed, the final design of everything which plays a part will be completed.

The concept of design to a structural engineer has been defined as simply a matter of determining what is to be built and preparation of the instructions necessary for building it (Addis 1990, p. 31). Design of any structure with any material, small or large, attempts to satisfy three criteria: Strength (will not fracture within its useful lifetime, remains stable), serviceability (functions as expected, does not excessively deflect or vibrate), and economy (overall material and labor costs). A common sequence of design begins with an appropriate statement of the problem at hand and continues along a well-defined course of action toward a solution, using the correct tools and a comfortable environment, being justified by independent review or supervision.

Stages of analysis and design often become intermingled in the mind of an untrained observer. A professional engineer is expected to understand the difference—proper analysis, research, and understanding of the system must be used to direct the engineer toward an appropriate design. All suspension bridges, for example, rely on the same principles that have worked since the beginning of time itself, though design has kept pace with modern theories, methods of construction, and new building materials. Analysis has been streamlined over the years, but it is still based on this same history and leads to a very similar design with each new bridge. Figure 1-5 shows two successful suspension bridges in New York, separated by time and tool, yet representing the same innovation and creativity in accomplishing a common goal. Sound principles of engineering are truly timeless.

## 1.3.2 Uncertainty and Error

Although the analysis and design of structures is based on logical, scientific principles, these products exist in a world that introduces elements of uncertainty, even randomness at times, and an engineer must be aware of this. When recommending

**Figure 1-5** Two types of suspension bridges in New York, designed in different time periods using timeless principles of engineering mechanics.

use of testing results or observations for practical engineering purposes, scientists apply statistical analysis to account for variation in material quality and environmental conditions, inherent limitations of any testing program, imprecision in methods, and error.

The load carrying capacity of building materials is understood in terms of two important limits: Yield and ultimate. A structural member's behavior is easier to understand and define if it is not loaded beyond the yield limit, after which that member will not return to its original shape when load is removed. Methods of structural design used in the profession are partially based on an effort to prevent the member from supporting long-term gravity loads beyond this yield limit, which can be easily defined with some ease in the laboratory for ductile materials, but is a bit more complicated for brittle ones. Almost every test imaginable has been run on the more common building materials, so that limits of acceptable performance can be dictated with a reasonable level of confidence. By comparing volumes of experimental data, material performance becomes less mysterious and variability of mechanical properties becomes easier to predict. Testing programs, however, are naturally limited by available technology, funding, quality and quantity of supporting and research staff, public interest, and perceived need of the engineering community.

**Imperfect Methods**   Part of being a successful structural engineer involves not only recognizing limitations of tools and methods, but also understanding the positive aspects of things being worked with. Completed structures are assemblies of individual elements that naturally create a level of redundancy as they function together to resist imposed loads. This is a positive quality that cannot be easily calculated, but it is nonetheless present. Methods of analysis and design may not be perfect, but the ability to predict structural behavior with reasonable accuracy prove to the engineer that a sound, robust building or bridge can indeed be made to stand up against the worst of elements when proper care is exercised.

A perfect structural model would be impossibly complicated and take far too much time to create with every conceivable variable in the design process. Random variables and unavoidable errors will always affect results and analysis, therefore a simplified, useable model should be developed which properly considers known and expected loading patterns and phenomena, and is accurate enough for the task. In fact, Priestley (1997) comments that although 3-D modal analysis for seismic design through the ability of high-powered computer programs is certainly useful for structures with unusual or irregular geometries, it is doubtful that better results are produced than would have been discovered by a simpler method, such as a lateral analysis on the basis of assumed force distribution.

Although limitations and some uncertainties exist in methods developed to help engineers accomplish their goals, there are crucial abilities acquired and nurtured over time that make it possible to sort through what can be known or discovered. Many methods of analysis and design begin on a global scale, introducing basic answers to the problem, and a trained engineer can close associated gaps and move the project into more fine-tuned detailing that properly considers the work of all other disciplines.

**Error**   Uncertainties in a building's resistance to load are generally based on difficulties with building materials, products of fabrication, errors in analysis and judgment, and human inexperience. Professor Henri Petroski reminds us that "a safe structure will be the one whose weakest link is never overloaded by the greatest force to which the structure is subjected" (Petroski 1992, p. 41). This presents the concept of *load path*, where imposed external forces set up internal reactions that will be traced throughout the structure until arriving at the external restraint, which is the earth itself. The greatest force must be determined as best as can be done by analyzing a series of combinations of load types most likely to occur at the same time, having been estimated by history and observation. Materials that might be part of the weakest link of a system must be reviewed with an element of safety so that they will not fail when subjected to the load combinations previously analyzed. Therefore, the reliability of structural analysis cannot be expected to exceed restrictions of current engineering knowledge, nor to exceed practical limitations of numerical accuracy: a tenth of a pound in structural engineering computations is nonsense.

An engineer learns that measurements, as careful as they may be, are subject to a certain level of error and uncertainty. This cannot be avoided, but it must be understood and mitigated if sleepless nights are to be avoided. Measurement to a structural engineer includes not only physical dimensions, but also complete equations intended to measure the effects of imposed forces. A complicated loading diagram can lead to errors of execution if not given the appropriate care. After an engineer determines the magnitude and position of external force, mathematical equations are developed to define the relationship between these points, which may be plotted in the form of a straight line, a parabola, a circular segment, or some other geometrical arrangement. A common understanding of error in measurements indicates that values used in constructing the appropriate diagram may not be exact, but should lie within a certain margin about the exact shape being defined, as indicated by error bars in measurements shown in Fig. 1-6. Once an appropriate equation of the applied load pattern is discovered, it can be integrated numerous times using basic calculus to determine the design effect of a member or an entire structure, depending on the level of study.

As careful as one might be in creating an appropriate mathematical model, certain errors are unavoidable. Estimations are necessarily made due to incomplete statistical data and an inability to accurately describe material and environmental variability. Models are simplified to describe and visualize complex phenomena and errors may be made as a result of oversimplification or due to a lack of understanding of those phenomena. These types of errors can be reduced, however, by improving or increasing the relevant state of knowledge.

Errors that a structural engineer may make include those of intention, concept, or execution (Nowak 2000, pp. 294–296). *Intentional errors* result primarily from an abuse of the engineer's moral obligation to the client, the company, and to society as a whole. These are commonly executed in the form of skipping design steps or not giving consideration to different aspects of a structure for the sake of

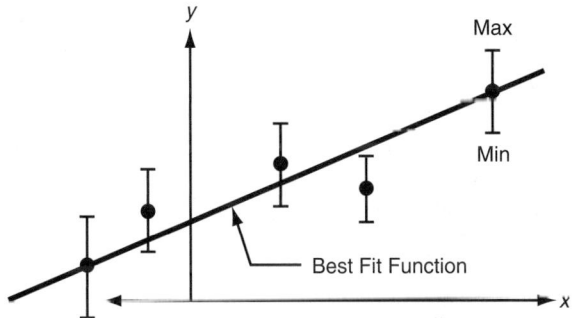

**Figure 1-6** Data points with range of error from testing and "least squares" function.

time, a limited budget, or an inability to perform the task. *Mistaken conceptual procedures* are not usually done maliciously, but can be more difficult to recognize and correct. Errors of this type include a failure to use correct methods or models without properly understanding or appreciating the consequences, and making analytical simplifications that are neither valid nor correctly assumed.

Even with good intentions and a solid understanding of mechanical principles involved, an engineer may still fail to *execute* the problem's solution properly for a variety of reasons including poor judgment, lack of communication with other members of the design team in order to be certain parameters have not changed, an inability to properly detect errors, or inexperience. Technological tools, such as 3-D building modeling programs that design both member and connection, are welcome innovations, but they can easily divert a practitioner from actually *experiencing* the behavior of structures. It is this experience that transforms a graduate from a technician into an engineer who understands the limitations involved in structural design and can adequately judge and check the accuracy of results delivered by an inhuman machine. Engineering problems deal with imperfect states of knowledge and part of an engineer's responsibility is to recognize this reality with human compassion and work diligently to mitigate associated risks or hindrances.

### 1.3.3 The Experience

Structural engineering is much more than just a career—it is a lifelong experience, meant to be passed along to future generations. A certain joy can be found in mentoring others by instilling pride in the realities of job security, earning potential, endless opportunity for growth, and important service to an international society. A whole new world of responsibility opens up to a graduate who begins working with peers and clients from a variety of industries, but confidence may be found through simple acceptance of this position: Responsibility for services rendered is unavoidable, others depend on that responsible charge given to a licensed professional, and it also opens the doors to great opportunity. To fully experience the profession of structural engineering, an individual must have a willingness to work with others, to accept certain directives from society, and to appreciate the importance of personal and technical growth.

Although there are inherent uncertainties in the work of a structural engineer, there are also volumes of solid scientific data to prove that a completed building or bridge is far more than just a sum of individual parts. There is robustness and redundancy, lending to incredible stability when subjected to even the strongest of external forces. An engineer's schedule can be complicated by many different tasks, but it is that wonderful variety of duties that keeps the profession exciting for a lifetime. A final encouraging thought can be found in the importance of responsibility: High as it may be within any professional career, a structural engineer will find pride and a sincere sense of accomplishment in creations that have a profound benefit to the international society of which we are all members.

# 2 The World of Professional Engineering

Because a professional engineer spends a great deal of time doing work without interpersonal contact, it is easy to lose sight of the fact that engineering decisions and discoveries have ramifications for societies around the world. For example, Fig. 2-1 shows a building constructed in China, made possible by engineering research and testing of seismic design theories performed in New Zealand, material specifications created in Germany, and design standards published in the United States. The international languages of math and science help engineers understand and encourage one another toward excellence and advancement of this global profession. It is this encouragement that leads to confidence, but it is important to remember that professional growth is not always pleasant—it requires hard work and facing certain realities, but the right attitude through it all can help engineers hold onto the joy and excitement of experiencing structural engineering. Responsibility to society need not be feared in the least, and a healthy acceptance of such can lead to tremendous growth in creativity and competence.

## 2.1 THE ROAD TO LICENSURE

Licensing of engineers and land surveyors began in 1907 when the State of Wyoming enacted legislation to help prevent incompetent and unqualified individuals from performing surveys affecting property ownership and water rights. California formally adopted the Professional Engineers Act in 1929 after collapse of the St. Francis Dam and the State of Texas enacted a similar Act following a public school gas explosion in 1937. The State of Montana Legislature was one of the last governments to formally enact professional engineer licensing laws, doing so in 1947.

| **Figure 2-1** Elegant skyscraper in China.

Though operational details of licensing laws vary from state to state, legal and technical provisions are often quite similar. All states will license an individual who has earned a 4-year bachelor's degree from a university with a civil, structural, or architectural engineering program accredited by the Accreditation Board for Engineering and Technology (ABET); passed an engineer-in-training examination (Fundamentals of Engineering); gained 4 years (typical) of professional experience under the direct supervision of an engineer licensed in that discipline; passed an 8-hour professional engineering exam (Professional Practices of Engineering); passed any supplementary tests required by the licensing state body (special seismic or surveying); and paid appropriate fees.

Being presented with the title *professional* is an honor that should not be treated lightly, as any career in a field by which service is offered to society brings great reward. Financial benefits are certainly helpful, but reward is also found in challenging and interesting work, in close associations with others of common interests and in formal exercise of the brain. A professional in any field of practice is commonly identified by the following characteristics:

1. A core of basic knowledge and some advanced techniques are mastered before offering service to society
2. Recognition and approval by society as *minimally competent* to exercise this knowledge in a practical way
3. Adherence to a code of ethics published by leading representative organizations
4. Subjection to a governing body of peers who enforce provisions of an engineer's practice act
5. Accepts a high level of responsibility for work completed personally or under direct supervision

*Licensing* is the process by which a government agency grants an individual permission to engage in an occupation after confirming that a level of minimal competence has been attained, as defined by a recognized gathering of peers. *Certification* for a variety of practices and skills, on the other hand, may be obtained through any association where the authority to use a particular title is granted based on a set of predefined qualifications. *Registration* is often a general term, indicating a requirement for all individuals who wish to practice in a specific field of expertise to pay fees and be added to a list maintained by some governmental consumer authority. Engineers authorized to practice within a state's boundaries are required to affix an identifiable stamp and personal signature to work for which they are directly responsible, as there is intended to be a value placed on such activity and the presence of a seal usually carries that significance.

## 2.2 AFFILIATIONS AND SOCIETIES

It is not only important for a professional engineer to promote unity within a design team, but to also be associated with a group of peers who are committed to excellence and to the honest execution of duties. Some of the ways that an engineering society brings members together to promote a unified front include publishing a standard code of ethics (validates their contribution to society), providing representation at the federal or state government levels (justifies their existence among regulating agencies), and offering technical publications or business-related assistance (substantiates their commitment to members). Association with other engineers in the same practice allows for an opportunity to compare business or technical procedures through networking, provides a forum for sharing knowledge in a particular area of expertise, and makes it easier to stay informed on current news events related to the profession. Currently, the three national structural engineers associations include the National Council of Structural Engineers Associations (NCSEA), which serves practicing structural engineers through education, advocacy, and advisory; the

Structural Engineering Institute of the American Society of Civil Engineers (ASCE/SEI), which serves the profession by developing formal standards, publishing books and journals, sponsoring conferences and continuing education programs, and advocating improved licensing and business and professional practices; and the Council of American Structural Engineers (CASE), which is organized into a risk management program and a business practices program.

## 2.3   ETHICS

Ethical behavior in any realm of life involves two main actions: Recognizing the difference between right and wrong actions to take in a given situation, and choosing to do what is right. Since engineers are ultimately responsible to society, consistent adherence to personal and professional ethics helps to alleviate some of the stress associated with such an important burden. Most practicing engineers are familiar with the ethical tale of the Citicorp building in New York (see Fig. 2-2), where repairs were made to strengthen the lateral force resisting system in response

| **Figure 2-2**   Citicorp building in New York.

to a student's question. Studies done on the promotion of ethical behavior by professional engineering societies and licensing agencies of countries associated with the North American Free Trade Agreement (NAFTA) revealed that common doctrine is found in three principles: Truth, honesty, and trustworthiness (Murdough Center for Engineering Professionalism, 1995). Some other common ethical elements in the dictates of NAFTA nations regarding the conduct of professional engineers include the following:

1. Engineers shall hold paramount the health, safety, and welfare of the general public in the practice of their profession.
2. Engineers shall practice only in their areas of competence, in a careful and diligent manner and in conformance with all applicable standards, codes, and regulations.
3. Engineers shall examine the societal and environmental impact of their actions and projects, including the wise use and conservation of resources and energy.
4. Engineers shall sign and take personal responsibility for all engineering work which they prepared or directly supervised.
5. Engineers should strive to enhance society's awareness of the profession's responsibilities and unmistakable value to the affairs of the world.

Unethical behavior of an engineer is not necessarily the same as negligent behavior, though the two may be closely related in some situations. *Negligence* is not commonly associated with technical inability, but rather with carelessness. If committed without appropriate regard for the safety of a client or of the public, a negligent act can also be described as unethical. However, sometimes an engineer does not give appropriate attention to decisions made or procedures followed because time or funding ran short on a project, and though the intent may not be to endanger or cheat someone else, carelessly cutting corners on engineering work in the interest of brevity is still negligent.

State regulatory agencies also govern those in the profession by penalizing actions of misconduct, gross negligence, incompetence, irresponsible delegation of design responsibility, or improper use of a professional seal. Claims of *misconduct* typically arise in response to violation of a law or established regulation and can also include acts of misrepresentation, where a design professional claims to have completed a particular task and either was not qualified to do so or did not perform the duty. *Gross negligence* is defined as a reckless act of willful disregard for professional responsibility and the determination of such requires experience, technical competence, and specialized knowledge of the profession and related standards. *Incompetence* is the demonstration of a lack of ability to perform functions of the profession, which may include actual design of structural members or a lack of understanding of basic principles such as load path and element stability. Incompetence cannot be rigidly defined by a single infraction, but is typically borne out through a process of repeated mistakes.

These disciplinary elements within the profession are statutory, being investigated, and decided upon by an appointed board of professional peers rather than attorneys, arbiters, or a pool of random jurors, though circumstances could lead to broad public litigation. Experts in the same field of engineering who are not acquainted with the accused are called upon to review evidence and offer opinions regarding the breach of state licensing board laws in order to provide fair and unbiased treatment.

## 2.4 CIVIL LIABILITY

Human beings make mistakes, sometimes due to incompetence, but many others are made because of unforeseeable factors. Some errors are serious enough to cause loss of life or property damage, and those extreme cases are always investigated and heavily scrutinized. Sound ethics does not demand perfection, but rather that errors are to be avoided as best as humanly possible. When a disaster occurs, such as the infamous Kansas City Hyatt Regency walkway collapse of 1981, investigators and attorneys sift through evidence and testimonies to determine what happened to cause the event in search of an answer to important questions: Could the disaster have been avoided? Should someone be held liable for the results?

Liability is a word that haunts structural engineers and insurance companies. However, there are important legal and practical safeguards in place that help bring everyone back to the understanding, uncomfortable though it may be, that engineers are capable of making mistakes simply because of their humanity. Business and profession codes of most states, for example, contain time limitations for the discovery of design defects in different types of engineering structures. California limits action brought against a design professional for patent deficiency (an obvious flaw in a product or document) to within 4 years after substantial completion of such improvement and within 10 years for any latent defect (a hidden flaw, weakness, or imperfection in an article which cannot be discovered by reasonable inspection). Other states have similar restrictions, protections, and definitions. In many states, the Economic Loss Doctrine limits claims by third parties that might be injured as a result of a design professional's act or omission to personal injury and property damage rather than arbitrary tort settlements.

### 2.4.1 Standard of Care

Structural engineers are expected to deliver a highly competent product, which may be a set of construction drawings and structural calculations for a new building, or a report completed after investigation of a building collapse. Society recognizes that no one is perfect and that perfection cannot be expected in every work and every instance, since all professions are equal in this regard. Engineers are therefore not held to the impossibility of perfection, but rather to a practical, somewhat universal standard of care. A person who brings a claim against a design professional generally has the burden of establishing what the standard is, that the professional owed a duty of care to the claimant, that it was materially breached, that a loss had indeed been suffered, and that a negligent act of the engineer was at least a contributing cause for any real damage or injury reported in the suit.

A working definition of care has both a legal and a practical background. The expectations of a structural engineer's competence become legally defined through court cases and defined in more practical terms by committees who are given the task of developing professional licensing exams. In both areas, licensed design professionals are called upon to give their opinions, based on experience and training, as to what a person is expected to know or do in the process of carrying out the duties of their profession.

**Legal Definition** The common religious and societal phrase, "do unto others as you would have them do unto you," is the indisputable backbone of every civil law and ordinance. Any structural engineer would not want his own office to be designed in such a way that it is too weak to stand up during a reasonably sized earthquake, therefore that engineer must execute his duties with the same attitude being expressed for fellow citizens: Design a building that will perform as well as you would want your own building to perform. Society must punish those who choose to offend, but society must also accurately define what an offense looks like.

Because there are so many differing opinions in the construction industry regarding what should be expected from whom, a standard of care that engineers should be expected to live by is often defined in the court system, from local offences to the nation's Supreme Court. Standing legal treatises on responsible care related to structural engineering practice include the following:

1. *Paxton v. County of Alameda* (California, 1953): The standard of care for a design professional was defined as "that level or quality of service ordinarily provided by other normally competent practitioners of good standing in that field, contemporaneously providing similar services in the same locality and under the same circumstances" (Ratay 2000, p. 95).
2. *City of Mounds View v. Walijarvi* (Minnesota, 1978): When an engineer provides professional services, he exercises judgment gained from experience and learning, and is usually providing those services in situations where certain unknown or uncontrollable factors are common. Therefore, some level of error in those services must be allowed (Kardon, 1999).
3. *Black's Law Dictionary*, 6th Edition, p. 1404–1405: "In performing professional services for a client, a professional has the duty to have that degree of learning and skill ordinarily possessed by reputable professionals, practicing in the same or similar locality and under similar circumstances."

**Practical Definition** Practical definitions for an expected standard of care typically have a stake in what qualifies an individual to be minimally competent, also being a critical milestone in an engineer's development. A set of basic requirements are developed by state agencies that must be met to sit for a licensing exam, which are first steps in determining who is at least qualified to become an engineer. A person who passes the exam is then recognized as minimally competent based on an agreed-upon set of definitions and expectations of ability. All of these requirements for professional licensure add to the understanding of what level of care an engineer is expected to meet. In realistic terms, the standard of professional responsibility is commonly exercised care (proper and serious attention to detail), absolute minimal competence in abilities and training (not godlike powers of perfection), and consistent ethical behavior.

### 2.4.2 Managing Risk

Awareness of risk is important for keeping sight of responsible care. Risk is often best handled in manageable pieces. Generally, risk in the building industry is defined in terms of economics, operations, and construction (Chen and Duan 1999, p. 4–12). *Economic* risk may be realized in the form of inflation, interest rate hikes, improper budget projections, or changing traffic patterns in bridge work. *Operational* risks for buildings and bridges are found in environmental impact, loss of life or limb due to some form of accident, or losses in value, assets, or income. Risks that can occur during *construction* include failure to meet budget or schedule, lack of quality required by project specifications, unforeseen underground conditions or installed facilities, delays or stoppages directed by governing authorities, and accidents.

Management of risk begins with an evaluation of the people who will be working on a project. First-time clients come with surprises and it is helpful to learn something about their history before agreeing to complete work for them, such as penchant for lawsuits, ability and desire to pay for services rendered, other associations or business connections, and personality. If it has been decided that the benefits of securing a new client are worth the risk, the next areas of possible risk include scope of services, compensation, and execution of a contract for a project of any size. All parties must be certain of who is doing what and how much is to be paid for it. What is decided then gets placed into a legally binding contract, properly executed by three basic elements: Mutual assent, which occurs when two or more parties agree to some type of arrangement; consideration, which is the goal or benefit earned from the arrangement (money or other goods); and legal capacity defined by several factors including age, authorization to enter into an agreement, and a proper understanding of the nature of that agreement.

Problems that begin as small opportunities can turn into hulking monsters if left unattended. Everyone working on a project has a schedule to keep and may be penalized by their suppliers or the owner if progress is unreasonably delayed. Once a project begins, the best form of risk management is clear, concise, and kindly communication. Thorough documentation of a project's events should be filed away in case a problem occurs and it becomes necessary to establish who was in charge of what and when services were to be performed. Records to keep include transmittal letters, logs of telephone conversations, factual e-mail messages or instructional letters, project documents (calculations, drawings, specifications), approval sequences, meeting notes, submittal schedules and milestone dates, site visit reports, and close out paperwork.

Risk is also managed in part by insurance. Common forms of insurance held during the design and construction of a project include (not necessarily limited to) bodily or personal injury, title, property, liability (professional errors and omissions, public, or otherwise), fire, loss of use, driving or equipment operating, special hazards, and workman's compensation.

### 2.4.3 Who's Responsible?

Responsibilities are moral obligations to perform duties honestly and responsibly, which automatically include an element of accountability. An engineer who is called upon to justify actions or decisions must be answerable for meeting obligations imposed by the public through education and licensure. Most states agree to a general definition of what describes an engineer who is in responsible charge of design work. Section 6703 of the State of California Business and Professions Code defines *responsible charge of work* to be the independent control and direction, by use of initiative, skill, and independent judgment, of the investigation or design of professional engineering work or the direct engineering control of such projects. The intended understanding does not necessarily involve financial responsibility.

Supervision of construction and site visits by an engineer-of-record for the purpose of observing completed work, as specifically related to the design intent expressed in the construction documents, sometimes raise issues of responsibility as well. For example, project structural drawings display the completed product and it is supposed to be understood that any temporary shoring, bracing, or other site safety issues are to be determined by the contractor in accordance with accepted procedures, unless the construction is of such an unusual nature that the structural engineer's expertise is needed for design of bracing or rigging to install the work. An engineer should not be in a position of approving means and methods of ordinary construction, or even specialized construction unless an agreement can be drafted, otherwise responsibility for site safety in implementing recommended procedures becomes an added liability.

## 2.5 DESIGN REGULATIONS

Regulations provide the roadmap that a structural engineer must follow while navigating by sound application of design theory. This direction defines the quality of the final product and puts all competitors on the same playing field, insofar as they are properly and consistently enforced. Regulations for design include building and bridge codes, mechanical or other energy conservation requirements (dictating the use of different building materials or types of construction), material testing standards and specifications, common state laws or statutes, local zoning ordinances (which may require additional services or protection of a building over a certain size), and the dictates of federal, state, local agencies or ordinances such as the Occupational Safety and Health Administration (OSHA) or federal Americans with Disabilities Act (ADA), or of property owners themselves.

A design regulatory system can take different forms depending on what goals are to be met and how those goals are defined. A *performance-based system* is that which identifies the ability of a regulated element to perform according to expected parameters, such as percentage of deformation or ductility. The viability of this approach is closely related to technological and risk-based criteria, each of which has their own identity as a form of regulation. *Risk-based* criteria include a qualitative element, such as the definition of an acceptable level of safety, and a quantitative one that is based on probabilistic analysis for the recurrence of a certain design event (IRCC 1998). *Prescriptive-based* design solutions include construction details, dimensions, and element specifications derived from a collection of codes and standards that describe how buildings should be designed, built, protected, and maintained in order to provide a minimum level of public safety.

As an engineer takes on different projects for a variety of clients, public and private, he or she is introduced to a plethora of codes, standards, and design guides and it can be confusing as to which governs what activity and how the law is going to define professional responsibility. In simple terms, the following hierarchy of regulatory documents for the design of structures, arranged in order of importance, applies across the United States and will be discussed in some detail throughout this chapter:

1. United States Constitution
2. Code of Federal Regulations (CFR), though some may dispute its position within this hierarchy because of state powers guaranteed by the U.S. Constitution
3. State Constitution
4. State Code of Regulations, which typically include a section, or "Title," that formally adopts a state-created or state-adopted model construction code
5. State-created building or bridge code (as applicable)
6. State-adopted model building or bridge code with all state amendments or revisions thereto
7. Voluntary consensus standards of design that have been developed according to the rules of the American National Standards Institute (ANSI) and are specifically cited or referenced within the state's Code of Regulations

8. Code-referenced standards of design that have not been developed according to the rules of ANSI
9. Nonreferenced standards of design, developed per ANSI rules
10. Nonreferenced standards of design that have not been developed per ANSI rules
11. General recommendations for design and construction issued by departments or branches within federal, state, or local government (in that order)
12. General recommendations for design and construction published by engineering societies, research agencies, publishing houses, or universities
13. Unpublished, but generally accepted engineering practices that are recognized locally or regionally
14. Individual or organizational research that has not yet been published

## 2.5.1 The Role of Government

In the United States, the hierarchy of government control and administration begins at the federal level, where congresspersons make and pass laws intended to be disseminated throughout the country. State legislatures create laws that are pertinent within their own boundaries, to the extent allowed by the U.S. Constitution, and other regional duties may be assigned to special branches, such as port authorities to operate airports, waterway or coastal ports, and develop property. Local governments regulate immediate land use through master planning, issuing or denying variances, and creating ordinances that govern local activities. All levels of government hold public meetings in an effort toward transparency, but there are certainly many improvements that can be made in all activities.

**Federal Versus State Government** The federal government also issues resolutions, orders, or acts to dictate the direction of engineering research or interest, usually in response to public reaction over a natural disaster. These proclamations do not typically affect written codes for design and construction, since regulation over construction lies in the hands of individual states by authority of the U.S. Constitution (Olshansky 1998, p. 4), but they have an indirect affect through funding of research or advertisement. For example, the Earthquake Hazards Reduction Act of 1977 (Public Law 95–124) was issued by the United States Congress for the purpose of reducing risk to life and property from future earthquakes, by which the Federal Emergency Management Agency (FEMA) was given primary responsibility for planning and coordinating the National Earthquake Hazards Reduction Program (NEHRP), established by the Act. Objectives of the program, which were delegated to different government agencies and departments, included educating the public through identification of vulnerable locations and structures; developing design and construction procedures in areas of seismic risk; development, promotion, and publication of model building codes, in conjunction with states and professional organizations, to encourage consideration of earthquake-resistant construction and land-use policies; develop ways to increase the use of existing scientific and engineering knowledge about how to mitigate earthquake hazards; and develop ways to assure the affordability of earthquake insurance (42 U.S.C. 7701 et seq.). This program has great influence in the direction of earthquake engineering, as methods of design have been based on these provisions since 1992.

The Code of Federal Regulations (CFR), updated on an annual basis, is a set of general and permanent rules created by executive departments and agencies within the federal government, divided into 50 titles representing specific areas. It is a daunting thought to consider what elements relative to structural engineering may be found within so many pages and it is critical to determine what regulations apply to any project prior to laying down the first line onto paper. A general idea of the subject matter of these regulations can shed light on when something will be enforced. Engineering and architectural societies, who send lobbyists to congress on the state and federal level, can help designers determine which regulations apply for different structural uses (educational, industrial, and the like) or locations of installation.

The 10th Amendment to the U.S. Constitution was ratified on December 15, 1791 and contained an important provision for state power that essentially left those governments in charge of most affairs within their borders. It asserts: "The powers not delegated to the United States by the Constitution, nor prohibited by it to the States, are reserved to the States respectively, or to the people." Many arguments have been presented over the years against federal intrusion in matters such as taxation and interstate commerce, citing the 10th Amendment for support, but the text was never intended to restrict Congress from anything and everything. *United States v. Darby* [312 U.S. 100, 124 (1941)] has been quoted as clarifying the power of the amendment: "The amendment states but a truism that all is retained which has not been surrendered. There is nothing in the history of its adoption to suggest that it was more than declaratory of the relationship between the national and state governments as it had been established by the Constitution before the amendment or that its purpose was other than to allay fears that the new national government might seek to exercise powers not granted, and that the states might not be able to exercise fully their reserved powers."

In fact, the 1981 case of *the Nevada State Board of Agriculture v. United States* served to clarify legal restrictions of the 10th Amendment in a dispute over the federal government's right to hold public lands within state boundaries. The state argued that federal lands could only be held in a temporary trust pending eventual disposal and that the retention of said lands violated equal footing doctrine rights. The court concluded that limitations on holding lands that were ceded by the original eastern states did not apply to western lands acquired after the U.S.

Constitution went into effect and that these newer states were not entitled to public lands: Congress had the authority of deciding which federal lands to keep or to sell. The equal footing doctrine only applied to rights of a political nature or those of sovereignty and not to economic or geographic equality (Prescott 2003, p. 14).

An Executive Order (EO) is given by the President of the United States to direct a federal agency or officer in their execution of congressionally established laws or policies. They do not require congressional approval, but have the same legal standing as though passed by Congress. The source of authority is found in Article II, Section 1 of the U.S. Constitution, which grants executive power to the President. These orders have relevance to any branch of government and can influence a great deal of areas responsible for society's growth, such as the creation of FEMA in 1979 (EO 12148) and of the National Science and Technology Council in 1993 (EO 12881).

**State Versus Local Government** Although states are responsible for adoption and enforcement of construction standards within their boundaries, codes are implemented and regulated at the local level through building, planning, and engineering departments. The keys to successful enforcement of code provisions are adequate funding to pay for the activities of government departments and appropriate staffing with qualified personnel.

State building, construction, or safety codes include both itemized and seemingly random provisions that dictate how structural elements are to be designed or detailed and it is difficult to know how broadly the requirements are spread out in published documents. For example, the section on construction safety orders found within Title 8 (Industrial Relations) of California's Code of Regulations, which is the section itemizing OSHA requirements, calls for all columns to be anchored by a minimum of four bolts except when columns are braced or guyed to provide stability to support a load of 300 lbs (136 kg) at an eccentricity of 18 in (460 mm). If this and similar measures are not properly coordinated prior to completing the construction drawings, there may be resulting architectural or project budget ramifications. Local governments may have their own regulations that need to be incorporated into structural design for industrial safety or other, depending on how much regulatory power those entities actually possess.

Typical land use planning begins with careful analysis of a region's current condition and trends for growth that seem to be occurring. Zoning ordinances are instituted to project residential, commercial, public use, and other areas for the direction of future growth according to assessed needs, resulting in a graphical master plan for an associated region. Once a layout is developed and refined, however, zoning changes, amendments, or alternate use provisions petitioned for by individuals, or as a result of further government survey, are continually reviewed for adoption by statute. These conditional use petitions to develop land for purposes other than what was originally intended usually involve a lengthy and costly approval process. Some of the more common zones of development include agricultural, commercial, industrial, preservation, recreational, and residential. Each zone has further subdivisions based on square footage of structures allowed or specific uses (i.e., light vs. heavy industrial). Master land use plans of local regions evolve with the election of new city council members or other government officials, but it is difficult to make significant revisions to a plan that has been set into motion.

The purpose of local zoning is common throughout the world: To promote the health, safety, and welfare of the public. In practical terms, some of the most important objectives include a need to prevent excessive population densities, promotion of a safe and effective traffic circulation system, provision of adequate parking and access to businesses, and protection of agricultural or rural land areas from urban sprawl. Zoning of land use was originally viewed as an extension of nuisance laws, being presented as a legitimate excuse for exercise of the state's police powers since their enforcement has a direct linkage to the protection of society's health, safety, and moral values. Older nuisance laws were often misunderstood and not universally enforced, so the establishment of specific zones of construction and land use was seen as a legal step in the right direction in terms of being able to equally apply the restrictions. Private land could not be developed in any way the owner saw fit, since dumping grounds should not be placed in a position that would cause harm to nearby residents (spread of disease, rodents, smell, and sound) and lot sizes could not be so small as to promote the spread of fire or block out adequate light and air flow around surrounding properties. Similar lot sizes were also required to be grouped in order to keep property values somewhat uniform and stable. To help protect the appropriate use of zoning regulations according to the needs of those who are affected, their administration was granted by the state to local agencies. Abuses of power within regulatory agencies, political pressure applied by involved parties, and the practice of payoffs and bribes have been uncovered over the years and will likely continue to plague the system.

There may also exist special districts within city boundaries not subject to construction (or other) regulations of that region. For example, the Port Authority of New York and New Jersey, established in 1921 under the title Port Authority of New York, is a self-supporting public interstate agency that is not subject to the local laws of jurisdictions where its properties are built. The Port Authority can create its own regulations, conduct its own inspections, and enforce its own rules without having to report to other regulators. It was originally created to administer the common harbor interests of these two states under a constitutional clause that allows for such compacts.

**Dillon's Rule or Home Rule?**  Since the U.S. Constitution does not mention local government, it is up to individual states to define their creation and associated powers by charter. In 1868, Iowa Supreme Court Justice John F. Dillon wrote an opinion on the limitation of regulatory powers of local governments that eventually became adopted by the U.S. Supreme Court as the default legal code for many municipalities across America. Dillon's Rule in this regard is embodied in the justice's statement: "It is a general and undisputed proposition of law that a municipal corporation possesses and can exercise the following powers, and no others: First, those granted in express words; second, those necessarily or fairly implied in or incident to the powers expressly granted; third, those essential to the accomplishment of the declared objects and purposes of the corporation—not simply convenient, but indispensable. Any fair, reasonable, substantial doubt concerning the existence of power is resolved by the courts against the corporation, and the power is denied."

Dillon wrote these words in response to local corruption and economic irresponsibility, as political abuses were rampant in municipal bodies. It is difficult to clearly define what specific functions should be handled by any level of government, however, since people differ in opinion on many important issues such as population growth, tradition, geographical protection, and economic function. Since Justice Dillon's decision, growing complexities of industrialization and increasing demand for public services on a local level brought difficulties of state power to the head of debate, as corruption became more evident within state land use regulatory agencies. In response, *home rule* was formally adopted and defined in 1875 by the efforts of Michigan Supreme Court Justice Thomas Cooley. In its most basic form, home rule simply means that local governments are granted any and all powers that are not explicitly prohibited by a state's constitution or statutory law. *Self-executing (mandatory constitutional) home rule* is deemed to be the strongest form, by which localities are granted powers of self-governance in all local affairs (unless specifically prohibited) through special recognition within that state's constitution. Some state constitutions, such as that for Illinois, automatically grant home rule to all communities over a certain population density. *Permissive constitutional home rule* merely authorizes state legislative bodies to delegate powers to local entities, but does not impose an obligation upon the legislature to follow through. *Legislative home rule* is doled out in bits and pieces by individual statues and does not carry the important authority of a state's constitution for support. In general, the sentiment of law is that Dillon states, and those with legislative home rule, entail weak local governments and strong state oversight, whereas strong home states offer greater freedom to local bodies with little state interference.

Courts in all 50 states continue to resolve controversy over state versus local rights, ordinarily in favor of the state through benefit of doubt. Some of the reasons given in support of strong Dillon's Rule include the following: Control from state level ensures more uniformity, resulting in fair and efficient governance; benefits local government officials by allowing them to use the rule as an excuse to not undertake certain tasks; and state oversight may prevent exclusionary and provincial actions by local governments. Proponents of strong home rule, on the other hand, claim that it empowers local citizens to select the form of government that they prefer, allows citizens to solve local problems in their own fashion, and reduces the amount of time state legislature has to devote to local affairs.

## 2.5.2 Codes

Codes and design standards dictate criteria necessary to assess the adequacy of a structure or its individual components and systems in terms of strength, serviceability, and functionality. Things can be complicated, however, as an engineer is often expected to know or be familiar with all codes and standards relevant to practice, including those written for site safety (OSHA). It is important to be well-informed, but it often takes the whole design team to share knowledge of applicable codes, standards, and regulations.

**Buildings** In general terms, a building code represents minimum acceptable standards for regulation of the design, construction, and maintenance of buildings for the purpose of protecting the health, safety, and welfare of those who will occupy or visit the structure. Buildings must not only be safe, but also predictable in terms of their performance during their service life or when subjected to extreme loading, such as an earthquake or hurricane.

The International Code Conference (ICC) is recognized worldwide for producing quality model codes related to all aspects of building construction. It is a nonprofit organization established in 1994 as a collective effort of three previous model code groups, the Building Officials and Code Administrators International, the International Conference of Building Officials, and the Southern Building Code Congress International. The process of developing a model code is consensus and private sector-based, open to all interested individuals or groups who submit a code change proposal and participate in open debate regarding all inquiries and submissions. ICC published the *International Building Code* (IBC) in the year 2000 with the intent of combining model building codes originally produced by originating members. It is updated every 3 years. Other standards and codes are referenced directly by the IBC to cover more ground, giving them the same legal credence, yet the IBC retains rights of preeminence when similar regulations are in conflict (2006 IBC, Section 102.4). Appendices are not usually mandated unless specifically adopted by the governing jurisdiction by ordinance.

The National Fire Protection Association (NFPA) is a nonprofit organization that has published fire safety codes since its founding in 1896 in New York, originally restricting membership to stock fire insurance organizations. Two of the most influential early documents produced included the *Report of Committee on Automatic Sprinkler Protection* published in 1896 (eventually came to be known as NFPA 13) and the *National Electrical Code of 1897* (becoming simply the *National Electric Code* in later editions). ICC and NFPA attempted to coproduce a national fire code in the late 1990s, but broke off the effort in 1998 because of disagreements over the code writing process and some of the contents. NFPA announced their intention to publish a separate model building code from ICC's IBC and NFPA 5000, entitled *Building Construction and Safety Code*, was born in 2002. It became part of the Comprehensive Consensus Code (C3) package of NFPA documents, which compete directly with ICC's family of codes. The IBC is currently adopted by a far greater number of jurisdictions than NFPA 5000, but building officials still tend to line up behind their favorite.

Design and construction codes have a great deal of prescriptive content intended to achieve predictable structural behavior, but much work is being completed to drive philosophies toward a performance-based set of codes with different criteria for measurement. The prescriptions and design measures in existing model codes are already intended to produce acceptable life-safety or collapse-prevention building performance when subjected to severe, infrequent earthquakes (SEAOC 1999, Section 1A-1.1; ASCE/SEI 7-05, Section 1.4), but these are minimum standards for which greater protection should be promoted as viable options. Section 5.2 of NFPA 5000 lists a variety of categories from which performance criteria can be measured, including safety from fire, safety from structural failure, safety during building use, surface water entry, control of contaminants, function, cultural heritage (additions, relocations, special regions), mission continuity (protected operation of community welfare facilities), environment (control of damaging emissions), and uncontrolled moisture.

**Bridges**  Bridge design is generally regulated with standards produced by the American Association of State Highway and Transportation Officials, including the *Standard Specifications for Highway Bridges* and *LRFD Bridge Design Specifications*, the latter of which has been mandated by the Federal Highway Administration for all projects using federal funds after October 2007. The design of railroad bridges is commonly regulated by the *Manual for Railway Engineering*, published by the American Railway Engineering Maintenance of Way Association and promoted by the Federal Railroad Administration. These organizations will be discussed at greater length in Chapter 5.

Any model bridge design and construction code adopted for use by an individual state, or mandated by the federal government to secure the nation's infrastructure, will be subject to review by the state department of transportation, which is given wide latitude in making revisions, interpretations, and adding amendments. This department usually provides all specifications and typical construction details to be used for projects of any size and a design engineer will work closely with that organization, even come to know the personnel at a more familiar level which is helpful during the process of review and approval of documents.

## 2.5.3 Design and Construction Standards

The American Society for Testing and Materials International (ASTM) has published specifications for test procedures used to determine the quality of building materials, construction methods, systems, and services since 1898. Technical committees of ASTM represent users, producers, manufacturers, designers, consumers, and the general public, for whom they work out and agree upon details that individual standards require in order to assure appropriate quality for whatever it is intended to control. All standards include metric equivalents, denoted by an "M" in the title, sometimes combined to form a single document. For example, the original standard ASTM A325-04b (specification for structural bolts; steel, heat treated) was last updated in the year 2004 under a third revision "b" (the letter identifies when more than one revision has taken place during that year, therefore an "a" would indicate a second revision), but the metric equivalent is denoted separately as ASTM A325M-05 and was last updated in the year 2005. ASTM A36 (specification for carbon structural steel), however, includes the metric equivalent within the same document and is therefore officially denoted as ASTM A36/A36M. Some ASTM specifications do not denote a particular product, but rather a process, tolerance, or characteristic, such as A6 (general requirements for rolled structural steel bars, plates, shapes, and sheet piling of a variety of steels), A370 (test methods and definitions for mechanical testing of steel products), and D2486 (scrub resistance of wall paints).

ASTM standards are developed in response to a need identified by a technical committee within ASTM or made known to them by any other interested party. Task group members prepare a draft standard that is reviewed by its parent subcommittee through a letter ballot. After the subcommittee approves the document and any related changes, it is submitted concurrently to the main committee and the entire membership of ASTM. All comments and objections must be given timely and thoughtful consideration before the document can be finalized.

The American National Standards Institute (ANSI) was founded in 1918 by five engineering societies (American Institute of Electrical Engineers, American Society of Mechanical Engineers, American Society of Civil Engineers, American Institute of Mining and Metallurgical Engineers, and American Society for Testing and Materials) and three government agencies (U.S. Departments of War, Navy, and Commerce) to coordinate and administer the voluntary standardization system of the private sector in the United States. It remains a private, nonprofit membership organization supported by both private and public sector financial sources. In order for a standard to maintain ANSI accreditation, the developers must adhere to a set of requirements and procedures that maintain a status of consensus, where the process remains equitable, accessible, and responsive to requirements of various stakeholders. All interested and affected parties, therefore, have an opportunity to participate in the development of any standard, which are written for products, processes, services, systems, or personnel.

Model building or bridge codes reference many standards and other supplementary codes produced by different agencies, giving them the same legal standing. In order for this to happen, certain criteria must be met: The intended application of the standard must be clearly identified within the body of the code; the standard itself must be written using mandatory language and cannot have the effect of requiring proprietary materials or methods; and the standard must be readily available and produced by a consensus process. Design standards referenced by model codes are also written by engineering societies, such as ASCE, in an effort to unify important procedures and methods based on the latest research. Material research or promotional groups, such as APA—The Engineered Wood Association, perform their own tests (see Fig. 2-3) according to protocols formally established by ASTM or other group, and disseminate the information to the public for practical acceptance.

The ICC also administers an evaluation service, abbreviated as ICC-ES, which is a nonprofit, public-benefit corporation that performs technical evaluation and issues approvals for specific construction products, components, and methods according to historic principles that have been developed for more than 70 years. Reports directly address compliance with model codes and have detailed instructions on how a component may be designed or used, which is of tremendous assistance to those who are charged with enforcing adopted codes. The process of

**Figure 2-3** Plywood roof diaphragm testing. (*Courtesy APA –The Engineered Wood Association*)

obtaining an official report from ICC-ES begins when a company submits an application, product information, test reports, and appropriate fees to cover the cost of evaluation. Testing is to be done according to accepted methods and results must be reported according to established procedures.

ICC-ES is a subsidiary of ICC, but is separately incorporated and has its own board of directors, staff, rules, and procedures. The reports can be quite cumbersome and difficult to understand because a large number of notes and specifications are included for design purposes. As long as the engineer systematically reviews all notes, relevant equations, and assures that the applicable model code is referenced as applicable to the report, this approval can be quite powerful in quality assurance.

### 2.5.4 Other Regulations and Considerations

International Organization for Standardization (ISO) is a network of national standards institutes of 157 countries and is a nongovernmental body. Though the organization's activities essentially began in the electrochemical field in 1906, its current designation and official duties were created in 1947. All standards are voluntary and every participating institute has the right to take part in their development, which is typically done in response to the needs of the free market. Some standards related to safety, health, or the environment are adopted by regulatory authorities or governments who then administer and enforce their use.

Underwriters Laboratories, Inc. (UL) was founded in 1894 as an independent nonprofit organization for testing and certifying product safety. It is primarily concerned with the safety aspect of hazardous products including life saving devices, fire suppression and containment equipment, and those used for analysis of harmful vapors. Electrical, heating, air conditioning, and refrigeration products are also tested. UL has three main forms of service: Listing, component recognition, and classification. The *listing* service is the most widely recognized, by which the UL-mark indicates that samples of a complete product have been tested by UL to nationally recognized safety standards and have been found to be free from reasonably foreseeable risk of fire, electric shock, and related hazards. The *component recognition* service covers testing and evaluation of individual components that are incomplete or restricted in performance capabilities and will be used later in a completed end product. *Classified* products have been evaluated for specific properties, have a limited range of hazards, or are suitable for use under limited conditions, generally including building materials and industrial equipment. Manufacturers producing listed products must demonstrate that they have a program to ensure each copy of their products will similarly meet the criteria for which it obtained recognition.

FM Global, often known simply as Factory Mutual (FM), is an international property insurance and loss prevention company with roots that can be traced back to

1835, when a proactive textile mill owner created a mutual insurance company to provide coverage for factories that made improvements to minimize the chances of loss due to fire. The knowledge base of how losses were caused, built up from inspections and research, became vital in identifying industry hazards. FM continues an aggressive research and development program for products and techniques that will help mitigate property loss, addressing electrical equipment, fire detection and signaling equipment, fire protection equipment, functional safety and assessment, different building materials, and roofing products. Of particular structural interest, roofing products are tested and certified according to wind uplift performance, hail-damage resistance, accelerated weathering, and corrosion resistance.

The National Institute of Standards and Technology (NIST) is a nonregulatory federal agency that promotes innovation and industrial competitiveness by advancing measurement science, standards, and technology. It functions through four cooperative programs: The NIST Laboratories, where research is conducted to improve products and services; the Baldrige National Quality Program, which promotes excellence among manufacturers, service companies, educational and healthcare providers; the Manufacturing Extension Partnership, which is a nationwide network of local centers offering technological and business assistance to smaller manufacturers; and the Advanced Technology Program, which accelerates the development of innovative technologies for broad national benefit by cofunding private sector research and development. Some of the ways NIST positively affects the construction industry include establishment and protection of basic weights and measures; ongoing research of the Building and Fire Research Laboratory, which presents findings relevant to the safety and performance of constructed facilities; and advancement in science and technology to measure and predict the service life of construction materials.

**National Flood Insurance Program**  Another regulation intended to reduce damage during an extreme event is the National Flood Insurance Program (NFIP), directed by FEMA. The program was created by the U.S. Congress in 1968 to provide federally backed flood insurance coverage and to reduce future losses by identifying floodprone regions and ensuring that new development is adequately protected from damage. In order to administer the program, FEMA maps all land into regions of low, moderate, high, and special flood risk, identifying hazardous areas across the nation by conducting studies and publishing Flood Insurance Rate Maps (FIRMs). FEMA makes reduced-rate flood insurance available to residents of communities that have adopted the program through their floodplain management ordinances or laws and requires the purchase of this insurance by owners of structures located within regions of high or special risk identified on the FIRM(s) for the adopting community.

Special Flood Hazard Areas (SFHAs) are those in which there is a 1% probability of equaling or exceeding a 100-year flood ("Base Flood") and structures built in these regions are subject to special floodplain management regulation affecting design

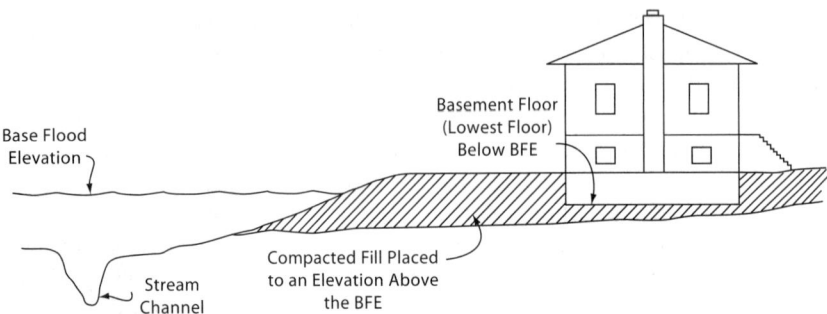

**Figure 2-4** Relationship between basement floor and base flood elevation. (*Courtesy Federal Emergency Management Agency*)

and construction. These include areas of high risk (identified on the FIRM as zones beginning with the letter "A") and high coastal hazards that are also subject to storm wave action (zones beginning with the letter "V"). The basic requirements for construction in these regions is found in Title 44, Part 60, Section 3(c) of the Code of Federal Regulations (CFR), including provisions for flood proofing and buoyancy.

As land is developed by placement of engineered fill or other topographic sculpting, owners or communities participating in the NFIP may request a revision of the FIRM to indicate that changed areas are no longer classified as special flood hazards. FEMA issues a Letter of Map Revision to officially declare the change in status of a previously affected region and once the FIRM is revised, special floodplain management standards no longer apply to structures built on this land and the owner is no longer required to purchase flood insurance.

## 2.6 RESPONSIBILITY TO SOCIETY

Public safety is the most important goal of every design and an engineer uses personal knowledge and experience to create safe buildings, while also giving careful consideration to economics, aesthetics, and constructability. Society will ultimately judge an engineer's work, whether in accusation after a disaster or in silence after hundreds of years of flawless service. Engineers create products that integrate people from all walks of society, from the little child who occupies a school building to the chief executive officer who makes decisions atop a multi-story office building, and this relevance is not easily forgotten.

### 2.6.1 A Healthy Workforce

One of the most significant ways that engineers can support society is to train the future workforce and encourage an increasing number of students to enter the profession. Engineering knowledge must be continually passed along throughout each generation if the profession is to survive, strengthened with the integration of new technologies and carried forth by enthusiastic professionals who are willing to teach and train others to eventually take their own places. Balancing the need to train a future workforce with tight project timelines is a challenge, but society requires it for health, longevity, and prosperity. Schools and professionals must work together to improve interest in a technical path of education and encourage kids to believe in their own abilities to do well in math and science, the foundational subjects for an engineering career.

In 2003, American College Testing, Inc. (ACT), a not-for-profit group that produces resources for education assessment and program management, outlined a plan addressed to school districts, colleges and universities, professional engineering organizations, and policy makers to help them understand the need to attract and develop a well-prepared and diverse engineering workforce. Some important findings and conclusions of the report include the following (Noeth, Cruce, and Harmston 2003, p. vi):

1. Continued growth in national productivity requires a continuous supply of engineers who are highly competent in mathematics and science and who are adaptable to the needs of a rapidly changing profession.
2. The number of potential engineering majors in college has wavered, resulting in a net decrease since 1991. Changes in academic interests and occupational preferences in high school students are likely explanations for this decline.
3. Since 1991, gender representation has remained relatively unchanged, with females making up only 18 to 20% of all potential engineering majors.

## 2.6.2 The Ring Ceremony

The Order of the Engineer, an independent American organization that consists of members who have participated in the Engineer's Ring Ceremony, was derived from Canada's Ritual of the Calling of an Engineer. It was initiated in the United States to encourage pride and responsibility in the engineering profession (to students and to practitioners), to bridge the gap between training and practice, and to present a visible symbol that identifies an engineer's commitment to a moral and ethical code of conduct. All students and practitioners who have graduated from an ABET-accredited engineering program or hold licensure as a professional engineer are eligible to participate.

The first ring ceremony in the United States was held in 1970 at Cleveland State University in Ohio, by which a candidate formally adopts the Obligation of an Engineer and receives a stainless steel ring to be worn on the fifth finger of the working hand. Canadians receive a wrought iron ring, following (what was thought to be) a rumor that the first rings to be used in the 1925 Canadian Ritual at the University of Toronto were forged from debris of the Quebec Bridge which collapsed in 1907. The purpose of the rings, therefore, is a reminder that engineers are not infallible and that there are serious consequences to failure. The Obligation of an Engineer is a formal statement of responsibility to the public and the profession and includes the following recognitions:

1. An obligation to serve humanity by wise use of nature's resources
2. The incomparable value that engineers bring to society and quality of life
3. The importance of fair dealings, tolerance, integrity, and dignity of the profession
4. A need for Divine guidance (in the original text of the Obligation)

## 2.7 INTERNATIONAL ISSUES

If a business chooses to explore an international market, the same basic strategic fields that create success domestically will still be active, but may function under different parameters. These fields include finance, technology, human resource management, marketing, and company operation. There are some advantages in each field when a business does not have to rely solely on the health of a single country's economy and political condition. Internationalization can also expand influence into regions where competition may not be as fierce and add a unique blend of interesting projects to a company's repertoire.

It should be obvious that conducting business across geographic boundaries also involves risk, related in part to a number of unknown factors. For example, a country's economic and political condition is found to be beneficial at times, but can also swing from profit to punishment in such a way that a firm cannot manage what is happening. A foreign legal environment may be lopsided in favor of the hosting country and international incentive to change this perception (or reality) often moves slowly. Cultural and technological differences involve special training for employees, requiring patience and persistence. Some of these risks are managed through diversification of influence, maintaining little company debt, and having a high level of asset liquidity to cushion adverse activity (Yee and Cheah 2006).

The National Science Foundation recognizes that a blend of domestic and international engineering talent drives the United States to a position of global leadership (NSB 2003, p. 7), therefore a broad practice across geographic boundaries can be seen as improving this unification and making it easier to integrate the best minds in the world to advance the needs of a much bigger society than one's own backyard. Balancing human resources and paying close attention to quality and consistency of work, complying in every regard with applicable codes regardless of which borders surround the practitioner, becomes an important, sometimes frustrating, activity in a broader format.

Integration of foreign-born workers into the domestic workforce is an issue that has existed with the birth of any nation. Limited term visas for nonimmigrant status are issued to address an immediate demand for skills in the job market, but do not serve as an effective long-term strategy, both for individual business or for the health of the country. The H-1B visa program was created by the Immigration Act of 1990, enabling U.S. employers to hire temporary workers in specialty occupations where a bachelor's or higher degree is necessary for entry. The number was initially capped at 65,000 entrants per fiscal year, but was raised to 115,000 in 1999 and 2000, and 195,000 between the years of 2001 and 2003. It reverted to 65,000 in 2004. This nonimmigrant visa is generally valid for 3 years of employment and may be renewed for an additional 3 years.

## 2.8 ADVOCACY

Engineers should be understood by society as adding irreplaceable value to daily living, working, commuting, entertaining, and relaxing. Recognition is driven by the federal government, sometimes being the best form of advertisement. The 109th U.S. Congress (2005-2006) formalized their sentiments in House Resolution 681, which was passed on March 7, 2006, affirming that engineers use their scientific and technical knowledge and skills in creative and innovative ways to fulfill society's needs; and that engineers have helped meet the major technological challenges of our time and are a crucial link in research, development, and demonstration in transforming scientific discoveries into useful products. Federal research agencies recognize the need to attract talented undergraduates to science and engineering majors with an emphasis on fostering skills, mobility, occupational and geographic migration, and demographic characteristics in such a way that a dynamic, global workforce in technical fields will strengthen the international leadership of the United States (NSB 2003, p. 29).

Engineers who display innovative and creative thinking not only provide an important benefit to society in resulting structures or technologies, but give strong impetus to other engineers to also do their best. R. Buckminster "Bucky" Fuller reintroduced architects and engineers to the strength and beauty of geometrical shapes, reminding the industry of its responsibility to develop structures that serve the surrounding environment in terms of efficiency and conservation. He introduced the idea of "tensegrity," which describes a closed structural system composed of compression struts and tension elements arranged in such a way that results in a stable, triangulated unit, and also invented the geodesic dome.

One of the greatest ways of educating the public on what engineering is all about is to involve elementary school kids in fun, engineering-related activities that stir the imagination and encourages them to share these experiences with others. Working with kids who represent a wide variety of socioeconomic, religious, and cultural backgrounds can open up an engineer's mind to the humanity which defines the profession. Many engineering outreach programs that are currently in operation across the United States use common themes and strategies, including active learning through hands-on activities, inquiry-based learning, curriculum supplements, engaged role models, younger student focus, and K-12 teacher involvement (Jeffers, Safferman, and Safferman 2004).

# 3 The Business of Structural Engineering

During the course of time spent at college, an engineering student becomes more familiar with research and development than the practical aspects of an engineering business. Even work on a campus directed solely toward advancing the profession is funded by public and private sources, therefore practical business skills (maintaining a budget, timely delivery of goods and services, effective people skills) apply to all areas of engineering practice.

## 3.1 PLACES OF EMPLOYMENT

There will always be a need for professional engineers around the world, from large cities to smaller suburbs, and businesses of all sizes and specialties can be found close to their own clientele. Engineers who work in close proximity to a downtown metropolitan region, such as is shown in Fig. 3-1, easily find inspiration embodied in a tall structure or a particularly creative bridge, but may discover less opportunity to develop client rapport than those who do business in smaller, more remote regions of the country. Many times it is this personal level of interaction that causes tremendous growth in one's professional career, as engineering work truly becomes service oriented and takes on far greater significance.

In addition to a nearly limitless range of geographical choices for employment, a structural engineer can find satisfying work in many different blocks of society. These include federal, state, or local government organizations, consulting firms of any size (from sole proprietorships to international corporations), industrial groups or other specialized companies that produce goods, and at the university

| **Figure 3-1**  Downtown Winnipeg, Manitoba, Canada.

level through teaching or research. Success and happiness in one's profession depends greatly on choice of geography as well as client base, perhaps requiring change in the course of time, but it can be refreshing to call to mind the great variety to choose from.

### 3.1.1 Typical Hierarchy

A basic outline of business hierarchy is shown in Fig. 3-2, which can be expected to exist in one form or another within all industries. The president, chief executive officer, and/or director of operations define a philosophy that all employees follow toward a unified end. This headship sometimes comes at a cost, as those who fill these positions of leadership work long hours and make difficult decisions that can have a profound effect on the lives of many, to an extent that family time is often sacrificed and friendship with coworkers may be looked upon with suspicion. These possibilities need not be realized, however, if interpersonal skills, powers of delegation, and time management strategies are aggressively practiced.

The next tier of leadership is given to business partners or supervisors with operational oversight, such as treasury or staff. Subdividing responsibilities makes it easier to funnel company plans and protocol to all members of the team and goes a long way toward fostering a sense of identity within each department. This identity is important if company philosophy is to be known and honored, where everyone

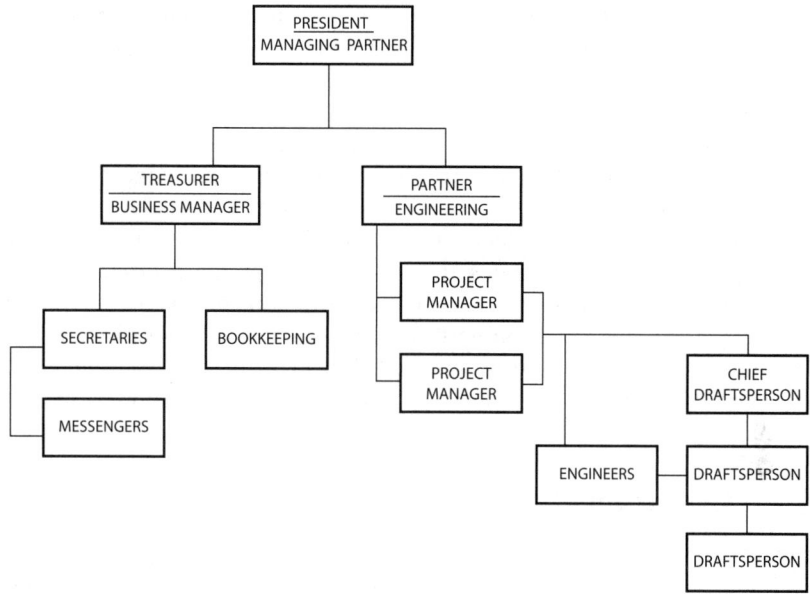

| **Figure 3-2**  Common engineering business hierarchical structure.

feels as though they have helped during prosperous times and are willing to evaluate their own contribution when things do not go as planned. Individual department heads must have the same interpersonal skills and creativity in leadership that a company president models. Some of the more common business skills that should be mastered include an ability to define and communicate the scope of a project, financially appraise business commitments, estimate product and labor costs of construction projects in order to have an idea of the company's associated risk, understand the basic legal structure of working contracts, schedule and allocate labor or resources, mediate over disagreements, and to honestly evaluate project success. An engineer-in-training will quickly realize that these skills in business leadership apply to his or her own activities, as designs are directed to supporting staff and other project team members. This need not be feared, but rather viewed as exciting career development.

Responsibilities of leadership extend to all areas of an engineering company, though some positions may not be formally recognized by established protocol, but rather through a general understanding. For example, bookkeepers often are afforded some administrative control over the secretarial staff and receptionists are well-suited to govern the activities of messengers. There is generally a lead draftsperson who is looked to for advice and guidance regarding company protocol for construction documents and this person may, in fact, be recognized with official project management responsibility.

Whatever leadership structure is established, the hierarchy must be acknowledged and respected by all employees: This requirement is obvious and is communicated at the time of interviewing or hiring. Once an employee is brought on board, a period of probation begins for the purpose of evaluating technical and personal abilities to do the proscribed work, as well as to reveal willingness and ability to advance. This period is a valuable time for both employer and employee to consider their relationship and to effectively communicate expectations regarding submission to the hierarchy. A contract or handbook for employment includes the terms of probation and instructions on how individual progress (or lack thereof) and administrative actions will be documented. Positive- and negative-work habits need to be discovered during this period (though some take time to surface), with appropriate reward or reprimand, to help avoid legal issues related to termination if things go sour. Company leaders are responsible for creating an environment that allows these discoveries to take place.

### 3.1.2 Government

Engineers who work for a government agency are also called civil servants, though the whole realm of engineering involves service to society regardless of employer. Compensation through public funds is usually maintained toward being competitive with salaries and benefits that are provided to engineers employed by other institutions of a similar nature in the same region. Work in the public sector includes not only oversight or approval of the work of private sector engineers whose projects are to be constructed using public funds, but also design work for government facilities when costs can be shown to be competitive with that of outside service. Administration of both people and projects tend to be the most important duties, as government work often involves enormous budgets, sensitive end users, and a large staff with a variety of specialties.

Government work can be quite varied or somewhat predictable, depending on level and venue, and will not suit everyone (though the same can be said for any place of employment, public or private). Frequent travel tends to disrupt family life, causing a need for major change or understanding of how things will be. Government agencies with a large staff of engineers may find it best to subdivide the work in such a way that each individual is somewhat "pigeon-holed" into performing the same tasks with great frequency, which is a great opportunity for some, but drives others to a state of insanity. On the other hand, work for a local government agency within a vibrant community may be far too heavy and varied for a limited staff to handle.

### 3.1.3 Private Consulting Firms

Consulting engineering businesses are found in many sizes and organizational structures including sole-proprietorships, partnerships, or corporations. *Sole-proprietorships* are the simplest to run and understand, as the company owner exercises complete control over operations, selection of clients and personnel, and vision for the future. Limited financial resources affect business expansion possibilities and a typically small staff of people can be cause for limiting project size. Professional liability insurance may be out of reach financially, but some consider this an unexpected benefit in the interest of avoiding the appearance of "deep pockets" if a lawsuit were to arise. Many legal advisors, however, consider this attitude to be poor business strategy.

A *partnership* is a business led by two or more professionals who combine talents and financial backing, which opens the possibility of serving a larger and more diverse clientele than if each were to operate independently. The basic legal structure is similar to a sole-proprietorship, except that all assets and liabilities are shared by several individuals. One significant disadvantage, however, is that personal assets of each partner are liable in proportion to their ownership percentage for prosecuted actions of the others related to business operations. Some legal structures include protection to a certain extent, such as limited liability partnerships.

An engineering company with a large number of employees and business volume may consider structuring as a *corporation* to have certain advantages, from federal income taxes to employee benefit packages (including ownership possibilities). Ownership and management of a consulting engineering corporation, or at least of the major company interests, are required to be in the hands of licensed professionals so that the public is protected from unqualified practitioners hiding behind a corporate shield. Although the assets of a corporation itself may be attacked for negligent acts of one of the principals, personal assets of the other principals are not held up for grabs. Other legal protective titles and arrangements are available, depending on business and licensure laws of states in which services are offered.

## 3.1.4 Industry

Depending on size, scope of business, and type of good produced, some companies in the industrial sector employ their own engineers. Tasks include monitoring and maintenance of equipment, designing the layout and structural support for piping and associated machinery (see Fig. 3-3), or overseeing the overall facility operation. If the company has its own engineering department, there will also be a staff of support to complete scheduling, place orders or manage inventory, and prepare cost or productivity estimates. Staff engineers can expect consistent work for a single entity and become well acquainted with operational procedures and expectations, which is preferred by many over having to cater to a multitude of clients with different nuances and demands.

The nature of work demanded in an industrial facility usually makes it necessary for an engineer to have some expertise in a variety of disciplines, including structural,

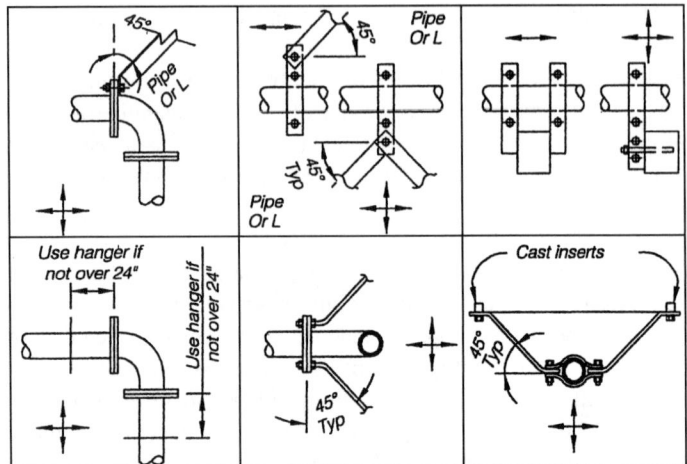

**Figure 3-3** Common methods for bracing piping systems. *(Courtesy U.S. Army Corps of Engineers)*

electrical, and mechanical. Corporate management recognizes that staff engineers cannot be fluent in all duties and a budget is usually maintained for retaining the services of outside consulting engineers if a need arises. An ability to solve problems quickly and efficiently will be paramount to success, as time is money in production. An industrial engineer will therefore also serve as head mechanic (to a certain extent) in order to give proper direction to supporting staff on mechanical matters.

### 3.1.5 Colleges and Universities

Historically, the term *college* was used to denote a portion within a *university* that offered a specific degree, though the ability to grant said degrees was retained by administrators of the entire university. Junior or community colleges offer associate's degrees and a variety of certifications and are the least expensive for a student to attend. Whereas universities commonly offer advanced degrees like doctorates and master of science, independent colleges may only offer more than a 4-year bachelor's degree without equivalent opportunity for laboratory work. The need for quality engineering professors who can instill a sense of importance and creativity in their students will continue to be strong in all levels of academia.

Teaching is probably the most noble of professions, where lessons learned from history are presented so that students do not make the same mistakes as their forefathers and the training up of a future society toward leadership has a lasting effect on affairs around the world. Structural engineering professors find employment in a variety of different schools, large or small, assuming positions of headship or simply blending in with the rest of the staff.

Advancements made in structural engineering begin with ideas of how to design or understand material behavior better, which are then put to testing in laboratories typically based on an educational campus. These facilities offer employment for directors, researchers, technicians, and other support staff to assist in fleshing out the truth in a hunch or hypothesis that may lead to a profound discovery. Facility staff members work closely with professors and graduate students in pursuing objectives.

## 3.2 HOW DOES AN ENGINEERING BUSINESS SURVIVE?

Engineers work closely with clients from all walks of life under a variety of conditions. Every decision made in the course of this interaction has an effect on future success of both employee and institution and must be made with care. Some choices are related to personal growth, where actions taken without careful thought can rob an individual of the ability to solve problems. Other decisions have a broader affect, possibly even to the destruction of a company's good reputation or financial stability. Survival of both individual and company depends primarily on effective management in two areas: Stress and risk.

*Stress* is often no more than a bad attitude. As an employee approaches the work day, choices need to be made on a variety of things, but the first choice should be regarding how problems will be handled, with a smile or with a scowl. Each problem should be viewed as an opportunity to learn and grow, as any issue is approached first by gathering information, then seeking assistance if it is needed. Sometimes it is important to simply consider how a career is valued: If it is a real pain just to make it to work every day, then perhaps you have chosen the wrong career path. Because so many different things can arise in a single day, and because engineers are responsible for public safety, this career should be one that instills pride and confidence in such a way that much stress during a period is managed.

*Risk* in every field within the construction industry is an unavoidable factor that cannot simply be ignored: Miscommunication between members of a design team can lead to costly repairs or change orders to the original contracted price for delivery of a constructed facility; structural engineers use a model code which is based on knowledge to date (subject to change due to new discoveries), use load approximations and methods of analysis based on assumptions (plane sections remain plane during flexure), and specify building materials that contain defects or physical damage. Business world survival does not mean the avoidance of risk or failure, but rather how these conditions are dealt with. A healthy acceptance of natural risk helps business managers to focus their efforts on controllable circumstances and, when things go as poorly as they possibly can, to learn from lessons that are always presented.

### 3.2.1 Management

The first line of survival is established by business managers who promote company philosophy and train their staff to work according to certain dictates. Company leaders work toward a common goal: Advancement. Everyone participates in this forward momentum and the management team, whether owners or supervisors, recognizes the contribution of employees on all levels to the success of the family. Knowing the value of each position within the company, a leader is able to properly motivate, encourage, and admonish those under their care through words as well as physical reward.

### 3.2.2 Employees

Though technical staff generates most of the revenue, an engineering company cannot survive without the help of support staff such as receptionists, financial managers, draftspersons, couriers, or interning students. A project is completed in a satisfactory manner only when communication with a client is maintained by all acting employees, bills are paid, invoices are honored, staff members are increasing in competence and growing to appreciate the business as a family, and tasks are appropriately delegated to allow workers to focus on respective areas of expertise. A thorough job description and employee manual will help formulate identity and remove confusion of duties that plagues disorganized companies.

Work in the profession of engineering is best when coming to the office every day is not a chore or headache, but rather an experience to look forward to. Management is partially responsible for fostering this type of friendly environment, but employees themselves also play a significant role in bringing this to realization. Problems of workplace safety or personnel must be made known to company principals or supervisors in a timely manner and all employees should have the freedom to suggest improvements in protocol or business matters. A company will only survive, however, as subordinates offer respect and honor to those in positions of management or ownership. A sense of family is the best environment to strive for, which can be difficult to attain because of obvious human intricacies, but not impossible.

### 3.2.3 Projects

A successful project is one that is well organized and understandable, even years after closeout, which requires the keeping of meticulous notes. Records that should be committed to storage on a project of any size often include the following: A comprehensive list of important contacts, including phone numbers and addresses; scope of services, including changes and fee adjustments related thereto; fee arrangement; construction costs, material deliveries, or recorded delays and reasons therefore; personnel scheduling, function, and cost; meeting and field inspection notes; and all telephone, fax, or e-mail correspondence. Storage of project information may be done electronically or in paper form, as long as it is easy to recall.

New projects often share common elements with existing ones and culling bits of data from historic records can save time and money. Always be aware of special conditions that will trigger additional funding or other requirements, such as when a project needs to be billed out according to prevailing wages (or some other financial schedule), whether other consultants need to be scheduled or contacted to complete another portion of the project, and be aware of time-saving steps that can be followed in the execution of project duties (i.e., computer spreadsheets, typical calculations or detail sheets, using less-expensive personnel or materials).

### 3.2.4 Procedures

The scope of work for which an engineer is hired to perform will certainly be project-specific, but there should be some consistency in type and cost of services offered to the same client, the same project type (based on owner, building occupancy, bridge type/style), and for similar levels of risk. Office procedures are difficult to streamline between multiple project managers who bring in work at different times with the knowledge of only few others. General company directives are easier to coordinate, such as deciding to refuse a certain type of work (for whatever reasons), though managers do need reminding from time to time.

All businesses have standard operating procedures for starting new projects and for closing them out. These usually involve recording the project's vital statistics (location, owner, client data), itemizing a scope of work and price or payment schedule that has been negotiated and agreed upon by all parties, and coordinating work with available staff. When a project is completed, generated documents, such as calculations and drawings, need to be properly stored for easy retrieval according to company protocol. An employee manual is the best place to store and manage procedures related to dress and behavior, but other workable documents should be available for the purpose of identifying and maintaining consistent quality in the product, such as drawing line weights and colors, general detail appearance, and drawing layout or numbering.

### 3.2.5 Communication

An ability to communicate clearly with colleague and client alike always leads to success, provided all parties are working toward this end. Some people have an uncanny ability to speak volumes without really saying anything at all, yet others are adept at getting to the point of a matter without being distracted from that path. Although every minute spent in communication should have some purpose, and brevity is often a virtue, there are times when many well-planned words are helpful. For example, when arguing for the validity of a controversial position (such as code interpretation), evidence should be presented from a variety of sources in great enough detail to answer objections or simply for the purpose of absolute clarity, which requires ample time.

Business leaders can take control of a conversation for the sake of guiding how others are to interpret what is being said, but this is a skill that only comes with experience. Some is learned through school team projects but the skill is best discovered and practiced through interaction with business associates and clients. In a meeting of different disciplines who are gathered to discuss the direction of a project, for example, it is not always the person who called the meeting that steps forward as the leader, but this position will often be assumed by a party that either has the greatest stake or simply the one who speaks loudly and frequently. These pseudoleaders muddy up the subject by introducing personal or unimportant concerns, or issues that are not on the table at that time, and clarity is sacrificed. A true leader can take appropriate control of a meeting and steer any conversation toward the initially stated goals. It is important to spend some time at the beginning of a meeting to simply observe whether the person who organized everyone together actually assumes a leadership position or if that person is derailed by some attending party. When multiple players dictate the direction of a meeting, chaos is following closely, and attending engineers must carefully record decisions affecting their scope of work and speak up to ask questions for confirmation of what has been heard.

Critical information that must be delivered to other parties cannot be done in laziness. The ease of sending impersonal messages via fax or e-mail, not knowing whether the recipient received and understood the message, can lead to trouble if the content is time sensitive. In 1995, one of the light towers erected in the Atlanta Olympic Stadium in Georgia collapsed, killing one worker and injuring another. A structural engineer discovered an error in the design almost 2 weeks before the collapse and sent a memorandum to the project manager with specifics of the discovery, but apparently did not follow up on the one-way message in an effort to avert possible tragic consequences. The Georgia State Board of Registration for Professional Engineers suspended this engineer's license after finding that the lack of urgency in his response to the discovery was negligent (Murphy, May 2000).

### 3.2.6 Qualifications

Design firms inevitably become involved in a bidding process to procure new projects for some types of work, but this is by no means universal and there are many owners who understand the value that structural engineers bring to a project. Qualifications-based hiring may sound like heed is not given to cost, but that is not the case: Part of what it takes to be a highly qualified design firm is the ability to deliver an economical product. The Brooks Act (H.R. 12807), which was ratified by the 92nd United States Congress on October 27, 1972, states: "The Congress hereby declares it to be the policy of the federal government to publicly announce all requirements for architectural and engineering services, and to negotiate contracts for architectural and engineering services on the basis of demonstrated competence and qualification for the type of professional services required and at fair and reasonable prices."

This promise, however, does not always follow in the process of selecting a contractor to build structures that are funded by public money, where the winning bid is typically the lowest responsible bid and it cannot be rejected unless there are legal problems with the bond or other paperwork. That company cannot be rejected on the basis of subjective opinion regarding competence unless gross misrepresentation of a company's qualifications, history, or ability to perform the work can be proven, which is often difficult to do. Most state- and local-government building, planning, and engineering departments have also adopted these requirements for projects using public finance, sometimes entitled "The Little Brooks Act" when published in a state's business and professions code.

## 3.3 CLIENTS AND CONSULTANTS

Clients that seek engineering services include government agencies, school or hospital districts, industrial plant directors, other design professionals, developers, and ordinary people from all walks of life. Many times, someone enters the office for the purpose of retaining engineering services because they were sent by the building department or recommended by another colleague and that person either does not know the importance of a structural engineer's work or simply does not care. Such is public perception, thanks to a lack of volunteers in the profession attempting to spread good news and a good name. However, that client who respects the engineer as an incredibly valuable asset to a project, and not simply a name on a piece of paper or a legal shield to hide behind, is worthy of pursuit and high regard.

Depending on the scope of a project, expertise needed to get a project completed may require the help of a trusted consultant. For example, a simple structural project like a metal-framed, cloth-covered canopy over the entrance to a building may require consulting work done by an electrical engineer to design and detail a lighting system pursuant to the owner's request. A single project can involve numerous consultants who may not have chosen to be on the same team, hired at separate times by the owner. Design firms in any given area are typically familiar with one another's work ethic, business philosophy, and reputation, and if they are to be part of the design team it is important to factor expense for time and frustration into the scope of services if warranted. All participants will have other projects going on at the same time and these must be properly managed according to priority and resources.

Though the realm of financial compensation is often delegated to an office manager or business partner, engineers within all company levels should understand that business is still business and the company will only survive on positive collections of money owed. Different strategies have been employed in the past to assure payment of fees, including the imposition of a retainer (money collected from the client before work is begun, held as final payment due for work completed) and refusal to issue work until all bills have been paid, including interest for outstanding invoices beyond a time period prescribed in a written contract.

## 3.4 ENGINEERING SERVICES

A proposal for professional engineering services is the first step in negotiating an acceptable agreement between two parties. A structural engineering consultant, for example, will explain to a project architect who has expressed a need for services precisely what is understood about scope of work (including what will and will not be done), limitations, compensation, or when a subconsultant will be needed. Although documentation of the project is fairly limited in the beginning stages, the outlining scope of a structural engineer's services is fairly simple: Provide design drawings, calculations, and specifications necessary to obtain a building permit and to adequately define the project to a contractor. Bidding and construction phase services, where the engineer assists the architect with questions or problems, may be negotiated within the same fee or under a different umbrella depending on the complexity of the project or the requirements of state and local codes. During the creation of a proposal, an engineer should give due consideration to time-saving tasks, such as the generation of typical calculations or detail sheets, in a streamlining effort to account for unknown factors that always seem to come up.

Limitations realized at the beginning of a project are set deadlines, availability of information about the site, knowledge of the behavior patterns of other consultants (including the architect), internal resources of staff and tools, and volume of current projects and their individual deadlines. Time often does equal money on a construction project, so milestone dates cannot be taken lightly. If the architect hasn't provided at least a preliminary schedule to the consultants during the proposal stage, it must be requested to properly consider the risks that might be involved. The engineer is to base a design using site specific data and to request any reports or evaluations, such as geotechnical or other physical conditions, to justify placement of the proposed structure.

## 3.4.1 Contracts

A written contract for engineering services is not only required by law in many states, it is also a powerful means of risk management. A properly executed document defines expectations, rights and obligations, indemnities and protections, and mechanisms available to accommodate changes for both parties. Architectural and engineering societies produce contracts that have been used successfully by companies for years, and they have also had enough time and exposure in courts to be thoroughly interpreted and understood. Even the most basic contract can be used many times for a variety of applications, provided the terms include consistent measures of protection.

An engineering firm's professional liability insurance carrier will often create or supply the standard contract to be used in all work. Sometimes terminology can kill a deal because one party asks the other party to assume more risk than is commonly attributed, needed, or possible. A design professional can be held liable for damages caused by their own errors and omissions (negligence), but *hold-harmless* clauses may dictate responsibility for more than that, including site safety and many other issues that the engineer has no control over. One party should not be allowed to escape liability due to their own negligence at the expense of a different party and responsibility for particular situations should always be left in the hands of the one best able to control them.

Contract negotiations may involve revisiting a defined scope of work. This is to be handled carefully with patience, confidence, and a commitment to fairness for all parties involved. Patience is necessary to review work from a new perspective, one that may not have been realized when the scope of services was originally thought through. Confidence serves to protect work that has already been completed, since a previously defined scope of work should not be completely abandoned. An initial expectation of what is necessary, when carefully thought out, will be very close to reality and only minor additions or deletions should be necessary. If a scope changes dramatically, then there was either a fundamental misunderstanding of what the participant was being hired to do or deleted functions are being delegated to a different party. The quality of fairness is not ambiguous because everyone deserves to be compensated in relation to risk assumed, level of participation, and quality of expertise. If something is added to a list of duties, there should be an equivalent adjustment to the fee.

Owners need to understand that an engineer is hired specifically for measurable competence, reputation for quality, experience, ability to solve problems, trustworthiness, or other service-oriented factor, not to *guarantee* the final result. There are too many other players, from multiple subcontractors to building inspectors, and professional liability insurance carriers typically will not cover guarantee claims outside of errors and omissions. A lawyer does not guarantee a win in the cases that he tries, but rather explains the risks involved in a legal proceeding so a client better understands the reasons for hiring an attorney in the first place. Likewise, a physician will sometimes present an informed-consent agreement before a complicated or serious medical procedure to make sure a patient fully recognizes that the doctor cannot guarantee success and that there are well-known risks or side effects that may arise, even if the physician performs to the best ability possible.

An engineer should not be expected to guarantee the performance of a building or bridge, or that a project will not run over the intended budget or schedule. The only guarantee a design professional truly offers is that services will not be performed in a negligent manner that either may cause or at least contribute to future damage or loss within a period of time after construction, as prescribed by law.

## 3.4.2 Scope of Services

In order to keep track of work, it is helpful to arrange promised services in categories such as contract administration, design, bidding, and construction. The work of needed specialty consultants, including provision for the cost of markup on a subconsultant's work or reproducible elements related to contracted duties, is also to be included and properly defined after discussion with staff that will be providing these services.

An individual's scope of work fits within the details of an assignment, but remains distinct from a fellow employee's tasks. A company hired to perform structural engineering services for a public high school building, for example, is best prepared to complete the work as individual tasks are assigned to different employees depending on size, complexity, and schedule. The project manager must completely understand what the company is responsible to complete and every employee within must also have a clear understanding of their own scope of work. The person serving as structural engineer-of-record begins distributing duties with himself, possibly handling the lateral force resisting system of the building, whereas an engineer-in-training may partially or completely design the gravity load-bearing components. Another engineer within the company may be assigned equipment anchorage design, and yet another is given the task of assembling typical construction notes and details. A lead draftsperson is responsible for making sure plans are completed to the satisfaction of the structural engineer-of-record, but responsibilities can also be delegated to other draftspersons.

The overall scope of the company's work must be carefully coordinated with the architect-of-record and other consultants and questions must be asked: One of the worst mistakes a project manager can make is to assume something is being taken care of without evidence to prove that assumption true. Some "forgotten" design responsibilities that can fall into the lap of a project structural engineer, as a result of unclear communication or expectation, include:

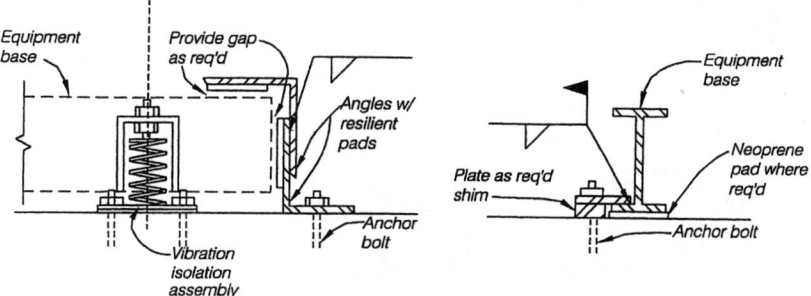

**Figure 3-4** Common methods for anchoring mechanical equipment. (*Courtesy U.S. Army Corps of Engineers*)

1. *Equipment anchorage.* Often specified by the mechanical engineer or a specialty contractor, the system selected and detailed might not be best for the project. There may be other types of anchors or manufactured products specified elsewhere for the building and it is important to coordinate all similar products in order to avoid confusion and to ease the contractor's work in assembling a budget. Figure 3-4 shows anchorage systems that have been used successfully for different types of equipment, where it is clear that anchorage considerations do not only cover bolting, but associated plates, stiffeners, clearances, and weather protection. Equipment may include air conditioning units, electrical transformers, chillers, and a whole host of others, even large satellite dishes or control buildings.

2. *Window framing mullions.* Glass, glazing, and storefronts resist wind (exterior) and seismic (interior and exterior) forces perpendicular to their plane and these forces must be transferred to the building's main lateral force resisting system. Figure 3-5 shows framing at the interior of a building, into which a glass panel will be placed, presenting a hazard if not properly anchored (with a complete load path) against the effects of a large earthquake. Glazing systems produced by reputable manufacturers are usually delivered to a project with anchorage specifications to the structure itself, but the building designer is responsible for completing the loadpath of gravity and lateral forces to the diaphragms and foundation. The architect is not in the best position to handle this duty, but a requirement for the structural engineer's assistance is not always verbally communicated, and someone winds up with a surprise at some future time.

3. *Anchorage of miscellaneous architectural features, including cladding.* Different factors may come into play to bring design or anchorage requirements of various premanufactured elements into the scope of services to perform. Enforcement agencies review project documents for evidence to suggest that someone has coordinated submittal of all items that have anything remotely related to ability to support structural load. A manufacturer supplies details to indicate a suggested method of attachment to the building, but the structural

**Figure 3-5** Indoor mullion and steel framing with block out for glass panel, requiring consideration of solid anchorage against seismic shaking.

engineer of record must review and approve them. Sometimes there may be a difference of opinion regarding the suitability of details provided by others and a new solution suggested; sometimes that solution calls for justification by appropriate details and calculations, but responsibility for the work is not immediately clear. If anything is provided at all in terms of anchorage, a manufacturer stops at the building surface and the building designer completes the loadpath.

Guardrail ornamentation, such as that shown in Fig. 3-6, is another example of a nonstructural element that often requires a structural engineer's review, as the anchorage details proposed by the manufacturer may not be suitable for the base provided. Loads that these elements impose onto the building, including weight, wind, seismic, or code-dictated live load, may necessitate a special system built into the main structure that the manufacturer does not typically design. Careful coordination is important, whether required by the enforcement agency or not.

4. *Staircases, elevator support and other means of circulation.* Because stairs are often embellished with artistic beauty and imagination, they are sometimes avoided by engineers for fear of treading where they are not wanted. However,

**Figure 3-6** Architectural railing system in need of structural anchorage against code-required horizontal forces, New Orleans.

sometimes an architect assumes that stair construction is being detailed on the structural consultant's plans and the scope of work must be very clear on this. It is not uncommon for an engineer to assume that the architect has a particular nonstructural design element handled, when in fact the architect is either not planning to address it or is expecting to receive something from the structural engineer.

The stability of elevators is dependent on support provided by the building itself and a manufacturer typically relies on a structural engineer to design guide rail supports or hoisting beams that will accommodate the system. It is also common for a model building code to require consideration of the effects of earthquake motion on elevator cars, guide rails, and their anchorage.

5. *Penetrations for services.* Footings occasionally bridge across large, compacted trenches where piping has entered the building, but the elevation of intruding elements must be confirmed with footing heights. If placements coincide, extra reinforcing of the foundation system may be required. Coordination of structural elements must be made with anticipated placement of services. Mechanical drawings indicate the layout of ductwork, but what is shown on the plans does not always match the contractor's placement due to interfering elements or other reason. The structural drawings must include strengthening details to cover a variety of scenarios in order to avoid costly errors during construction because something was not immediately clear.

### 3.4.3 Estimating Your Work: Time and Cost

Regardless of fee arrangement, all employees must keep careful track of time. It is not uncommon to work on several projects in a single day and the work needs to be properly billed. A contract that is negotiated on a *time-and-materials* basis may be less of a burden to manage, but will be an unpleasant surprise to the client if it is not clear what to expect. Project managers need to work with an accurate estimate of labor and resource expenditures to create profitability. An efficient way to approximate time and cost anticipated for engineering services is to break duties down into small portions and associate individual commitments with each. Afterwards, add the total, increase by a certain percentage depending on the type of work (or client), and be sure to introduce a fixed or reimbursable sum for cost of copies.

A multitude of different fee arrangements can be stipulated within a business contract. A *cost-plus-fee* agreement includes compensation for direct and indirect costs associated with services rendered (i.e., reproducibles, shipping, delivery), in addition to a paid fee for those services either on an hourly basis or as a percentage of construction cost.

### 3.4.4 Estimating Your Worth: Fair Compensation

Fair compensation can be somewhat elusive when business is slow and luxury of selecting clients is not available. Structural engineers perform critical duties, but this importance can be quickly hidden by competition, laziness, disrespect, or a misunderstanding of the value that they bring to a project. If the negotiated fee is too low, work can become tedious and the danger of cutting corners in an effort to stay within budget changes attitudes toward everyone else involved.

Part of estimating the worth of engineering services involves determining the constructed cost of a facility. The International Code Council (ICC) does an excellent job of keeping the public current on national figures that can be used to place an average dollar value on a permitted project for the purpose of calculating code-related fees. Currently, 6 months is determined to be a reasonable interval between issuance of new figures, likely because valuations used for the purposes stated are not necessarily affected by a project's bidding results or volatility of the market. Valuation is based on both occupancy group and type of construction multiplied by a building's square footage, including all elements and systems to be permanently installed prior to occupancy of a permitted structure, including structural, mechanical, electrical, and architectural systems.

As far as a client is concerned, an engineer's worth will be directly related to the satisfaction of needs or emotions. The less an owner needs to worry about during the course of a project, the more value is placed on services of those who make life easier. Do not extol personal worth, but help the client decide by persuading him to recognize that you understand his needs and have the resources to accomplish projected goals.

## 3.5 CRISIS MANAGEMENT

It is important to understand that a disagreement between two or more parties does not always involve conflict. A disagreement can be regarded as a simple difference of opinion over a matter to which neither of the persons involved feels as though they are being attacked or ridiculed into abandoning their position. Because of individual tastes and preferences, human beings are prone to disagree with one another: It is a natural part of our species. Conflict, on the other hand, involves a more militaristic approach to problem solving and is created from a sense of insecurity or entrapment felt by a participant who is in a subjugated position, such as an engineer trying to persuade a government official that a completed design meets the requirements of adopted codes. Resolution begins at a point before a conflict can even ignite, where one or all parties involved in disagreement behave or interact in a conscious manner that will avoid unnecessary escalation.

### 3.5.1 Philosophy of Conflict Resolution

A professional engineer deals with disagreement over a variety of issues, ranging from simple matters of company procedure to design failure and allegations of incompetence. These are defining moments: Will a responsible person give up under pressure or will he cut through the frustration and strive to bring equitable solutions to the table? Conflict is never fun, no one seeks it out, but a healthy attitude toward troublesome situations will help those involved to learn valuable lessons that will be remembered with greater vividness than morsels picked up at a continuing education seminar. A healthy approach to unexpected conflict is one that builds maturity, endurance, and an appreciation of the truth and service to our fellow person.

Therefore, all situations that lead to conflict must be approached with appropriate respect for the problem itself and for those involved. Respect involves a conscious decision to avoid lashing out at an opposing party with unfounded accusations, or even to deliver truth from a seat of anger or resentment. Begin with the assumption (or expectation) that an opposing party is worthy of respect and honor. Speak gently, not out of anger or frustration. An acknowledgement of respect immediately leads to an attitude of humility, regardless of the positions of those involved. Everyone is entitled to their opinion and it makes an impression to admit that your own ideas or positions may be completely wrong or insufficient, even if you are the boss, supervisor, or plans examiner. In fact, a truly humble person is not prideful, arrogant, or blindly assertive. Even when stating positions that are firmly based on indisputable fact, dialogue is carried through with kindness, patience, and an unswerving (gentle, but firm) commitment to the truth. If a disagreement is first approached with an eye toward humility, it will be easier to admit that you are wrong because you've already established the fact that this outcome is possible.

Approaching a disagreement with respect and humility may not be enough to avoid conflict if the persons involved cannot (or will not) see the problem from another person's perspective. Reasons for this are numerous and can only be partially discovered due to the complexity of human behavior. The ability to coexist in the midst of differing cultures, temperaments, and loyalties is one of humanity's greatest qualities, but it is also responsible for a lion's share of conflict. A person's attitude and opinions certainly have a basis in upbringing or culture, but one cannot completely understand another person's history. An engineering education is completed with some study of the humanities and other nontechnical subjects which serve to form a student's general approach to problem solving and opinion of worldly matters, but not in total (and not usually enough). Religious affiliation also shapes a person's perspective. In fact, there are so many things that contribute to one's world view it is impossible to completely understand someone else's perspective. Understand it to the best of your ability and make up the difference with a clear position on the true facts of the matter in dispute. Ask questions to gain as much knowledge about the other positions as possible, in order to put everything on the table. All valid opinions are to be given consideration and each participant should strive to see things from other perspectives, regardless of how clearly reasons can be articulated.

The final position within disagreement or conflict must be taken by a firm commitment to the truth. Fortunately, engineering is based on science: Theories are tested and retested in a laboratory; physical phenomena are easily observed in the field; observations are made as to what happened to a structure after a major environmental event. Societies established for the advancement of knowledge within the field of engineering publish findings, standards, and opinions that bring cohesiveness to information that can be referenced in support of a certain opinion or perspective. Opinions that cannot be supported by existing knowledge must be tested for validity—that's how science works. A hypothesis leads to a theory, which leads to established fact under careful procedure and protocol. The truth of a particular matter must be established by solid factual data and can be introduced into a discussion at any time, which may take place during the stage of disagreement or interjected in a more formal way when things have turned into conflict. Gather as much information as you can to establish the truth of a matter, especially the truth of what you are actually hearing from the other parties: did you understand what they actually said or did you hear them wrong?

Once a disagreement has escalated to conflict, it may be necessary to take a step back from everything and determine what caused things to sour. Was it something you said and can immediately correct, or was it something another person said that might require gentle correction or firm confrontation? Some claims must be confronted if they are blatantly false or demeaning so that they do not fester in the minds of participants, casting a fog over the entire situation. No one benefits from lies, deceit, laziness, or negligence. Do not hope that a conflict will resolve itself because it won't. Be proactive in working toward an atmosphere of mutual respect and understanding.

## 3.5.2 Working with Difficult People

Though we may want to live at peace with everyone around us, it is not always easy to do so. The qualities of humanity that are so cherished—different cultures, styles of humor, personalities, levels of education—can also bring down the most cheerful of hearts. Most of the time, one discovers that the best way to deal with difficult people is to first search for issues or problems, then to honestly evaluate elements causing the difficulty. Many times a "difficult person" is no more difficult to get along with than a cherished family member if the causes of strife or discomfort can be properly discovered and personally or cooperatively dealt with. Some suggestions that will help bring about a more congenial relationship with anyone include the following:

1. *Do not get hung up on minor offenses.* Others may be impatient, rude, or combative toward you for a variety of reasons. If there is no specific dispute causing this behavior that needs to be resolved, do not hold the uncomfortable reaction as a reason to end a relationship.

2. *Check your own attitude.* Your impatience or anger toward another can stir up conflict and close doors to peaceful coexistence. Open your mind to differences of opinion, strange habits, and unexpected reactions with humility and patience. Always ask yourself, "Am I treating this person in the way that I would like to be treated?"

3. *Always respect authority and cling to the truth.* Some of the most difficult people to get along with are those in positions of authority who do not know how to handle or administer such power. Arbitrary decisions and accusations are frustrating, but you must offer respect to those in higher positions without compromising the truth. Know your place within the hierarchy of a company and all of society.

4. *Seek help if communication is constantly being misunderstood.* Sometimes two different personalities come together like oil and water, requiring another person to help sort out what is causing the difficulty. Sometimes it is best to avoid meaningful contact with a difficult person if business can be handled by others.

5. *Ask forgiveness for your own contributions to the difficulty.* Recognize problems that you bring to the table yourself, freely admit them, and work to eliminate them from the equation. Most people find it hard to admit when they are wrong, but this action goes a long way toward smoothing out the bumps in any relationship.

Sometimes difficulty may arise from an employee's inability to perform necessary functions. A short period of probation does not always reveal whether this will be a problem, as delegated work is appropriate to the skills a new employee brings to the table. Engineering is a profession that requires knowledge of how to apply building blocks collected at the university and it cannot be learned overnight or in a matter of several weeks. It takes months for even the brightest student to learn nuances of navigating through codes and standards, which include an important ability to select and apply appropriate sections while passing over others.

Evaluation of an employee's ability to pick up new concepts and to mature in the profession is often subjective and difficult to administer, but there are a few procedures that can help make this determination reasonable and productive:

1. *Make sure the employee understands and accepts what is expected.* Growth in the profession, or movement into more advanced abilities and new challenges, should be communicated as terms of employment. An employee shows acceptance through an expressed desire toward growth, commitment to scheduled presence in the office (arriving and leaving on time), offering suggested seminars, or opportunities for further training, and asking technical or business-related questions.

2. *Show an employee the attitude or areas of work that need improvement.* Behavioral problems should be monitored and disciplined immediately—they cannot be allowed to continue without warning because of potential disruption to company ambiance or client perception. Technical performance expectations should not be presented from a vague list of ideas, but only what is pertinent to work at hand in such a way that an employee can see the necessity.

3. *Provide appropriate training and support to correct deficiencies.* Focus attention on specific areas of improvement, such as writing ability or work with a particular building material, and demonstrate how the work is to be done correctly. Supervisors should have some awareness of subjects that keep coming up for correction in case something is not settling in, which is the source of difficulty. Close scrutiny should also be paid to training methods, as perhaps an employee does not understand something because the company is not teaching it effectively.

4. *Determine whether other avenues are available to help overcome or compensate for an employee's limitations.* Sometimes a simple change of environment, such as relocating the employee closer to a supervisor or into an office with a window, can provide the needed boost toward competence. Only assign work that can be completed with reasonable assurance of accuracy. Provide more advanced work, or portions of projects, only when an employee is ready, then line out all necessary tasks in specific order and demonstrate how each can be completed, either through reference or example problems.

5. *Give sufficient time for an employee to correct or show improvement.* Probation trials of employment may be increased sequentially, but they do not usually extend beyond a 12-month period. When a worker is not growing in one area of expertise, there are usually other areas which show promise, and it may only be a matter of time and patience for other skills to develop.

### 3.5.3 Legal Means of Resolution

Different ways of settling conflict have been used for many years and decisions that result from each one are legally binding if all parties enter into a contractual agreement to accept the conclusion of the matter. When there is no independent party to help each side work through the issues or to render decisions, the process is called *negotiation*. Because of the simplicity of this method, it may be chosen first to try to bring about a solution to the problem. It is typically faster, less expensive, and offers a greater level of privacy than means involving outside assistance. Each party will often be supported by their own legal council or material experts to try to persuade the others, and occasionally stronger voices will dominate the meetings as well as the terms. It is important for everyone involved to strive for equity, fairness, and patient treatment of the issues.

If those involved cannot reach an agreement, the next stage may involve *mediation* or *arbitration*, which involves the direct assistance of an independent party. A mediator's primary duty is to make certain that all quarreling parties are communicating clearly and that they understand all points being presented. Advice is offered and different solutions to the problem may be explained, but no one is legally obligated to accept the mediator's suggestions. This person may be a professional in such matters or a mutually respected friend or colleague, whose patient intervention is critical in calming nerves and tempers. Mediation still has the possibility of weighing in favor of the stronger party, but it offers better control than simple negotiation.

Sensitive issues require *legally binding arbitration* in which a neutral party is chosen by those involved in the dispute to sit in a position similar to that of a judge, where that person does not work to improve or clarify a party's argument but rather listens to all of the evidence presented and renders a decision. An arbiter may be a legal professional, like a judge or an attorney, or another individual who has been properly trained. The process still offers better privacy than formal litigation, but may not follow typical rules and procedures that can be critical for maintaining order in a court of law. A party that disagrees with the decision of the arbiter is given opportunity to appeal under certain circumstances as allowed by the state in which the matter is being considered, but options and venues for such an appeal are limited.

In some cases, it may be necessary to define lien rights and invoke them to be paid for services rendered. A *lien* is basically a claim against property and can take many forms. A trucking company, for example, has a lien on goods they are carrying and can refuse to surrender them until shipping costs are paid. When imposed on real estate, a lienholder has a right only to the value of the property, not to possession of it. *Mechanic's Liens* are created by statute and have different applications in each state, but they essentially represent a right to secure payment for work or materials used in the improvement of real property. Most states allow an engineering company to file a Mechanic's Lien against property for which construction documents were prepared (other than public property) after the process of making improvements has begun in order to receive payment for their work, but there is often a period of time before which claims need to be filed.

## 3.5.4 Litigation

Judges, juries, attorneys, and well-established court procedures define the process of *litigation*, the most public of all methods of conflict resolution, whereby extensive records of the event are kept so that a proper appeal can be filed if necessary. They are easily more adversarial as each side has far more to lose when a decision is rendered. Though the decision may follow established precedent, less flexibility is offered in the solution and a court is not at liberty to attempt to clarify misunderstanding between the parties, resulting in frustration and bitterness. It is costly to administer, but the court has the power and the means to enforce resulting decisions, which may have great value under certain circumstances.

This process of settling disputes before a court of law can easily become an emotionally-charged experience, though there are steps any party can take to help ease the burden. Having a well-defined goal and plan of action is one of these important steps, rather than simply charging forward with a disruptive, vindictive spirit: What are the real interests at stake and how can they best be met? To help keep a financial burden from fueling spent emotions, attorneys may agree to handle some cases on a contingency basis, which means legal fees are not paid unless (and until) they win that case, but this is a practice generally limited to accident or personal injury litigation. A lawyer will not only consider cost and fees in deciding the best course of action, but will also consider available evidence of a case, the need for an ongoing relationship between parties, available resources of both sides, time the case will occupy, and which court will hear the arguments. This careful consideration of all aspects of litigation can help each party to make it through and return to their business practice a little bit wiser.

# 4 Building Projects

In its most basic form, a building can be described as an assembly of construction materials that are fastened to one another, as well as the ground itself, in such a way as to provide shelter for living things, machinery, possessions, or other elements. A building may be fully enclosed to keep out the effects of the environment (temperature fluctuations, snow, rain, wind) and should be provided with a well-defined system for resisting lateral forces from hurricanes, earthquakes, floods, or soil pressure. The form it takes is limited only by the imagination, which is directed through principles found in regulatory codes and standards, and its function may be that of an elegant residence (see Fig. 4-1) or an important place of business.

## 4.1 BUILDING SYSTEMS

All building systems and functions are regulated by locally enforced codes adopted by state legislatures. The general scope of the major model building codes, including the International Code Council's 2006 *International Building Code* (IBC) and National Fire Protection Association's *Building and Construction Safety Code* (NFPA 5000), is described in the following terms:

1. Code provisions apply to the construction, alteration, movement, enlargement, replacement, repair, use (operation) and occupancy, location, maintenance, removal and demolition of every regulated building or structure.
2. The main purpose of the code is to establish minimum requirements to safeguard the public health, safety, and general welfare through structural strength, means of egress facilities, stability, sanitation, adequate light and ventilation, energy conservation, and safety to life and property from fire and other hazards.

**Figure 4-1**  Oak Alley plantation residence, Louisiana.

A building designer must correlate an owner's needs with restrictions associated by the intended use. This requires consideration of maximum area and height based on a building's occupancy or use, type of construction material (combustible or noncombustible), setback from property lines through codes and zoning ordinances, quantity and character of hazardous material storage, fire protected separation between interior spaces, special considerations within the means of egress system, accessibility for physically disabled persons through adoption of model codes and federal Americans with Disabilities Act of 1990, and other identified conditions. In fact, model building codes may be thought of as dictating two primary characteristics of all structures used by humans: Structural adequacy, confinement of damage that may occur during a disastrous event, and safety of occupants as they exit during an emergency. Occupancies are generally classified for uses associated with assembly, conducting of business, education, industry, sale of goods, special hazards, residential, and storage or other miscellaneous needs.

There are a variety of resources available in the construction industry encouraging a whole-building approach to design, including the Leadership in Energy and Environmental Design (LEED) Green Building Rating System. LEED promotes a design approach which focuses on human and environmental health in terms of the sustainability of a site's development, efficient use of energy and water supplies, inclusion of durable and environmentally friendly building materials, and maintenance of a high-quality indoor environment. The rating system is intended to give industry professionals a consistent set of standards for what a *green building* should look like. It is defined and continuously developed through a principle of consensus, much like the pattern of other building standards, and architects monitor different projects to see if subject buildings could be made to qualify.

### 4.1.1 Structural

Building codes define different types of construction based on a level of fire protection needed and some of these forms prohibit the use of combustible framing, of which the structural engineer must be aware. NFPA 5000 defines *combustible* construction as that which ignites and burns (Section 3.3.388.1) and commonly includes wood framing. A *limited-combustible* material is defined as one having a potential heat value not exceeding 3500 BTU/lb (8141 KJ/kg) when tested per NFPA 259 (NFPA 5000, Section 3.3.388.9), and a *noncombustible* material is one that will not ignite, burn, support combustion, or release flammable vapors when subjected to fire and heat (NFPA 5000, Section 3.3.388.11).

Construction types allowed and defined within the model codes discussed in this chapter have different restrictions on the use of combustible materials, fire-resistive assembly construction requirements, and allowed storage of hazardous materials. A structural engineer needs to be aware of the classification an architect seeks (or requires) in order to be certain to design with the right material. It would be a shame to work on a wood-frame system and realize at some later time that only noncombustible or limited-combustible materials were allowed (or desired) for the project.

**Vertical Load Resisting Systems**  Roof framing systems support the covering and overall drainage layout, being of lightweight or heavy materials, and include insulation, membranes, and associated piping. Mechanical and other equipment, as well as special appurtenances like antenna-supporting towers or satellite dishes, provide special challenges that are to be incorporated into the structural support system. Live loads required in model building codes are usually stipulated to cover construction loading and reroofs or other future work.

Floor framing systems must not only be chosen to support the weight of materials and live load imposed by a building's use, but also to handle deflection and vibration when those sources are present, such as for aerobics or dancing events. Live loads tabulated in model building codes are based on historical information and have actually been under development for a long time. ASCE/SEI 7-05 was originally created in 1924 by Herbert Hoover's Building Code Committee, then entitled *Minimum Live Loads Allowable for Use in Design of Buildings*.

**Lateral Force Resisting Systems**  A building's lateral force resisting system defines behavior during an extreme event, such as a hurricane or an earthquake, and provides for general stability and interaction with site soils below. Any configuration may be selected as long as sound engineering principles, coupled with appropriate testing data, are used to justify needed ductility and strength (ASCE/SEI 7-05, Section 12.2.1). The most common structural systems are described in Table 12.2-1 of ASCE/SEI 7-05 and volumes of data are available to substantiate good performance, provided members and connections are detailed as described in other areas of the code. These systems include *bearing walls* (those which provide primary support of gravity load in addition to resisting lateral forces), *building frames* (gravity loads are carried primarily by a system of beams and columns, whereas lateral forces are absorbed by shear walls or braced frames), *moment-resisting frames* (providing support of both gravity and lateral loads), *dual systems* (frames, shear walls), *cantilevered columns*, and even steel systems that are not specifically detailed for seismic resistance.

Though these systems are described in the seismic section of the standard, they also provide resistance to wind and other forces that will cause the structure to rack, overturn, or slide in relation to the ground below. The importance of having an adequate lateral force resisting system can be seen in Fig. 4-2, as prevention of building collapse prior to escape of occupants during a significant loading event is commonly accepted as an absolute minimum level of performance.

**Joints** Construction joints are built into a structure to accommodate changes in length, volume, and planar alignment because of a number of conditions that may occur during service or construction. The decision to include joints in a building's design is based on aesthetics, where cracks in brittle materials would be unsightly;

**Figure 4-2** Residence damaged in the Northridge Earthquake, California 1994. (*Courtesy UCLA Department of Earth and Space Sciences*)

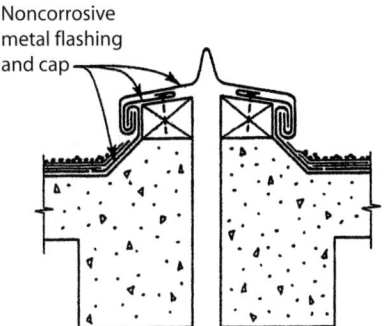

**Figure 4-3** Seismic joint and flashing in building system, requiring notation and detailing on project drawings. (*From Gaylord, Structural Engineering Handbook, McGraw-Hill, 1990*)

weatherproofing, where uncontrolled cracks in exterior walls allow moisture to enter the building envelope; structural relief, where expansion of adjacent materials would impose undesirable internal forces upon building elements and lead to damage; or seismic movement for adequate performance (see Fig. 4-3). Not all buildings require construction or expansion joints and the need for such should be properly coordinated with the architect and owner.

Coefficients of thermal expansion for different materials are tabulated in ASTM C1193, which also includes procedures for calculating the required joint width and range of potential movement that can impose forces significant enough to cause cracking in building elements. In cold regions, building materials will contract with falling temperature and enlarge the joint width, whereas hotter climates cause member expansion as temperature rises. A good, waterproofed joint includes a priming material, backer rod, bond breaker, and sealant. Masonry, concrete, metal, and wood surfaces require a primer to improve adhesion of the sealant, depending on limitations imposed by the manufacturer. A backer rod is installed at a certain depth to control thickness of the sealant, in addition to serving as a secondary waterproofing element. Finally, a sealant that is allowed to flex with structural movement, yet remain intact, is placed to fill the joint.

## 4.1.2 Plumbing

The *2006 International Plumbing Code*, *International Fuel Gas Code*, and *International Private Sewage Disposal Code* are used in the design and installation of piping systems and related fixtures for use in buildings and the surrounding site (2006 IBC, Sections 101.4.2, 101.4.4, and 101.4.5). When codes produced by the NFPA are adopted, plumbing systems are designed according to the *National Fuel Gas Code* (NFPA 54) and the *2003 Uniform Plumbing Code*. Most piping systems are designed by the use of simple tables within these codes considering a variety of influencing factors. For example, gas piping is sized according to system pressure, allowable pressure drop, maximum gas demand, and the specific gravity of the fuel gas itself.

Water main service piping (supply for public use, owned by a municipality or public authority) and distribution piping (conveying water from service to fixtures within a building) may also be determined conventionally based on a knowledge of pressure range, meter size, development length of piping, and quantity of fixtures to be serviced. Water piping is constructed of acrylonitrile butadiene styrene plastic (not commonly used, but approved), asbestos-cement (predominantly used for street mains, where asbestos fibers are bonded and not considered a health hazard), brass, galvanized or stainless steel, cast or ductile iron, polyvinyl chloride plastic, chlorinated polyvinyl chloride, copper or copper-alloy, or cross-linked polyethylene plastic.

Other plumbing that is commonly found in a building consists of sewage disposal, water drainage systems, fire suppression, and ventilation pipe runs in order to remove odors. Fixtures are provided to supply and collect water or waste, and those for use in waste lines must include a liquid-seal trap designed such that the passage of air and gases is prevented during the flow of liquid. Figure 4-4 shows a basic wastewater removal plumbing system that is common in buildings.

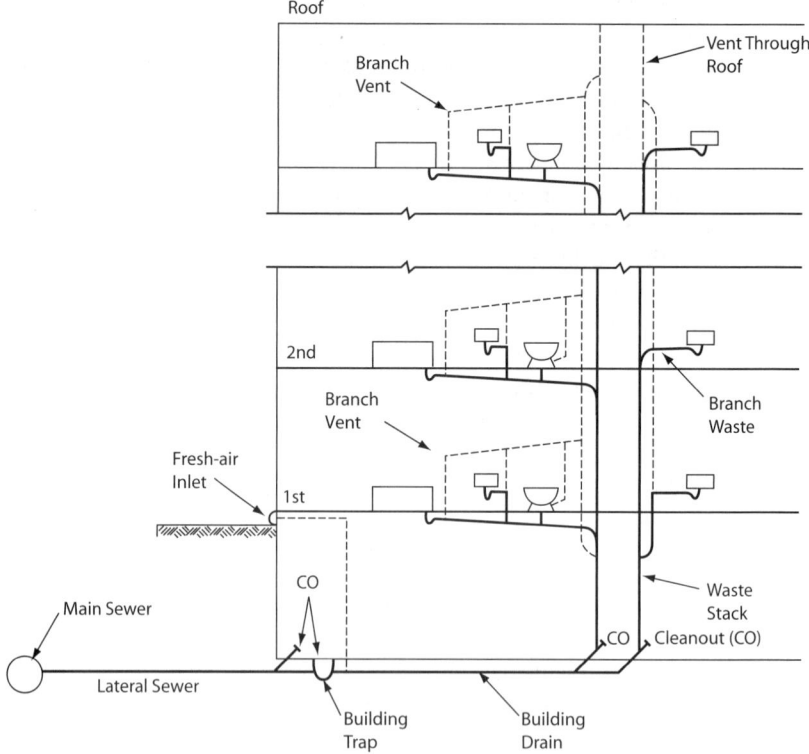

**Figure 4-4** Common commercial building waste plumbing layout, identifying simple fixtures and pipe runs.

### 4.1.3 Mechanical

Section 101.4.3 of the 2006 IBC requires work related to equipment and appurtenances used for ventilating, cooling, heating, air conditioning, refrigeration, incineration, or other energy-related system to be in conformance with all requirements of the *2006 International Mechanical Code*. Many different codes are also available through NFPA's collection when adopted. Equipment may be installed within a structure, anchored to a foundation at grade, or fastened to a roof and the design of structural elements needs to take vibration effects into account where they occur. A platform is often created for a unit to sit flush atop a sloping surface, though convenient premanufactured curb systems are also available for most installations. Lateral bracing of ducts and piping should be considered in medium and high Seismic Design Category (SDC) regions as defined by ASCE/SEI 7-05.

Heating, ventilating, and air-conditioning systems (HVAC) are critical to the health and comfort of a building's occupants. The work of air conditioners is to move heat from the interior environment to the outside and requires the use of air handlers, compressors, evaporators, condensers, pumps, and cooling towers (for large systems). Each component has its own weight and center of gravity that dictate how the mechanical system will be anchored to the structure against wind and seismic forces. Larger heating systems require heat pumps, boilers, furnaces, heating elements, and forced-air equipment (where applies) to heat water, steam, or air and circulate warmth to all parts of a building. Different regulations and installation guidelines, as well as construction protocols, are published by the Sheet Metal and Air Conditioning Contractor's National Association (SMACNA). A common term used in the air conditioning industry to define the cooling rate of equipment is *ton of refrigerant*, which indicates the ability of an evaporator to remove 200 Btu/min or 12,000 Btu/h. Units are identified in terms of these tons (i.e., "1-Ton unit"). Living space within a building requires a ventilation system that exchanges air within the closed spaces with fresh outdoor air to maintain a healthy indoor environment. Other spaces that collect unwanted moisture, such as attics, crawl spaces, basements, kitchens, bathrooms, or indoor pools, must be ventilated by mechanical means or with sufficient openings to allow for a natural flow of air.

### 4.1.4 Electrical

Regardless of complexity, an electrical system is basically an arrangement of conductors and equipment that transfer electrical energy (the movement of electrons within a confined pathway) from a power source to functional devices like lights and motors. Power sources are either the distribution system of an electric utility company, a privately owned generating plant, or batteries. It is common practice for electrical plans to indicate the layout of parts and their interconnection within a circuit using simplified symbols along a single line, called a one-line diagram. Some of the more important definitions related to electrical systems are listed here:

1. *Circuit*: The path along which electrons flow, generally formed of insulated conductors. A basic direct current circuit may be described as a run of wire conductors (similar to pipes used to transport water) and a generator (similar to a centrifugal pump which supplies the work of moving water through the piping system) that is used to power a simple motor (similar to a turbine that operates in response to water pressure), regulated by a switch (similar to a water valve).
2. *Feeder*: All circuit conductors between the service equipment or power supply source and the final branch-circuit over-current device.
3. *Ground*: An intentional or accidental conducting connection between an electrical circuit or equipment and the earth or other conducting body. The electrical system must be grounded in order to limit voltage imposed by lightning strikes, line surges, or unintentional contact with higher voltage power lines, as well as to stabilize voltage to the earth during normal operation.
4. *Main*: The circuit from which all other smaller circuits are taken.
5. *Raceway*: An enclosed channel designed solely for holding wires, cables, conduit, and tubing.
6. *Voltage*: The greatest root-mean-square difference of potential between any two conductors of a circuit which causes electrons to flow in a predefined direction, establishing the current. In the water piping system used to illustrate the concept of a circuit, voltage is analogous to water pressure.

The *National Electrical Code* (NFPA 70) was first drafted in 1897 as a product of architectural, electrical, insurance, and business interests (Ayers 1984, p. 197). This code remains in broad use throughout the construction industry and is supplemented by the *2006 ICC Electrical Code*, which contains administrative text necessary to incorporate compliance of electrical provisions of all other I-codes with NFPA 70.

### 4.1.5 Fire Protection

Model building codes have consistently adopted NFPA provisions for all occupancy types and have defined all stages in the process of design that must be given attention, including a recognition of the importance of clear spaces around a structure to allow fire department access; careful selection of a fire suppression system and layout, such as automatic sprinklers used in combination with standpipes for taller constructed facilities. The incorporation of local and state fire marshal concerns into the overall plan is a vital component of a total fire protection system, as they are most familiar with the history and ramifications of regional fires. The model building codes regulate the use of combustible building materials for certain occupancies, areas, and structure heights, and also establish an effective means of communication through alarms and the like. Some barriers to the spread of fire, like special doors and dampers, are also triggered automatically to create fire resistive spaces within a building.

General building material and construction assemblies must be subjected to tests described in ASTM E119 if they are to qualify as *rated*, which defines the assigned time period based on duration exposed to furnace testing conditions, such as $3/_4$ hour, 1 hour, 2 hours, and so on. Roof covering materials are also assigned a fire resistance classification by Underwriters Laboratories (UL) per Standard UL 790 to define their performance when exposed to a fire source originating from outside the building onto which they are installed. Class A indicates that the material is not readily flammable under severe fire exposure, Class B will protect the underlying roof deck to a moderate degree, and Class C is not readily flammable only under light conditions of fire exposure.

In structural steel assemblies, there is a distinction made between *restrained* and *unrestrained* framing members when subjected to fire and this may dictate selection of a structural deck section. The yield strength of steel is reduced at highly elevated temperatures, and the first place where a plastic hinge will develop (where member stress exceeds the yield point) as a result of this phenomenon is the location of highest stress. For simply supported framing members, the end connections are not considered to provide bending restraint under ordinary conditions and would therefore be identified for fire resistance purposes as unrestrained (though we will see in Chapter 12 that even simple steel bolted connections have some restraint). When such a beam is considered as having end restraint by one mechanism or another, the formation of a plastic hinge at the middle does not immediately lead to failure because carried positive moment is redistributed to the ends where there exists reserve capacity.

Recognition of a restrained condition, in addition to other factors, may allow for a reduction in the amount of fire protective material that must coat, enclose, or encase primary structural elements depending on the rating required by code. Some of these materials can be quite expensive, such as spray-applied cementitious mixes or intumescent mastic coatings, and a structural engineer should at least be aware of cost-saving options depending on level of involvement dictated by the project architect. Other means of protecting primary structural steel elements include concrete encasement, masonry, or gypsum membrane enclosure, wrapping in mineral fiber blankets, or flame shields, all of which lead to a particular fire rating established through calculations according to Section 721.5 of the 2006 IBC.

Through-penetrations of fire-resistive barriers are to be sealed with a material meeting the requirements of ASTM E814. Because mechanical ducts provide a continuous pathway for fire, smoke, and heat, coverings, linings, and interruptive dampers must be provided and tested per applicable standards, such as ASTM E84. Movement of flame and gases is restricted in wood-frame buildings by lumber fire-blocking defined in Section 717.2 of the 2006 IBC, placed within concealed passages such as floors, walls, and stairs, forming an effective barrier between horizontal and vertical spaces. Draftstopping is also constructed and installed within floor/ceiling or roof/attic assemblies according to Sections 717.3 and 717.4 of the 2006 IBC to divide the overall area into manageable concealed spaces.

### 4.1.6 Egress and Circulation

All conditions required within a means of egress system are identified in Chapter 10 of the 2006 IBC or Chapter 11 of NFPA 5000 including rise and run of stairs or ramps (see Fig. 4-5), handrail or guardrail size and placement, areas of refuge for emergency rescue, hardware and performance stipulations, illumination and signage, doors, and travel requirements. When reviewing the provisions, a designer quickly observes that everything hinges upon something known as *occupant load* (OL), defined as the maximum number of persons within a building or room at any given time, calculated in terms of occupants per square foot (square meter). The entire structure will be assigned an OL based on a usage which produces the greatest number of persons and each individual room will also be assigned a unique OL used to calculate the number of persons who must exit directly from that space.

Continuity of every means of egress system is established by predetermined segments defined within the 2006 IBC. The exit itself is defined as the portion of a means of egress system separated from other interior spaces by fire-resistance rated construction, linking an exit access to an exit discharge (that which signals the

**Figure 4-5** Means of egress (exterior stairway) from second floor of a commercial building, which requires positive anchorage against a high live load under emergency conditions.

terminus of an egress system). It is typically described in terms of several different components, including exterior doors at ground level, enclosures or passageways, exterior exit ramps or stairways, and horizontal exits that provide passage through specially constructed area separation or fire resistive walls. Calculated exitway widths and areas are usually larger than required for emergency purposes in order to account for possible massing of occupants and movement or relocation of equipment.

Elements of an elevator system include the traveling car, hoistway, machine room, pit, and structural supports for vertical and horizontal movement. The car is guided by vertical rails and may be moved through the hoistway by a hydraulic cylinder or an electrical driving machine. Depending on type of cab drive and support, a hoisting beam may be required at the top of the shaft, designed to carry all weight and live loading of the elevator cab for emergency purposes. Pits or machine rooms located below the base flood elevation as determined by the National Flood Insurance Program of FEMA require special floodproofing to protect equipment. Other guidelines for elevator construction and maintenance are found in ANSI A17.1.

**Accessibility**   Chapter 11 and Appendix E of the 2006 IBC, as well as Chapter 12 of NFPA 5000, list regulations for buildings related to providing facilities and means of egress for people that have a physical handicap. Many states add further regulations to these base documents, but the general philosophy is that the public should be afforded reasonable access to building facilities, including travel across the site from a designated parking space. This is not to mention the Americans with Disabilities Act of 1990, which exists as a civil law, and is intended to set guidelines for accessibility to places of public accommodation and commercial facilities, including different standards than what appear in the model codes, so enforcement and compliance with all regulations can be somewhat tedious.

## 4.1.7   Weatherproofing

From a simple, energy conservation point of view, a *building envelope* is defined as the combination of elements that enclose conditioned spaces through which thermal energy may be transferred and includes exterior walls and finishing, the foundation and roof system, windows, and doors. There is a bigger picture, however, as the enclosure of a building is also required to meet performance objectives. The framing and fenestration system must be laid out and fastened in such a way that all weight is supported and lateral forces are properly transferred to the building's structural frame. The intrusion of water, air, and sound is controlled by various mechanisms while energy used in creating a livable environment is conserved.

Most reconnaissance reports written after major hurricanes concur that properly designed and constructed buildings can withstand these elevated wind forces and that the majority of damage that a building does experience occurs after the weatherproof envelope had been breached. Current codes include special fastener requirements at building regions where wind pressure builds up (roof eaves, wall corners) in recognition of this need to secure the envelope. Because of this importance, the National Institute of Standards and Technology (NIST) recommended that state and local organizations consider licensing roofing contractors, providing continuing education for contractors, and establishing field-inspection programs to monitor the construction of roofs (NIST 2006).

**Cladding Systems** Exterior cladding is defined in generic terms as the protective layer fastened to the outside of a building and includes a wide variety of naturally occurring and manmade materials such as stone, concrete, wood, and aluminum. Panels are subjected to environmental differentials between exterior and interior surfaces which can cause bowing due to uneven shrinkage (moisture) and temperature response. Structural elements within a building frame are designed to carry the weight of cladding supported above grade, but to also resist wind or other pressures applied normal to the wall surface, creating a multidimensional load path as may be seen to occur in Fig. 4-6. Cladding is also controlled by seismic movement, sometimes resulting in tricky anchorage details, and can affect weather

**Figure 4-6** Commercial building with stone cladding, subject to seismic shaking that may affect closure of the building envelope if not properly detailed.

protection of interior elements if not properly accounted for by structural design and appropriate joint sealing.

Curtain walls are supported by spandrel beams or directly from a building's foundation. In either case, it is important to control cracking to maintain the effectiveness of weatherproofing. Cladding reactions and magnitude of supporting element deflections are sometimes placed on the project drawings so that the supplier can properly design and detail the curtain wall system for strength and movement, though independent review of these systems is also important to allow for the professional opinion of material and system suppliers.

**Roofing and Flashing**  The artistic flavor of a building may call for a roof system to be broken up into different pieces and shapes, such as the residence shown in Fig. 4-7, leading to special structural and weatherproofing challenges that can be solved by a variety of products. The first layer of defense is the roofing material itself. Among the least expensive, and among the lightest weight, are wood shakes or shingles that are made from species naturally resistant to moisture, sunlight, and breakage from common hail stones. They may be installed over both wall and roof surfaces with a slope of not less than 4 in 12 (4 units of rise for every 12 units of run).

Asphalt-type shingles have been used successfully for roof slopes not less than 3 in 12, provided installation measures are carefully followed, and typically are surfaced

**Figure 4-7**  Roof diaphragms placed at different levels can present a challenge in tracing lateral forces to the foundation.

with embedded water-resistant mineral granules according to ASTM D225 (organic felt, manufactured with a base of cellulose fibers such as recycled waste paper or fibers) or D3462 (fiberglass felt). An asphaltic system used in flat or low-slope applications, known as built-up roofing, consists of multiple layers of adhesive/waterproof bitumen (asphalt- or coal-tar pitch) and reinforcing ply sheets, that are commonly placed over a layer of rigid insulation fastened to the structural deck. The finished system is overlain with some type of surfacing material to protect against energy transmission and solar radiation, being in the form of a roll-type asphalt cap sheet (ASTM D3909) or crushed aggregates that are prepared according to ASTM D1863.

Clay, slate, and concrete tiles are the heavier roofing materials to be expected in the industry and have found broad appeal because of their durability and long-lasting aesthetic value. These elements are manufactured according to the requirements of ASTM C1167 (clay), C406 (slate), and C1492 (concrete) and are recommended for slopes not less than 4 in 12. Mortar is used to fill gaps and seal lapping surfaces when necessary, giving the designer an idea of the weight that could be involved in these systems, although lightweight materials are typically used in modern construction to help keep placement and design manageable.

Single-ply roofing membranes are commonly used for slopes not exceeding 2 in 12, such as EPDM elastomeric sheets manufactured from ethylene, propylene, and diene monomer according to ASTM D4637. Sheets are laid loose or fastened to the structural deck, or a layer of rigid insulation below, either mechanically or by a bonding adhesive. A gravel ballast may be laid over the membrane for protection and insulation after joints have been sealed, though precast roof paver systems have also been used.

Metal roof panels are available in a variety of types (aluminum, galvanized steel) and have been used successfully on slopes as low as 1/4 in 12, depending on manufacturer's recommendations. The performance of standing seam-type roof panel systems is regulated by ASTM E1514 (steel) and E1637 (aluminum). A layer of waterproof felt is placed between metal panels and the structural deck below to protect against leaks due to construction errors, tolerances, or wind-driven rain.

In all roofing systems, light-gage metal flashing components play an important role in sealing the building envelope and are manufactured of copper, aluminum, stainless or galvanized steel, zinc alloy, lead, or some other type of coated metal. Any type of shape can be created to direct water away from joints, edges, corners, valleys, roof deck penetrations, or other areas where discontinuities exist (see Fig. 4-8). Flashing is also provided at seams, joints, and other areas of wall finishing assemblies to seal the envelope against water intrusion. To complete waterproofing measures on the roof, a system of drains is built to channel water away and this can be a tricky part of structural design as well. Some drain systems require a low profile in order to properly collect water, which may require cutting into the structural deck as shown in Fig. 4-9. Special care and attention should be given to the structural system at a

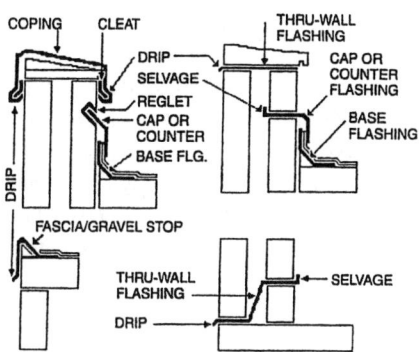

**Figure 4-8** Metal flashing systems and terminology. (*From Ratay, Forensic Structural Engineering Handbook, McGraw-Hill, 2000*)

building's perimeter, which could be the location of required lateral force collecting elements.

**Thermal Insulation** Insulating materials are intended to control temperatures of inside surfaces for the comfort of occupants, assist in the prevention of condensation or water-vapor transmission, reduce the passage of noise, and conserve energy. These include vapor retarders, reflective foil insulation, fiberboard, foam plastic, cellulose loose-type fill, or fiberglass blankets, all of which require a flame-spread or smoke-developed index depending on placement, determined in accordance with ASTM E84.

Thermal resistances of elements and assemblies are identified in terms of an R-value, which is the inverse of the time rate of heat flow through a building envelope from one of the bounding surfaces to the other for a unit of temperature difference between the two surfaces, per unit area. Higher R-values indicate better insulating properties (less heat flow through the material). Windows, skylights, and doors are rated with a U-factor, which is the reciprocal of the R-value.

## 4.2 THE BUILDING TEAM

The building team consists of those responsible for financing, designing, and constructing the facility to be occupied by the owner at the end of the project. Small building projects usually contain the same elements and players as much larger

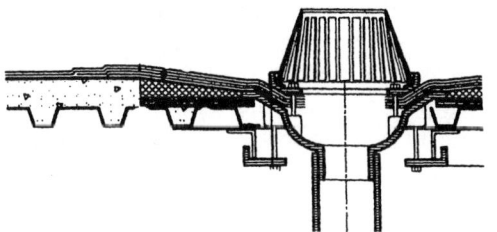

**Figure 4-9** A roof drain which is set down into the structural deck needs to be given special attention, as lateral forces may need to be transferred around the area blocked out.

projects, but the scope and intensity of work is reduced. Some members of a design team, however, can turn a relatively simple project into a complete nightmare if they are approaching it as a "black box," trying to apply everything they are familiar with to the work, and in the process they make things more confusing for everyone involved. Some of these things include the process of having meetings just for the sake of having meetings, developing volumes of specifications when requirements could have been summarized on the drawings, and hiring other consultants when they are really not needed. As with anything else, good, respectful communication is the key to success.

### 4.2.1 Owners

Most individuals or entities that sit in the seat of ownership do not fully realize what it takes to bring a project from imagination to occupancy. Although they rely on decision-making and problem-solving skills of other professionals to bring their dreams to reality, there are important powers that owners bring to a project, such as financing, dictation of a building's form and function, and the ability to hire or fire. The function of a structure is one of the first things projected from the dreaming stage of development, followed by a general idea of form that an owner would like the dream to be molded into. Sometimes a site is already selected and this will influence decisions on appearance and functionality of the structure, but owners typically do not have the education or training to see the beauty and artistic impression of blending a manmade object into its surrounding environment.

One of the first things an owner decides is how the project will be approached. There are different methods of integrating the work of the design team with that of the contractor, the most common being the *design-bid-build* approach, in which the owner selects a design consultant (usually an architect) at the beginning of the project to help develop schematics, determine restrictions due to codes or local ordinances, assist in setting a budget, hire necessary subconsultants, and complete construction drawings that are suitable for bidding by contractors. The design team assists the owner in selecting a construction company during the bidding phase and helps in the award of a contract to construct the building. Sometimes this process is carried out with the assistance of a construction management company.

The *design-build* process is often hailed as the most efficient system in regard to time of delivery. An owner begins with a schematic of what is desired, and when funding has been secured and the timing is right, a single consultant is selected to complete both design and construction of the building. One of the advantages to this system is that the owner only needs to deal with a single entity as opposed to a team of individual firms. There may be some cost savings as well. Government studies have indicated that design-build projects can deliver a product at about 6% less on average than what a traditional delivery system may have amounted to (Harrington 2000).

A possible disadvantage to the structural engineer working within a design-build firm is an apparent conflict of interest: A contractor may choose to build with less robust (thus less expensive) materials, whereas the engineer may decide that something of better quality is necessary for individual peace of mind. Who judges such a matter if both are being paid by the same principal? Because it is common for construction to begin on a fast-tracked design-build project before the design is even completed, there are also ethical issues that must be dealt with regarding code compliance and assurance of complete documents.

### 4.2.2 Architects

Architects form the first line of defense for everyone else on the design team. They are best equipped for making sure the owner respects and understands design team responsibilities for nurturing an atmosphere of clear communication and defining reasonable expectations for the design, bidding, and construction process. The architect determines what other consultants are needed to assist in areas that lie outside the parent company's expertise, such as landscape architecture or acoustical design.

Architectural firms come in different sizes with a variety of project tastes. Some specialize in a limited or exclusive range of building uses, such as healthcare-related facilities, yet other firms are familiar with all conceivable functions and operate with a large staff to support a broader vision. Some larger architectural firms include in-house engineering departments and can offer their clients a complete line of design services, even essential land surveying or soil testing.

Even for simple projects, an owner is usually in a better starting position if a qualified architect is selected early in the process of imagining and dreaming. Some projects, however, are primarily structural in nature and do not require the services of an architect, for which a responsible engineer will then shoulder more administrative responsibility for the work and must understand what steps are necessary to bring things to completion. In very broad terms, an architect considers the following basic segments of work for every project, regardless of size: Financing and budget, site selection, architectural harmony with the surrounding environment (visual and functional appeal), soil type and condition, structural form and function under external loading, interior comfort and usability, and continuing use through solid construction and maintenance.

The architectural licensing examination, developed by the National Council of Architectural Registration Boards, addresses eight different divisions of knowledge, skill, and ability required to provide services to the industry. Therefore, it is intended that a licensed architect will have a practical (though not necessarily a complete and thorough) understanding of the following subjects:

1. *Pre-design*: Project development is addressed in relation to architectural programming and analysis (assessing client needs, spatial and functional relationships); environmental, social, and economic issues (understand building surveys, alternative energies, and historic precedent); codes and regulations (building, specialty, and zoning codes); project and practice management (develop project delivery methods, budgeting and legal issues related to architectural practice); and site planning and design (incorporate the implication of human behavior with design theory, assess impact on the environment).
2. *General structures*: Identification of general structural principles in the design and construction of buildings, including general theory, codes and regulations, materials and technologies, and effect of the environment on the selection of building materials.
3. *Lateral forces*: Incorporation and appreciation of lateral force principles in the design and construction of buildings.
4. *Mechanical and electrical systems*: Evaluation, selection, and integration of mechanical, electrical, plumbing, conveying, and specialty systems. Content areas within the exam include codes and regulations, environmental issues, plumbing, HVAC, electrical, lighting, and specialties (communications, acoustics, conveying, fire detection and suppression).
5. *Building design, materials, and methods*: Selection of building materials and systems as related to environmental issues, codes, and different technologies. This division also covers schematic and design development phases of an architect's practice.
6. *Construction documents and services*: Project management and professional practice related to the preparation of contract documents and administration thereof. Focus areas include codes, standards, and environmental issues related to site planning and building design, preparation, and coordination of drawings and specifications, preparation of cost estimates, scheduling, coordination, and related issues.
7. *Site planning*: Integration of site requirements and conditions into a genuine solution, including topography, landscaping, climate, and geography.
8. *Building planning*: Schematic design development of interior spaces, furniture arrangement, handicap accessibility, and functional utility of buildings.

A practicing architect must be something of a master when it comes to defining a structure's form and function, the most basic of elements. A building's form is what people see and a building's function is what people experience. Function is what causes the form to work for an occupant. Architecture is the marriage of form and function in such a way that the occupants, public at large, and surrounding environment benefit in some way to the addition of this building. The form of a building is based on artistic talent and desire of the architect as he or she works with the client in creating a dream to be enjoyed for many years to come. In many regions, artistic form of a new building must be designed to fit into

**Figure 4-10** Aesthetic challenge of fitting new buildings into a region with older, historic structures.

an existing historical setting, as is shown in Fig. 4-10, using new materials and technologies. This can certainly be a challenge as archaic finishing materials are difficult to match precisely and handicap accessibility regulations cannot always be neatly incorporated into a remodeled form.

The function of a commercial building is quite busy with a multitude of different activities being performed at the same time. In a typical hospital building, for example, administrative duties are carried out in one area while medical procedures are practiced in another and patients are sleeping soundly in yet another. Residential buildings, on the other hand, are primarily used as a type of retreat from everything else, where families can grow closer together and refresh themselves for the following day's activities.

## 4.2.3 Engineers

An engineer is typically a specialist in one of the systems that are constructed into or around the building: Structural, mechanical, electrical, and civil are the most common engineering disciplines in the building industry. Some jurisdictions allow a professional civil or structural engineer to design and take responsibility for the construction documents of an entire building (including architectural, mechanical, and electrical), provided they can demonstrate that they are competent to do so. Specialists in one or more areas of building construction may have real weaknesses in pulling all of the respective pieces together to move a project through the construction phase, unless they have been trained by an architect to handle these duties.

### 4.2.4 Contractors and Subcontractors

The general, or prime, contractor is in charge of the complete building project, but may actually only participate from a labor standpoint in a particular area. For example, the general contractor for a building project will be in charge of making sure all requirements of the contract are completed, but their staff may only physically complete the concrete work. They would then retain the services of subcontractors to complete the rest of the work, such as framing, roofing, finishing, painting, landscaping, grading, and the like. The general contractor is in complete and sole charge (and responsibility) for completing all of the required work.

Financing a project can be complicated as the contractor often submits a lump-sum bid (though not the only means) for completing the work and will sometimes capitalize on areas where extra work can legitimately be billed to the owner, put into the form of something called a *change order*. It is the architect's responsibility to make sure the project documents are clear enough as to the intended construction and it is the owner's responsibility to clearly communicate his needs to the design team. It is one thing to have a structure drawn nicely on paper but a completely different thing to actually piece it together, so most projects inevitably involve some additional, unforeseen costs which should be included in the budget.

A construction manager needs an uncanny ability to foresee problems ahead of time to properly deal with delays or added project cost. It is difficult to picture a solution to a problem that has not yet occurred, so evidence will need to be collected from project documents, as well as the experiences of contractors that have been affected by some form of design or construction trade sequencing change, in order to head off a problem or make the project flow more smoothly. *Requests-for-information* (RFIs) can serve as tools for a look into the future progress of a project as they may identify weak links along the way. When problems occur, they will affect the cost or schedule of a project if there is an inefficient redirection of labor, extra equipment or material has to be brought to the site, portions require dismantling and reassembling, or there will be a time lag for approvals.

## 4.3 LAND DEVELOPMENT

City planners and councils have land use regulatory powers within their own established boundaries, whereas counties provide jurisdiction in unincorporated areas. Each group establishes a master plan by which all development is judged, providing a future of managed growth and proper incorporation of regional infrastructure. Every site deemed to be suitable for construction of a new building must be documented sufficiently to identify existing features and controls. Some of these features include boundary lines, existing right-of-ways that have been granted to transportation or other agencies, easements for utilities or other purposes, position and use of existing structures, location of fire hydrants and light sources, location of swamps or other water sources, geologic hazards, outline of wooded areas, and contours that define the topographic characteristics.

Information regarding the suitability of a site for proposed construction is often gathered through a topographic survey or aerial map, historical geological surveys or other record data, local knowledge of property owners or professionals in the building industry who have completed successful projects in the region, well records, reconnaissance surveys, or a subsurface geotechnical report.

### 4.3.1 Ownership and Legal Interests

Over the years, numerous laws have been enacted by the federal government to grant, sell, or transfer ownership to private individuals, which was done in abundance after the American Revolutionary War to raise revenue. Land ownership was also transferred to existing states, but some areas continue to be reserved for the benefit of the public (national parks, historical monuments, military or educational purposes).

For admission into the Union as a state, persons living within legal boundaries of a territory had to agree that, in accordance with the respective Enabling Act, all unappropriated lands be granted to the Union for disposition on a first-come and first-served basis to anyone with enough money to purchase it, having first been divided into sections and townships according to the Land Ordinance Act of 1785. The purchaser is then protected by the Fifth Amendment of the U.S. Constitution: ". . . nor shall private property be taken for public use, without just compensation." There have been cases brought against the government for confiscation of property through the practice of Eminent Domain without appropriate compensation, which has been cause for wariness and caution on the part of land owners with property susceptible to political maneuvers or environmental reassessment.

### 4.3.2 Surveying

A *boundary survey*, which is also known as a record of survey, is completed for the purpose of locating legal corners, boundary lines, and easements of a parcel of land and is completed in one form or another to define the working perimeter of a site plan. A *topographic survey* is necessary for the purpose of locating existing features across a certain region including buildings, improvements, elevations/grades, fences, trees, land contours, and much more. A *control survey* is one which locates horizontal and vertical positions of points on a site to determine boundary lines, develop maps from aerial photographs, lay out construction staking, or other purposes. Architects and civil engineers use data from these surveys to determine how best to make improvements on a site, such as grading land so that storm water will drain away from all new structures.

Title companies work to ensure that title of ownership to land is clear and provide insurance to cover the assessment, thereby requiring far more information about a parcel of land than would be needed by a land owner for improvements. An *ALTA survey*, developed by a joint effort of the American Land Title Association and the American Congress on Surveying and Mapping, is one which collects as much information as possible on the existing condition of a piece of property, including right of way lines, other legal easements, boundaries, services, and anything else necessary to justify issuance of title insurance or other sensitive needs.

### 4.3.3 Civil Engineering Work

Civil engineers design features that cause a parcel of land and associated structures to operate, including grading and drainage of the landscape for storm water, determining need for stabilizing embankments, planning of pipelines for a variety of different materials (waste disposal, water supply, runoff management), establishing subdivision features, designing wastewater treatment systems, and more. They often specialize in a particular area, such as structural engineering, surveying or construction management, hydraulic design, or land development.

Civil engineers determine how to make use of existing sources for water and transport of waste away from a site. Surface fresh water sources include rivers, streams, marshes, ponds, or lakes and normally require some type of treatment before the supply can be drinkable. Ground waters include springs and wells that are tapped into aquifers or other saturated region below the water table and can actually provide a fairly clean supply of water in many cases due to filtration through the soil above. Water is also collected and stored in manmade reservoirs and basins with some level of security to keep the public from littering and spreading disease into a region's water supply.

A storm drain installed below grade receiving water from the surface through inlets is a closed-type of drainage system and is used to convey rain water, subsurface water, condensate, cooling water, or other similar property discharges (except for sewage) with terminus at a lake, river, ocean, or drainage basin. An open-type of drainage system involves the sculpting of landscape into swales, with a possible addition of ditches or culverts, to direct rain water and runoff away from a site. Most lot drainage solutions include both open- and closed-type elements.

When sewer lines are not available for transporting liquid waste material off site, a septic tank and leach line are commonly used. A septic tank retains sludge that is in immediate contact with sewage flowing through for a sufficient amount of time to decompose organic solids by bacterial action while the liquid makes its way into a leach line or leach field. It is thus allowed to percolate into the surrounding soil. Obviously, there are many governmental regulations for leaching waste material, locally imposed and enforced.

## 4.4 PROJECT PHASES

The first stages of all construction projects are characterized by information gathering regarding the site (land development), applicable regulations (federal, state, local), and responsibilities of each member of the building team (owners, design professionals, contractors, regulatory agents). Some information is easily available and may be retrieved quickly, but there is always a certain amount of data that takes time to research or to discover through the natural course of experience and intuition.

### 4.4.1 Design Phase

Once a structural engineer has finalized a scope of work and preliminary questions have been answered, design begins with a schematic layout of the anticipated structural system (if this work has not already been completed as part of the contractual negotiations). Materials of construction and a type of lateral force resisting system may have been selected on a rudimentary level by the architect to fit a particular vision of the building's form and function, therefore one of the structural engineer's first tasks is to determine whether that system is valid and what changes are necessary.

An architect's drawings should have sufficient detail to determine design loading. Fire-resistive requirements may necessitate inclusion of spray-applied fire proofing to structural steel and installation of a fire suppression system, adding weight to the framing elements. Some walls also require additional layers of finishing, such as gypsum wallboard or a cementitious-type covering, and the roofing system is often dictated by fire resistive or energy conservation requirements. An aggregate-surfaced built-up roof weighs more than a mineral cap sheet-type surface, and a veneered wall weighs more than a plastered wall, thereby making the case for careful determination of design loads by review of the architect's intent.

A structural engineer may not be consulted until the next stage of design, called *design development (preliminary design)*, which occurs after the owner has approved the schematics. The main focus of this phase is to resolve any outstanding issues or clarifications in order to move forward with firm decisions on system and layout. Engineers make recommendations or corrections to the architect that may affect the preliminaries. Information regarding the use and installation of special constructions, such as premanufactured skylights or awnings, should be investigated at this stage so that any data can be forwarded to a supplier with plenty of time. All major components of construction should be identified at this point, though the nuances of connection and continuity may still be unresolved.

The *final design (construction document)* phase is the closing assembly of drawings, specifications, and calculations to the level of detail required to obtain a permit for construction, a collection of responsible bids, and to clearly identify the scope of a contractor's work. The architect will also provide bidding and contractual articles according to the owner's needs, but they may not be introduced into the set of project documents until a building permit has been issued.

On occasion, a project engineer is called to put together an estimate of construction costs for the structural system, but this can be difficult in an environment where prices fluctuate in response to availability and political climate. It has been jokingly theorized that engineers should double the value of their cost estimates and triple their expectation of construction time and then they may be in the "ballpark of reality." The architect is usually the one who sets the tone regarding budget and estimation of cost and decides who is responsible for such, but the general contractor, construction manager, or other specialized cost estimator (as applies to a particular project) will also play a significant role.

### 4.4.2 Approval Phase

Regulation is a necessary part of engineering and jurisdictions enforcing applicable codes do their best to be fair and impartial to all applicants. The intent of a code requirement may certainly be complied with by unconventional methods, but a regulator will require proof and a convincing argument for deviations from commonly seen solutions. These officials serve as society's front line of defense against incompetent designers and it is often a thankless job. In fact, design engineers often blame building or transportation departments for imposing strict regulations on a project, failing to understand that they themselves have every right to challenge a regulator's interpretation of the code, albeit patiently and professionally.

Review of construction documents is not only a necessary step to compare information to what is required by code, but the process also represents a valuable independent check of a professional engineer's work as a fresh look for errors, omissions, or other things that may have been inadvertently left off. This step in securing a permit for construction is seen by so many as unnecessary bureaucracy and a waste of time, but if it is properly and reasonably administered in close accordance with directions provided in the model codes, this review is of great benefit to the owner, the design team, and society at large.

**Planning Department Review**  A regional planning department communicates and administrates a vision for land use plans produced by planning commissions, zoning boards, commissioners, and the general public. It reviews subdivision maps, zone change requests, site plans, variances, or projects that may affect a region's aesthetic or acoustical setting. Occasionally, a site plan will require preliminary review and approval before construction documents can be submitted for a building permit. Most owners do not fully understand the global requirements of planning and how sensitive decisions can affect the master plan of a community, therefore an understanding of planning department duties and historical administration thereof can greatly reduce the time it takes to break ground on a project.

**Building Department Review**  A typical building department consists of a designated building official who delegates different responsibilities to other employees, such as plans examiners and building inspectors. The beginning chapters of state construction codes spell out the duties of these officials in great detail and also clarify that personal liability for decisions rendered during the process of administering code-prescribed duties cannot be assumed (2006 IBC, Section 104.8; NFPA 5000, Section 1.7.4.1). Review of building construction plans is often a complicated process, as personal opinion can easily creep into the realm of code interpretation. A building official or certified plans examiner reviews submittal documents for compliance with the adopted building code and generates a correction list with items relative to fire protection, means of egress, handicap accessibility, structural, mechanical, plumbing, energy conservation, and electrical disciplines. A written response to each correction item by members of the design team is important (typically requested) in streamlining the review process in a timely fashion.

Objections to any correction item may be made and they will be better received if supported with appropriate documentation of code references and accepted design works. Patience and a respectful attitude can make the difference between having a correction item overturned and causing additional issues to be raised. Additionally, the process of plan review and approval can be slowed or accelerated depending on how an official and a design professional choose to work together. Once everything is addressed to the satisfaction of these officials and other departments have given their approval for whatever scope may have fallen to their jurisdiction and the applicant has filled out forms and paid fees, the project is given a permit to proceed with construction and follow-up inspections.

**Environmental Restrictions**  A need to protect the surrounding environment has prompted government influence in geographies predetermined to be of a sensitive nature, with new regions being added continually. Planning agencies in charge of approval review a project's documentation and evaluate existing topographic patterns, giving consideration to potential for erosion, drainage, sources of pollution, protection of sensitive vegetation, and succession of growth within a changed area. Tolerance of indigenous wildlife species to a newly constructed environment, including evaluation of their movement patterns, is also given special attention. Unfortunately, regulation and response to it are often mired in politics and it can be difficult to determine the truth of a matter.

Protection of natural resources is driven by state and federal laws administrated by special government agencies with power over land development in various geographic regions, such as coastlines and lake basins. Cities and counties are charged with identifying resource-sensitive areas within their boundaries, although their focus may be geared toward management of their use for proper economic growth, such as the harvesting of timber for construction versus completely protecting systems against any use that would affect the environment to a prescribed degree. It is this prescribed degree of prescription that is often interpreted by the courts and the success of a building project within an environmentally sensitive region depends on early consideration.

### 4.4.3 Bidding Phase

Documents assembled for contractors to base a bid for necessary work on public projects are prescribed by various statutes designed to protect public funding, prevent fraud or favoritism, and obtain quality services at a reasonable price. When the owner wants to select from a number of qualified contractors for the work, bidding on privately funded construction projects is conducted according to that owner's rules and regulations, often based on the advice of a licensed design professional. The first step in bidding out a project is advertisement to interested and qualified parties, describing the nature, extent, and location of work as well as procedures for obtaining documents and submitting a valid bid.

Design professionals need to be aware of rules set forth in the bid documents regarding inquiries, receipt of bids, and award. All construction projects, especially those for entities who will be paid with public money, are scrutinized to a degree by contractors who want to make sure rules are being followed in order that a fair and competitive set of bids is collected. Those who believe they were treated unfairly or discriminated against may file a protest with the owner, requiring the assistance of legal council to review submittals and formulate an appropriate response to the protest. Needless to say, no one wants to have to deal with protests, so care in the assembly of bidding rules and thoroughness in the execution of those rules will be paramount to getting a project started on the right foot.

Bidding documents include a full set of approved construction drawings (those which have been permitted by the approval agency), project manual (including bid and contract documents), and any addenda or schedules. The owner designates whoever is best suited to control the issuance of documents, typically the prime design professional (architect), and also calls upon that same person to help advertise for bids, control the distribution of documents, advise on the acceptability of proposed subcontractors, and keep records of the process.

A surety bond is required on most construction projects by the general contractor to guarantee complete execution of the contract and all supplemental agreements pertaining to the project, as well as payment of all legal debts related to construction. It is not an insurance policy, but rather an extension of credit by the bonding agency as endorsement of the contractor. It is common practice in contractual agreements to call for two separate contract surety bonds, one to cover expenses for labor and materials and another which covers performance of the contract itself. Other bonds may be required by the owner or public agency to assure certain conditions or performance criteria are met.

### 4.4.4 Construction Phase

Construction of even a small building involves many different trades, materials, operations, and schedules that must be properly managed in order to meet a budgeted amount or time period expected for occupancy. A general contractor sets up a type of factory by which the products (building and site improvements) are created and installed to the owner's specifications. The construction process is often wrought with uncertainty and unpredictable events, but once everything is completed, a sense of satisfaction (and relief) is had by all.

The custom of most contract documents is to specify that the contractor is responsible for safety of the workers on the construction site, in addition to the appropriate selection of means, methods, techniques, and procedures necessary to complete the work. It is sometimes necessary for a contractor to hire a qualified, independent engineer to design and detail systems that will maintain safety of site appurtenances and working conditions.

If an unsafe condition is observed during a visit to the site by the design professional, it must be reported to the superintendent and followed up in writing for documentation. The engineer remains neutral in regard to actually specifying to the contractor how to fix the dangerous situation, which releases him or her from liability, but it should also be clear in the documentation that the intent of the site visit was not to assess safety issues. During the erection of elements, for example, temporary bracing is always necessary against intermittent wind, other environmental or accidental loading and a general contractor should have sufficient knowledge of the needs of each area of the structure.

When responding to a problem or question from the field, it is important to respond as quickly as practical, as a contractor may be required to pay liquidated damages if the agreed upon schedule is overrun. If there is a sense that the project engineer is not performing duties responsibly, or with appropriate attention to the urgency of matters, that engineer may be charged with paying damages for each day the contractor runs over time. These issues should be discussed and arranged with the owner and the architect. It may also be necessary to provide alternate contact information, and have someone in the engineer's office standing by to assist when the responsible professional is on vacation or is out with an illness.

**Shop Drawings and RFIs** Shop drawings are produced by fabricators or manufacturers of a particular product for the purpose of showing laborers how to install the pieces. They contain dimensions, materials, specific connections, conditions, and occasionally methods of construction. The scope of a design professional's review is usually restricted to simply comparing the information in the shop drawings with the design intent expressed in the project documents. An engineer with project design responsible charge, including the design direction taken for resisting applied forces, should be the one reviewing submittal documents. An engineer's review of the drawings created by others for coordination purposes, or for determining compliance with the project documents, does not relieve the contractors of their duty to perform according to the project documents, applicable codes, and construction industry standards.

Shop drawings are marked by the engineer-of-record as reviewed, rejected, revise and resubmit, approved (furnished) as corrected, submit specific item, or no exception taken. Other variations are possible and it is critical that the intended meaning of any statement is either understood from the project specifications or defined within the shop drawing approval stamp itself. *No exception taken* can be interpreted to include all aspects of the documents this statement is affixed to, whereas it usually only applies to the work of that particular design professional. If required corrections are minimal and the reviewing engineer is satisfied that the drawings fundamentally satisfy the conditions of the original design, they should be *approved as corrected*. Limited corrections might include sporadic, minor dimensional errors and shop notes that are not fully consistent with project documents. Instead of sending back a complete set, individual sheets may be rejected and asked to be resubmitted due to nonconformance, poor quality, or other problems that would otherwise lead to serious shop errors.

It is generally best to make corrections on the drawings themselves and avoid requiring a resubmittal which only delays a project's momentum. It is important to remember that *review* simply means that the submittal is being checked for information only, whereas *approval* indicates that the design professional may require changes in the submittal and analysis of submittal documents will be more thorough. When subcontractors have made plans available, the engineer-of-record needs to make sure the general contractor has placed their review stamp on the submittals before reviewing them in an effort to avoid a misallocation of duties required by the contract established between the contractor and the owner.

During the course of bidding and construction, anyone who has a job to do in relation to the plans or specifications may submit an RFI directly to the construction manager or the project architect, who will then follow up with the designer of whose system is being asked about. It is difficult to assemble a complete set of construction documents that are absolutely clear on requirements and intent to every person or entity who has a stake in the project, so RFIs are both common, necessary, and to some extent, welcome.

**Revisions** Changes can be necessary to the project documents for a variety of reasons. Sometimes a contractor frames a particular condition because it was not clearly identified on the plans and without consulting the project engineer, the resulting condition may be worse than imaginable, such as is shown in Fig. 4-11. The structure needs to be given proper attention during the generation of drawings, not delayed until construction is already underway.

Revisions that cause an increase in cost or delay the anticipated completion date require a change order to be signed by the owner, contractor, and sometimes design professional. All members of the design and construction team must work together to resolve issues impacting the project in a negative way and attempt to find a solution that causes the least hassle and requires little additional cost, if any at all. This is often the moment when a structural engineer can demonstrate real problem-solving skills and an ability to think creatively, leaving an impression not easily forgotten.

**Figure 4-11** Poor wood-frame construction detail with lack of proper load path and connections.

**Field Observation and Inspection**   Site visits or project inspection requirements by the structural engineer may be specified by the owner, model building code, specific project needs, or required by the governing body within a jurisdiction, and these visits can be afforded different levels of responsibility. The term *observation* is used to denote reviews made of the structural system at significant stages of construction for the purpose of confirming that work is being done in general conformance with the project documents—general conformance because it is impossible to note and review every element, fastener, and condition. *Special inspections* are also necessary for materials and installations requiring special expertise to ensure compliance with codes and standards, such as welding, masonry, concrete placement and testing, and drilled-in anchors.

Inspectors hired by the owner for general or special inspection needs must have proper training, education, and experience to understand code requirements and relevant standards of quality. A large construction project may have one or more general inspectors and a number of others who specialize in a particular building material or an entire system, such as HVAC or electrical. An inspection program identifies specific work and testing to be performed, citing a code or standard to which the contractor is expected to meet so there is as little ambiguity in the inspection process as possible. The program will be formally introduced at a preconstruction meeting in order that all players know what is expected, from the owner to the last subcontractor on the list whose work will be affected or influenced.

If the structural engineer of record is to perform periodic on-site inspections, the extent to which the work is being qualified must be properly understood by all parties. No warranties or guarantees of the final constructed product should be promised, or even understood, regardless of the intensity of the inspection process. Procedures and observation points discussed in this section are intended to give a general picture of an onsite inspection and there certainly may be other areas in need of review. Again, the purpose of this work is to determine that the contractor has performed the work in general conformance with the project documents.

Good communication is necessary between the owner and the design professional when it comes to the issue of contractor's work that does not conform to the requirements of the project documents. Only the owner can reject such work because he is the one who entered into the contract with the builder, but it is usually done on recommendation of the engineer-of-record. If nonconforming work does not violate any codes and merely represents a different way to accomplish the same purpose, it is important that a contract allows an engineer (with appropriate financial compensation) to exercise judgment and discretion in rejecting such work, advising the owner to do such, or recommend a fix in an effort to keep the project moving. If some material or labor savings from the original requirements would result, these should be credited to the owner.

Before performing a general foundation observation, the engineer of record needs to verify that observation of soils by the geotechnical engineer or other party has been completed and approved (when required by the contract documents). Basic elements of a foundation observation include determination of the following:

1. *Width and depth of continuous and pad footings.* Main load bearing or lateral force resisting footings should be given special attention, whereas other common footing types may only require a spot check.
2. *Type and placement of reinforcing bars.* Areas where atypical bends or lap splices are present should be reviewed in detail and tolerances, including clear distances from soil or forms as dictated by code, should be monitored throughout. Assure that sufficient chairs or dobies (small concrete blocks) are present to avoid sagging of horizontal bars.
3. *Position and reinforcement around through-penetrations, such as pipes or bundled conduit.* The project drawings should give general guidelines on maximum sizes and when reinforcement will be necessary. The structural engineer must determine the work of other trades that will affect the foundation system during the process of completing construction drawings and not wait until inspecting to begin dictating new requirements. All penetrations that exceed a reasonable set of typical standards will require special detailing.

During general observation of a floor or roof system, the engineer of record should consider the following installations according to what is prescribed on the construction documents:

1. Sizes of main girders and critical secondary framing elements
2. Spacing and support of selected joists or purlins throughout the project
3. Connections at the end of main elements
4. Connections at elements to the floor deck, such as in composite construction
5. Drag struts, ties, chords, or subdiaphragm framing elements
6. Connections that complete the lateral force resisting system
7. Support conditions present to assure vertical continuity, such as at columns providing support for an upper level or roof
8. Reinforcement for concrete slabs
9. Penetrations through framing members or the deck itself
10. Placement and condition of splices between members
11. Other special conditions that may have been shown on shop drawings

### 4.4.5 Occupancy and Continued Use

Once a facility is built, an owner needs assistance during the process of occupying the space. This activity is sometimes called the *postcontract phase* and includes a recognition of maintenance and operational programming, start-up assistance, the creation of record drawings, review of all warranties and product manuals, and postcontract evaluations. A specialty contractor instructs the owner on essential building or equipment operations during initial occupancy once systems are tested and confirmed to meet specifications.

*Record drawings* indicate changes made as compared with the original contract documents, whereas *as-built drawings* are intended to show how a structure was actually constructed, requiring extensive dimensions and surveys. The process of assembling record drawings may prove to be cumbersome, if not impossible, if paperwork has not been filed or organized during a project's history. Any changes made to the contract documents should be certified in writing at the time of installation in order to save time and frustration during closeout procedures.

Some owners will require a post-occupancy walk-through at a designated time after equipment startup and substantial use of the facility has begun. Attendees usually include the architect of record, construction manager (if any), a representative for the general contractor, and the director of maintenance or operations for the structure. There are different reasons why such an evaluation should be made. First of all, an assessment should be taken of how the occupants are performing within the structure: Are they comfortable? Does any particular function distract them from their duties? How does the equipment meet their needs? Secondly, the owner or facilities director can share information on how much energy the building uses, whether it has suited the company's needs and vision, and what things might have been better additions or features. Lastly, a post-occupancy walk-through gives the architect a better idea of what works well, what has not particularly met expectations, and what types of modifications to design philosophy can be explored (or pitfalls to avoid in the future).

# 5 Bridge Projects

Bridges are perhaps the most vulnerable structural systems in terms of laying bare all of their wonders, their failings, and being subject to criticism. If the public perceives that its form inadequately represents its function, the structural engineer may be saddled with a bad image. On the other hand, feats of great courage and imagination, such as the crossing channels shown in Fig. 5-1, can instill pride in the hearts of engineers around the world.

## 5.1 TYPES OF BRIDGES

Bridges are classified according to function or structural type. In terms of function, most bridges carry highways, railways, service lines (water, power cables, telecommunication), pedestrian routes, or a combination thereof. Though type of structure is planned according to budget and practical constraints of the site, selected form will also be dictated by requirements of strength, serviceability, stability, fatigue, durability, expected service life, and maintenance requirements.

### 5.1.1 Highway

Highway bridge construction is overseen by the Federal Highway Administration (FHWA), a division of the U.S. Department of Transportation (USDOT), in order to assure that minimum standards of quality are built into the nation's infrastructure. Though most of the nation's highways are owned by state, local, and tribal governments, the FHWA provides financial and technical assistance to support construction, improvement, and preservation. Title 23, Section 625.4 of the Code

**Figure 5-1** Canal Bridge Magdeburg, Germany. (*Courtesy Dillinger Hütte*)

of Federal Regulations (CFR) incorporates American Association of State Highway and Transportation Officials (AASHTO) *Standard Specifications for Highway Bridges* (Standard) and *LRFD Bridge Design Specifications* (LRFD) as national design standards, along with others addressing different aspects of highway and appurtenance construction like structural support of signs and luminaries, structural welding, and testing of materials. After October 2007, the LRFD specifications published by AASHTO became mandated by the FHWA for all federally funded projects.

Highway bridges are classified by individual localities according to importance, traffic type and volume, design speed, and other aspects. A median is provided on major highways to separate opposing traffic flows, but some bridges may be completely divided into unique structures for further isolation. Pedestrian walks are usually protected from traffic by a railing system designed to contain an average-sized vehicle, yet allow for freedom of view. It is important to remember that a median strip that is poured monolithically with the main concrete deck contributes to longitudinal stiffness and it may be necessary to model this into a structural analysis. However, it is acceptable practice to simply add the median's weight into the analysis, as these portions are often placed as a secondary pour over a cold joint in the bridge deck.

Design live loading for highway bridges is primarily based on a *design truck* as identified by AASHTO, but also includes that due to bicycles and pedestrians.

In 1944, five general classes of vehicle live loading were introduced into design specifications: H20, H15, and H10 (trucks with two loaded axles); HS20 and HS15 (truck and trailer with three loaded axles as shown in Fig. 5-2). Some state DOT's add a multiplication factor to these base load values and require additional consideration for military usage. One of the primary differences between the AASHTO Standard and LRFD specifications is the live load design truck and application. The new model for use with the LRFD specifications is known as HL-93 and was developed in 1993 as a combination of the old HS20 design truck and design lane requirements. Mathematical formulas that define vehicular live load distribution are based on bridge type and complexity and may be based on a longitudinal strip (slab-type), simply supported slab reactions according to the *lever rule* (beam-girder bridges), or three-dimensional consideration per the *refined analysis* method (truss-arch bridges). Such analyses may also require consideration of nonlinear effects (cable-stayed) or large deflection theory (suspension bridges). Horizontal live loads are taken into account as braking or centrifugal forces applied at a distance of 6 ft (1.8 m) above the roadway.

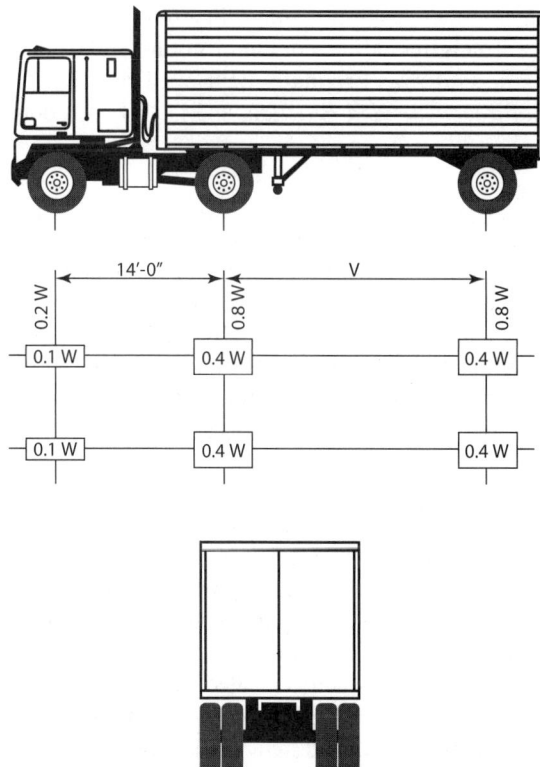

**Figure 5-2** AASHTO *Standard* truck weight distribution.

## 5.1.2 Railway

The Federal Railroad Administration (FRA) was created in 1966 as a division of the USDOT to promote and enforce rail safety regulations, administer railroad assistance programs, conduct research and development in support of improved safety and national policy, provide for rehabilitation of rail passenger service, and to consolidate government support of rail transportation activities. The Track and Structures division issues manuals, circulars, and organizes meetings in order to provide technical expertise related to structural safety.

The American Railway Engineering and Maintenance of Way Association (AREMA) was formed in 1997 as a merger between the following groups: The American Railway Bridge and Building Association, originally formed in 1891 for the purpose of exchanging information and discussing solutions to problems; the American Railway Engineering Association (AREA), first organized in 1898 to develop recommended practices through the study of engineering challenges related to rail, ties, track, buildings, yards, and terminals; the Roadmasters and Maintenance of Way Association, organized in 1883 as a gathering of maintenance officers to discuss problems and solutions and develop standard maintenance practices; and some functions within the Association of American Railroads including communications and signals. *The Manual of Railway Engineering* was originally developed by AREA in 1905, then entitled *Manual of Recommended Practice*, and is still considered the industry standard.

Railroad bridges are designed for a higher live-to-dead load ratio and require careful consideration of serviceability issues. Live load is specified by axle-load diagrams, or by an E-number of a *Cooper's Train*, consisting of two locomotives and an indefinite number of coupled cars. Design values for bending moment, shear, and resultant reactions are commonly tabulated for E10 loading, multiplied by a constant for proportional adjustment to achieve other loadings, such as E72 (multiplication factor of 7.2) or E80 (multiplication factor of 8), which happens to be the most common design load. Loading is distributed to each track according to historical assumptions.

Because a railroad system travels along a set of tracks, bridge construction tolerances must match those necessary for placement and restrictions of movement in service. Suspension-type bridges are rarely used for supporting railroad traffic, due in part to these strict deformation tolerance requirements. It is common in early literature to picture the movement of a locomotive and cars across a truss-type bridge, such as that shown in Fig. 5-3, which is created by fastening beams together in a triangular configuration: There will always be a top and bottom chord element, either of which may be arched, and individual web members that distribute carried forces between the chords by alternating tension and compression forces.

**Figure 5-3** Scotland's Forth Bridge, cantilever steel-type for railroad use.

## 5.1.3 Others

Pedestrian bridges deserve a great deal of artistry and imagination because those who use them will have sufficient time to admire the trip across. They are placed to cross such obstacles as interstate highways, other main thoroughfares, rivers, ravines, railroad switchyards, or other physical obstacles to keep people moving about their business. Placement of these bridges must be done strategically in such a way that the traveler is invited to cross rather than being forced to do so by some type of barrier. A careful study of the geography, normal behavior patterns of pedestrians, and traffic volume in a particular region guides the decision making process in terms of form and placement. In 1997, AASHTO published the *Guide Specifications for Design of Pedestrian Bridges* for assistance in designing bridges that will be used primarily for pedestrian and bicycle traffic.

Aircraft runways or taxiways require bridges to span across grade separations, transportation routes, or locations of weak soil. Width, capacity, and clearances are to meet requirements of the Federal Aviation Administration or other agency having such control.

## 5.2 SIZE AND FUNCTION OF BRIDGES

The width and height of a deck over water or a roadway influences the final shape and size of a bridge system to a certain degree. Columns cannot appear too narrow when compared with the rest of the profile and cable-supporting towers should be shaped according to the majesty being portrayed. Architecturally, the function of a bridge should also influence its form, expressed in conformance with structural requirements. An appropriate design is one that efficiently responds to the flow of forces after consideration of a variety of engineering issues as opposed to one that focuses too heavily on artistic appeal. Geometry, therefore, must closely follow structural function. The most common structural designs include the more common beam or cantilever forms and signature systems such as the arch, truss, suspension, cable-stayed, movable, or floatable.

The superstructure of a bridge has a geometry determined by route alignment, required clearances above and below, special restrictions in the area of construction (flight path, type of traffic, noise restrictions), environmental issues (high temperatures, the need for snow removal) and some architectural considerations. It is primarily this assembly of members that dictates behavior under load, though type of bearing, abutments, and supports from the foundation also contribute.

### 5.2.1 Reasons to Span

Bridges are critical elements in a region's infrastructure, as they serve to transport people and goods across barriers to sustain a healthy economy. The particular crossing influences choice of foundation and pier construction measures and difficult routes can control economic viability in some regions.

**Bodies of Water** Most early bridges were constructed to cross a waterway, especially for mobilization of the military. Some ancient bridges were even retractable to allow water traffic to pass through. As technology and commerce accelerated over the ages, conflict between water and land traffic became inevitable and more elaborate types of movable bridges were produced. The early industrial era marked a significant increase in the construction of swing-type bridges, which rotate horizontally on a pivot located under the center of gravity, because they were easier to design than draw-type bridges and commonly believed to be more economical since mechanical equipment did not have to lift an enormous mass high into the air. The U.S. Coast Guard determines elevations by which bridge clearances above navigable waterways are to be measured.

Floating bridges have also been used to carry traffic across a body of water, in which sealed, floating containers called pontoons are fastened together to support a roadway placed on top. These are particularly suited for crossings of unusual depth or with soft/unstable bedding that restrict the use of conventional-type foundations and pier supports. Though direct vertical support by a foundation system is not required, anchors against translational and lateral movement are necessary to keep bridge segments in proper alignment. Long-term projections of a particular waterway's usage become critical in the design and specification of a floating bridge.

**Grade Separations**   The most common type of grade separation is found between crossing highways or thoroughfares, such as is shown in Fig. 5-4, where a bridge must be provided with sufficient clearance to allow for traffic flow in any direction that commerce may dictate.

Arch-type bridges are particularly suited from a structural standpoint for crossing ravines or valleys with steep, solid walls because the force of the carried load is transferred directly to the abutments through compressive action. They can be constructed in a variety of forms, including deck-type (straight deck placed atop the crown of an arch), through-type (straight deck is located at the springline, or the base, of an arch), or a half-through form (intermediate placement of deck). The thrusting force at each end due to a curved compression element may also be taken up through a tension tie, thus eliminating or greatly reducing high abutment reactions depending on where the tie is placed relative to the bottom of the arch. Although a pure structural form is offered, other mitigating factors come into play to decide whether this type of bridge will be suited for a particular site, especially including economic review.

I **Figure 5-4**   Simple concrete grade-crossing bridges.

## 5.2.2 Scour

Scour in the vicinity of a bridge that crosses a stream occurs as a result of long-term changes in streambed elevation; removal of material from the bed and banks across the channel width; removal of material from around piers, abutments, spurs, and embankments due to acceleration of flow and creation of vortices where obstructions occur; and due to natural tendencies of a waterway to scour and fill during runoff cycles (Mueller and Wagner, 2005 p. 3). Larger bodies of water are subject to similar phenomena. Sound design procedures call for an arrangement of intrusive elements in such a way that the potential for scour is controlled to a reasonable degree. Adverse flow patterns can be avoided by streamlining column or foundation elements and through provision of sufficient abutment height or footing depth out of scouring regions.

In designing bridge systems against the effects of scour, different activities are defined in order to provide focused protection. Degradation occurs as a lowering of the stream or river bed due to movement of the water and elevation changes when sufficient replacement sediment is not delivered from upstream regions. Contraction scour occurs from increased flow velocity when a waterway's flow width is reduced naturally or by manmade intrusions. These phenomena are evaluated using accurate data of existing conditions and identification of potential problems by field observation, historic scour investigation, identification of problematic design or natural features, and projections of damage during an extreme event, such as a flood.

## 5.2.3 Elements of Bridges

The overall performance of a bridge is based on the contribution of each constructed element. The term *superstructure* is given to the combination of structural components above a foundation or pier system and includes a traveling surface (deck), supporting elements upon which the deck rests (girders), components that provide intermediate support of the girders (suspenders and cables), and main structural bodies that transfer all imposed forces to foundation walls (towers, arches, columns). Some bridge forms exclude one element or another, but the basic load path is always directed toward the foundation and abutment systems.

The *substructure* of a bridge system includes bearings, piers, columns, foundation walls, abutments, and foundations. Towers are sometimes included within this grouping depending on their position relative to the deck surface. Translational forces are applied to these systems from creep, shrinkage, traffic braking or centrifugal effects, and thermal changes. Rotational-type forces occur from normal effects of traffic, construction tolerance, and uneven settlement of foundations.

Bridge systems work together to not only support all imposed loading without collapse, but also to maintain some tolerance on deflection and vibration. Excessive deflection of a bridge not only builds up deformation stresses and fatigue effects due to extreme vibration, but people who travel across will experience great discomfort, mostly psychological in nature. Inbanathan and Wieland (1987) list seven factors that influence forces produced in a bridge by this dynamic load condition: Type of bridge and its own natural frequencies of vibration; vehicle characteristics such as mass distribution, vibration natural frequency, and suspension; speed of the vehicle; profile of the approach roadway and deck; traffic intensity; damping characteristics of the bridge and vehicle; and behavior of the driver. Using a mathematical model and the results of a statistical analysis, these researchers concluded that the maximum dynamic response of moving vehicles is produced by trucks traveling at high speeds and that the dynamic effects of passenger cars can often be disregarded.

**Decks** The structural deck of a bridge is critical in transferring all imposed forces, including environmental and live traffic loads, to supporting structural elements and different types have been used successfully over the years, including the concrete deck shown being placed in Fig. 5-5. Technological innovation has introduced the use of glass fiber-reinforced polymer (GFRP) panels that were shown by testing to act in a somewhat composite fashion with steel girders when properly connected (Stiller, Gergely, and Rochelle 2006), though the effectiveness of this behavior is not considered by many designers of great enough value to include in analysis.

An orthotropic steel-framed deck was introduced in Europe during the post-World War II years in an effort to maximize use of available steel. Engineers were

**Figure 5-5** Placement of concrete bridge deck. (*Reproduced by permission of the Virginia Department of Transportation*)

able to economize an assembly of framing members due to previous testing and great advances in structural analytical theory. Some of these tests were performed by scientists for the American Institute of Steel Construction in the 1930s on a layout known as a *battledeck floor* (see Fig. 5-6), consisting of standard structural beams with steel plates above. Plating served to transfer wheel loads to stringers and to increase the effective width of supporting girder top flanges, thus reducing section sizes.

The formal model is based on a simple orthotropic plate, which has constant thickness but different flexural and torsional rigidities in perpendicular directions. Simplified forms of analysis are valid under most circumstances since load distribution is mainly directed through flexure and torsion in the longitudinal and transverse directions, with shear deformation remaining negligible. Bridges are modeled as orthotropic plate-type systems for simplicity as long as certain limitations are observed: Deck width and depth remain constant and uniform along the bridge's length with a maximum skew angle of 20°; support conditions are projected as linear; curvature in plan is not unreasonably tight; where a concrete slab deck is used, a

**Figure 5-6** Original Severn Bridge crossing between Wales and England, using a steel battledeck floor.

minimum of four girders with equal rigidity are provided and the length of deck overhang is restricted to 60% of the spacing between girders; and a minimum of three cells exist where cellular decks are used (Bakht and Jaeger 1985, p. 81).

The soundness and suitability of a deck's wearing surface is based on its ability to completely protect against corrosion, resist penetration of water and chemical agents, expand and contract in such a way as to avoid cracking, and strength of bond to the structural deck below. It should be lightweight, yet thick enough to cover irregularities over the structural deck below, and provide a skid-resistant, even traveling surface.

**Abutments and Retaining Structures** The two most common abutment types are the *seat-type*, which allows for relative movement of the deck, and the *integral-type*. Design of any form of abutment according to DOT requirements typically assumes linear behavior when subjected to load, but nonlinear behavior during a strong seismic event may occur in response to cracking and yielding of the material, in addition to changes in stiffness of the supporting soil. This has an effect on the dynamic response of the superstructure itself. When a complete structural model is created for analysis, a properly reinforced concrete approach slab extending far enough beyond an abutment can be counted on for rotational restraint and considered in determining the dynamic response of the bridge.

Reinforced earth-retaining structures are essentially unrestricted in height, though lateral movement needs to be evaluated. Mechanical stabilized earth (MSE) structures are formed of multiple layers of fabric strips to reinforce fill material in the lateral direction so that the resultant mass is capable of holding back the associated abutment. The completed mass acts as a gravity-type retaining structure and can be evaluated as such in terms of overturning, sliding, and other conditions of soil support.

**Piers and Foundations** Piers that transfer all imposed loads to the foundation below are designed in a variety of different ways, as shown in Fig. 5-7, from

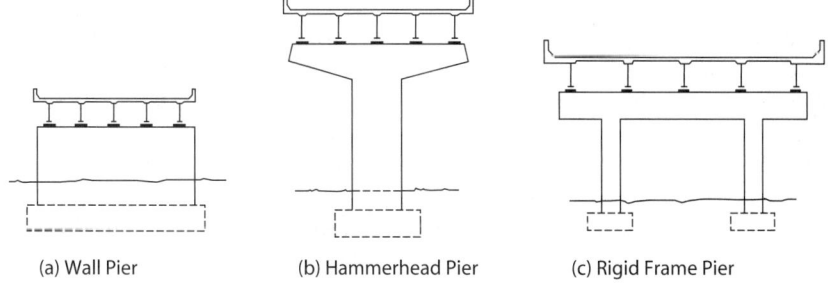

(a) Wall Pier  (b) Hammerhead Pier  (c) Rigid Frame Pier

**Figure 5-7** Different types of concrete bridge piers depending on site and economy.

simple columns to elaborate towers. Difficult conditions such as deep water or soft soils can greatly increase project cost and an engineer may choose to reduce the number of piers and foundations by increasing a bridge's span length. A simple, concrete mat may be placed below piers when the ground is strong and stable. If the ground is soft, it can either be excavated to a layer of bedrock below for direct placement of bridge piers, or weak layers can be penetrated by piles to the depth of bedrock which will then carry the piers above.

Underwater foundations require special construction procedures and administration. A *cofferdam* is a temporary watertight enclosure, constructed of reinforced steel sheets driven into the ground, placed directly onto the site where a bridge pier is to be founded. The chamber is pumped dry to expose the soil beneath, which can be excavated to bedrock or can provide a clear working area for driving piles. A *caisson*, on the other hand, is a large cylinder or box constructed ahead of time and sunk down to the river or lake bed. Foundation work is completed within the submerged caisson, which may actually be left in place and filled with concrete to form part of the total foundation system.

Other pier types are discussed in detail in Chapter 9.

**Deck Support Components** Suspension bridges are constructed using two main cables draped over towers and solidly embedded into anchorage blocks at the abutments. The deck and carriageways are suspended from these cables through a path of smaller vertical cables, though secondary diagonal cables are sometimes included for added stability. Some familiar towers used in suspension-type bridges are shown in Fig. 5-8. Dynamic characteristics of suspension bridges are based on superstructure stiffness of the system and on the mass of deck, piers, and main cables. Main cables provide the bulk of the vertical bending and torsional stiffness which can be boosted during construction by increasing deck mass near the piers. The deck contributes greater stiffness to the torsional response than that due to bending, indicating the need for a strategic erection sequence since the overall stiffness and stability of the bridge will be at risk until the final piece of deck is secured.

A cable-stayed bridge, however, does not use main cables or anchorage blocks, which aids the speed of construction. The deck is supported from towers by a series of diagonal cables. There are basically two arrangements in cable-stayed bridge systems: Those where the cables are separated into two planes and those forming a single plane. Cables of a single-plane system are normally placed along the median strip of a divided highway, whereas a double-plane system arranges the cables so that they support the outside of the roadway. These cables either fasten to the tower in separate planes or converge toward each other along the bridge's centerline (see Fig. 5-9). There are four basic cable configurations when viewed in elevation, called the *converging type*, where all cables are taken to the top of the tower; the *harp type*,

Bridge Projects | 119

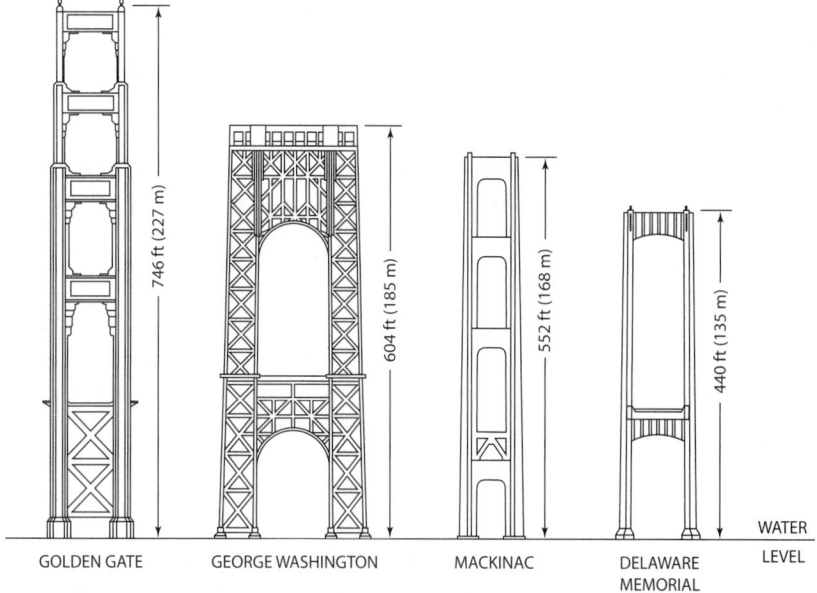

**Figure 5-8** Different steel suspension bridge tower configurations.

**Figure 5-9** Cable-stayed pedestrian bridge with added support for fabric shade canopy.

where cables are parallel to each other and equally spaced along the deck and tower; the *fan type*, where cables are equally spaced along the deck and tower but are not parallel to each other; and the *star type*, where cables are equally spaced along the tower but converge to a common point along the deck.

Bearings are basic structural elements that transmit forces from the superstructure to the substructure, classified as *fixed-* or *expansion-type*, and are subjected to dead, live, wind, thermal, and longitudinal seismic loads. Some common bearing types are shown in Fig. 5-10. An *elastomeric bearing* pad, by far the most common, is constructed of rubber reinforced with steel plates or fiberglass in order to control excessive deflection under compressive loading and is considered to be one of the least expensive (Mistry 1994). *Sliding bearings* are those where a steel plate attached to the superstructure slides over another plate attached to the substructure. Lubricant is used to reduce frictional forces and they can only be used for shorter spans. Sometimes guides are provided along the sliding parts for stability and to control the direction of expansion.

A *rocker bearing* is in the expansion-type category, consisting of a pin at the top for rotation and a curved surface at the bottom to accommodate translation, yet restrained from walking by a pin and steel keys. A *pin bearing*, on the other hand,

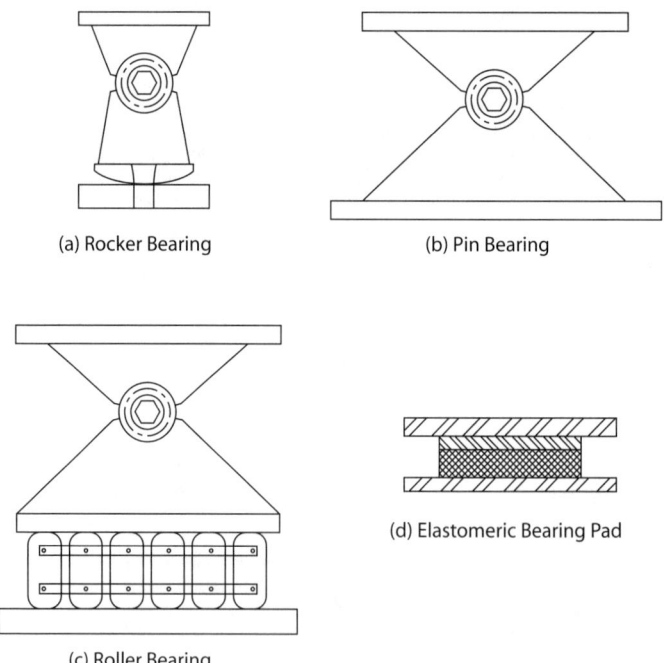

**Figure 5-10** Different types of bridge bearings.

is a fixed-type mechanism similar to the rocker, only the bottom portion is rigidly attached to the supporting surface. *Roller bearings* consist of one or more rollers placed between two steel plates, where a single roller is capable of accommodating both rotation and translation but multiple rollers need to be coupled with a pin bearing to carry the rotational component.

## 5.3 BRIDGE SYSTEMS

A bridge system can be visualized in terms of different parts integrated together by simple, rigid, semirigid, or composite connections. Simplified methods of design independently consider these elements—deck, secondary joists, main girders or cables for deck support, piers or towers, and foundation—with minimal sharing of stiffness or strength. In most cases, this leads to conservative results, but can also prove to be a dangerous simplification because the stability of longer span bridges depends greatly on their interaction. A conventional method of analysis, for example, considers deck support conditions to be rigid, whereas the system below actually deflects and twists to a certain angle that is difficult to accurately define without some use of advanced analysis, either by hand (cumbersome) or with the aid of computers.

### 5.3.1 Slab Spans

Concrete slab-type decks are constructed of either solid, voided (prestressed), or ribbed segments to serve as the superstructure spanning between abutments and piers. They may be designed as simply supported, cantilevered, or continuous across multiple supports and prove to be economical for spans of up to 80 ft (25 m) (ACI 343R-95, Section 2.5.3.1). Prestressing steel added in both the longitudinal and transverse directions assists in achieving longer spans. These greater span lengths can also be achieved by the use of composite action, where a concrete slab is connected to individual steel or concrete girders with headed bolts or some other form (see Fig. 5-11) in such a way as to share shearing forces over the length of contact.

### 5.3.2 Steel

Weathering steels are currently supported by ASTM A242, A588, and A709 and can be quite expensive. Mistry (1994) recommends that these steels be considered for applications that involve exposure to highly corrosive fumes, severe marine conditions with repeated wetting by salt spray or fog, burial in the ground, submerging without adequate protection, and details subjected to run-off containing deicing salts. Nonweathering or other unprotected steel I-girders have shown greater corrosive attack predominantly along horizontal surfaces (bottom flanges, gusset plates, stiffeners, and splices), in crevices, and at reentrant corners.

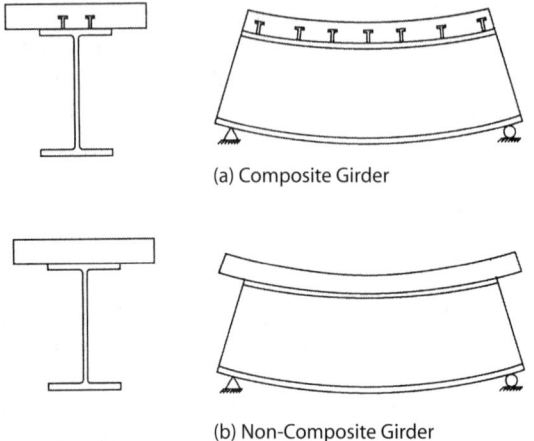

**Figure 5-11** Demonstration of composite action between steel girder and cast-in-place concrete deck through strain compatibility.

Steel is a material that lends itself well to the formation of horizontal curves, as the bending radius is controlled by parameters for travel and is not expected to be uncomfortably tight. Erection study tests have shown that providing even minimal radial restraint for curved I-girders during construction and removing the shoring in a nonuniform fashion have a beneficial effect on their response (Linzell, Leon, and Zureick 2004).

### 5.3.3 Prestressed Concrete

AASHTO has approved a variety of concrete girder cross-sections, including special beams (Types II through VI) particularly suited for spans up to 118 ft (36 m) and box sections that are suited to perhaps 590 ft (180 m). A concrete box girder-type deck consists of top and bottom flange elements connected together with vertical or inclined webs, forming cellular closed sections. They may be constructed with a single cell below two lanes of traffic, or multiple cells below several lanes, and include cantilevered arms to support walkways or provide additional roadway width. Torsional rigidity is one of the greatest qualities of these sections, making them perfectly suited for curved installations where torsional shearing stresses are developed. If the radius of curvature is large compared with its span, a curved box-girder section can essentially be designed as though it were straight, which is of particular value to an engineer seeking to simplify the design process.

Segmental concrete bridge construction has been used successfully for a variety of different systems, including those with simply supported or continuous deck spans, those erected using balanced cantilever methods, and those set in place by incremental launching. Cast-in-place segments may be made on the site, but it is a time consuming process and can occupy a great deal of space. Precasting is the preferred method of creating concrete segments, provided means of transportation and adequate preplanning can be secured. When sections are cast in a plant, better measures of inspection, protection against inclement weather, and control of materials can be expected.

### 5.3.4 Timber

Wood has been used with great success for short- and medium-span bridges over hundreds of years due to advantages of a high strength-to-weight ratio, predictable constructability, good energy absorption capability, reliable durability, and economics. Primary disadvantages, however, include potential for decay, insect infestation and related damage, and obvious hazards of fire. Preservative treatment is usually added to structural members for protection against each of these hazards, though the best material can be quite expensive. Waterborne solutions are used for surfaces receiving human or animal contact (handrails, pedestrian decks) and in marine installations, whereas oil-type treatment is used elsewhere for durability and has less effect on the cell structure of the wood.

Timber abutments have been specified when structural loading is not overly demanding, either driven into the soil as piles or directly bearing on an embankment. When designed as piles, a beam cap is anchored to the top for support of the superstructure against weight, lateral movement, and potential uplift.

### 5.3.5 Movable

A movable bridge deck can either be lifted (vertically or in draw-type fashion) or swung horizontally depending on the mechanism and needs or characteristics of a particular crossing. They are particularly suited for situations in which long approaches to a high level fixed bridge are not feasible (or even constructible) or other alternatives, such as blocking off a waterway to certain types of traffic by a lower fixed-height bridge, installing a tunnel which has its own set of issues, or continued use of a ferry system have been ruled out in the overall plan. Figure 5-12 shows the London Bridge in Great Britain as a unique example of a movable draw-type bridge combined with standard suspension-type sections. In general, bridges that are counterbalanced and opened by pivoting about a horizontal axis are known as *bascule* bridges, those which open by pivoting about a vertical axis are identified as *swing* bridges, and those that open by lifting without rotation or horizontal translation are called *vertical lift* bridges.

**Bascule** A double-leaf bascule bridge provides the quickest passage for waterborne vessels since structural components necessary for lifting are located on opposite sides of the channel instead of placed in the waterway. Structural members experience a significant change in dead load stress as a bascule leaf raises, which has been the cause of failure in some older bridges of this type. Wind loading must

**Figure 5-12** London Bridge, combination suspension and movable-type.

be carefully reviewed when leaves are held in the open position for design of the structure as well as the foundation because of load reversals, possible disengagement from the supports, and general instabilities related to this lifted position.

**Swing** A swing bridge is essentially viewed as two cantilever spans supported on a central pivoting point in the middle of the waterway. When the bridge is closed for land traffic, the ends either rest directly on abutments or on special piers. The ends of these bridges must be free to swing open, but they must also be rigidly attached to allow easy passage of vehicles. This has prompted the use of tricky bearing elements, such as retractable rollers, wedges, shims, and jacks, not to mention elaborate mechanisms necessary for keeping railroad tracks together.

**Vertical Lift** A vertical lift bridge, such as is shown in Fig. 5-13, contains a simple spanning deck, usually a truss-type structure, that is raised straight up with the help of counterweights installed in towers, placed at each end or at all corners of the deck. The entire dead weight of the deck is transferred to the towers by counterweight ropes mounted on special sheaves during operation and are susceptible to wear, corrosion, and fatigue, commonly requiring replacement after 40 or 50 years (Koglin 2003, p. 60). This type of movable bridge is best suited for very long spans because it is the most stable and least complicated to design. The lifting bridge is designed as a simple span for both dead and live loads, can be built to any width desired, and may also contain numerous girders. They are intended to seat comfortably at the corners under all loading conditions without requiring a complicated central alignment.

**Figure 5-13** Movable bridge with towers that have become city landmarks (Sacramento, California).

## 5.4 OTHER ISSUES

Economics and aesthetics are likely the two biggest public concerns when considering a bridge construction project. Appearance will have a lasting effect on residents and visitors alike and bridges often serve as symbols of a region's prosperity (or lack thereof). For signature bridges, the engineer makes a conscious selection of form in such a way that the most visible elements can be weaved together and connected to form a work of art, appreciated by those who purchase, observe, and make use of it, otherwise the form of a bridge is usually dictated by the one with minimum ongoing maintenance costs. Structural artists like Thomas Telford (1757–1834), Gustav Eiffel (1832–1923), John Roebling (1806–1869), and Robert Maillart (1872–1940) have created bridge forms that stand the test of time and criticism, serving to remind all engineers of their responsibility to human sensitivity. Although structural art is disciplined by material mechanics and issues of economics, freedom in creating pleasing forms can take many different paths toward elegance.

A necessary, though sometimes inhibiting first step an engineer must take in designing a bridge is to evaluate existing creations in order to provide scientific justification for any innovation that might be proposed. Precedent drives not only selection of structural form, but also becomes critical during fabrication and construction, as workers are not often paid an adequate sum for exploring avenues they have never seen before. The principle of aesthetics does not always follow brand new configurations, but many times is built on an innovative modification or combination of existing solutions. For example, a simple-span girder can be formed into a variety of profiles; a rigid frame bent may have straight or inclined legs with different choices of section geometry; arch-type bridge systems have been aligned as normally expected and inverted for visual effect; and towers that support cable-stayed or suspended decks are seen formed into unique architectural shapes that express a region's character.

The image of engineering must fit well into a natural setting, as all forces to which a bridge is subjected are derived from nature and delivered back into nature through a continuous load path. Patterns of behavior can only be understood as the natural flow of forces between members and connections is analyzed using mathematical, scientific, and even philosophical equations according to a rational expectation of what has been experienced for thousands of years. History represents continuity of thought and progress, which in turn determines the direction of aesthetics.

### 5.4.1 Drainage

Surface water that is allowed to sit on a bridge for any length of time poses a danger to traffic, as flow is disrupted by driver uncertainty and loss of visibility due to splash and spray. Hydroplaning, or fear thereof, has long contributed to accidents resulting in fatality and sometimes damage to a bridge's structural elements. During winter months, standing water quickly turns into a dangerous patch of ice, and the presence of deicing salts leads to corrosion of steel if not properly controlled. The depth of water that builds up along a roadway is based on pavement type and texture, rainfall intensity and duration, consistency of maintenance, and adequacy of transverse and longitudinal grading. Spreading of water across the deck will have a different effect in arid regions that are subject to thunderstorms of great intensity versus areas that experience weaker, more frequent events.

Adequate drainage of rain water or other runoff from the deck in the transverse direction may be accomplished with a suitable crown, placed strategically along the alignment in relation to the number of lanes and width. Longitudinal drainage is handled by camber or gradient added to the profile and water may also be directed into adjacent gutters. A bridge engineer is rarely given opportunity to modify a roadway profile that has been adopted and in use by a particular jurisdiction for years, since it is important for travelers to expect a level of consistency or uniformity over the course of their movement.

There is also a need to properly channel the flow of water running off the end of a bridge in such a way that erosion of the approach embankments is not exacerbated. A piping system sloped for gravity flow can be placed along the span of longer bridges to collect water that flows into scuppers or drain inlets and transport it away from the structure.

## 5.4.2 Joints

Creep, shrinkage, and temperature changes in deck length are accommodated by expansion joints. Some concrete bridge decks have performed adequately with expansion joints placed nearly 4000 ft (1220 m) apart (Chen and Duan 1999, p. 11–29). Joints have logically been placed at the tip of finished cantilever segments or over piers for span-to-span construction methods. It is always desirable to keep the number of joints to a minimum because they increase the initial cost of a bridge and must be properly maintained over the years to retain their function (see Fig. 5-14). Construction joints of posttensioned bridge structures require coupling of tendons for continuity, which is done either directly at or around the joints, and this also increases complexity of ongoing maintenance.

An integral abutment-type bridge is one that does not have thermal expansion and contraction joints. Rather, movement of the superstructure is accommodated by flexible pilings carefully designed to account for lateral movement in addition to vertical loads. Greimann and Wolde-Tinsae (1984, p. 83) suggest that lateral loading

I **Figure 5-14** Joint in concrete bridge deck.

may be simply distributed to individual piles within a group such as this if the spacing of piles perpendicular to direction of applied loading is greater than about 3 times the pile diameter or width. The impact of abutment movement due to thermal expansion and contraction on the approach slab and fill requires careful consideration for this type of bridge, in addition to axial stresses induced within the superstructure relative to this movement.

## 5.5 PROJECT PHASES

A bridge project, in a nutshell, begins with an effort to raise funds for a conceptual design that has received public approval after discussing the results of many studies, including functional or capacity expansion, environmental impact, economic development, and traffic safety. The owner then awards a design contract to a consulting engineering firm who is responsible for producing project documents, followed by an approval process directed through all agencies with an interest in public safety. Documents are then put out to bid where contractors propose a lump-sum cost for completing the work (most common) and oftentimes the lowest successful bidder is awarded a contract for construction. After construction is finished, the bridge is opened up to much fanfare and political stumping.

### 5.5.1 Approval Phase

Just as is necessary for building projects, establishment of new infrastructure must follow a predetermined sequence of design according to adopted regulations, careful analysis of environmental and social impact, obtaining adequate financing, and approval of public agencies, the public in general, and elected representatives.

**Planning, Financing, and Politics**  Planning and financing of infrastructure is based on respective visions of federal, state, and local agencies for economic and strategic development well into the future. Regional and local efforts are not always well coordinated and locally driven public works or capital improvement projects are unavoidably dictated by patterns established through regional infrastructure, such as location of airport facilities, route of freeways, and placement of rail lines. Funding sources for projects of any size include taxes, tax-exempt bonds, donations, user and developer fees (though these types are more locally based), or a combination thereof. Fees charged to developers can only be imposed in proportion to the effect of a new real estate project on surrounding infrastructure, which makes it difficult to collect revenue in this manner for remote bridges of any size. In many cases, capital from different developments is banked and combined with other resources to fund a future project.

The longevity of infrastructure involves not only new facilities, but monitoring and maintaining existing ones. Old bridges are replaced when the cost of maintaining

**Figure 5-15** Concrete piers for support of new bridge adjacent to existing bridge. (*Reproduced by permission of the Virginia Department of Transportation*)

them exceeds budgeted amounts or when traffic volume or weight exceeds original design limits, and because of deterioration (see Fig. 5-15). Maintenance operations can be expensive and are funded on many routes through the collection of tolls. There is also an operational cost absorbed by society during the time when an older bridge is replaced, as people adjust their behavior patterns and routines during this period.

It is common practice for state and local agencies to pay a portion of the cost for erecting new bridges, but most of the bill is funded by the federal government because of the impact on and interest in national infrastructure. Because planning of federal, state, and local infrastructure is grounded in economic development, it is also driven by political pressure. Public citizens and special interest groups can effectively promote or condemn plans for regional growth or rehabilitation due to the (mostly) open process of planning, the ease by which information can be obtained, and an often complicated legal system. Public interest helps add accountability to the planning process, but land use regulations are often exploited by well-financed groups large enough to gain leverage over city councils, supervisory boards, regional managers, and legislators in all levels of government. Overall, however, planning and financing of infrastructure is served well by the inclusion of a variety of interest groups that are able to balance one another and have appeal to a broader public.

**Traffic/Use Review** Regional and local planners study land use demands and transportation patterns in order to recommend policy and a master vision for future prosperity. The layout of regional transportation routes gives not only function to a community, but also dictates the form it will take and capacity for growth, for better or worse. Local projects that do not expand over jurisdictional lines are important in filling the gaps left over by larger projects to link citizens together, giving them a strategic solution to problems that accompany regional infrastructure, such as traffic congestion and environmental impact.

Federal and state transportation departments, local public works departments, port authorities and other land owners, operating organizations (railroads, airlines, ship and truck lines), and metropolitan planning organizations all have a stake in new bridge renovation or construction projects. New routes and expansion or rehabilitation of existing traffic ways are studied for future impact, regardless of jurisdictional control. Increased traffic through a particular region may call for new bridges to follow an existing route and disruption of daily activities in the surrounding area will be considered during the approval phase. Transportation planning studies usually begin with a *request for proposal* (RFP) by the sponsoring agency when an outside consultant is to be chosen, or by a detailed work program for in-house projects. Once personnel are on board, details of how the study is to be administered are worked out, then relevant statistical, economic, environmental, and demand projection data is gathered, analyzed, and summarized in a written report.

**Environmental** Evaluation of any transportation project relies a great deal on environmental impact assessment, which can be a deciding factor of project feasibility, priority given for funding, or selection between different alternatives. The National Environmental Policy Act of 1969 requires preparation of an environmental impact statement for any action taken by the federal government, including funding of state or locally regulated projects that have some effect on the environment. The report is to address probable impact on the environment, unavoidable conditions created by the project, alternatives to the proposed work, and resource commitments necessary for completion.

Different impacts requiring consideration include social and economic changes to neighborhoods, housing or business relocation, air quality and noise, energy consumption, potential harm to environmentally sensitive regions, land use, preservation of significant archaeological or historical sites, contamination of water sources, construction-related disturbance of wildlife habitats or landscaping, and aesthetic changes. According to Section 4(f) of the DOT Act (Public Law 89-670), special planning effort is also necessary for preservation of public parks, recreation areas, wildlife refuges, historic sites, and other lands under protection by the federal government.

### 5.5.2 Design Phase

The designer of most outsourced bridge projects (those not completed by government staff) is a consulting engineering firm responsible for producing drawings and administering the owner's bidding requirements and procedures. Specifications are usually provided by the state DOT. In order to select the best structural form of a bridge, engineers must anticipate type of traffic or load that is to be carried, maximum span, height and clearances necessary for avoiding obstacles, temperature and other environmental conditions, character and quality of abutments, geometry of approach paths, budget, and completion date. The quality and expertise of local labor, availability of construction materials, and local transportation to the site will also influence the design decision.

Imagining a schematic solution to a bridge problem is the first step toward discovering a suitable design. A bridge engineer must see the solution before committing it to paper. Preliminary concepts are developed prior to initiating any hard scientific analysis to establish an appropriate plan of attack. Existing bridge solutions are meant to guide a designer in the process of comparing different systems that meet the goals of a project and discovering creative solutions that may not have been used before. Preliminary structural analysis and further research into anticipated response to loading is necessary to weigh scientific advantages of different solutions. Alternatives are not only to be based on structural behavior, but also on construction time and cost, maintenance demands, availability and transportation of building materials to the site, geologic or environmental sensitivities related to the options under review, and the ease with which a solution can be justified through analysis and experience.

A successful design is judged to meet certain expectations of technical, functional, aesthetic, and economic suitability for the intended use. These characteristics are codependent and do not exist in isolation, therefore any final design solution should be chosen as the best combination possible. For example, a suitable location chosen for simplicity may initially be thought of as one requiring the shortest span, but deeper water also requires a more substantial and costly substructure. Another engineer finds that one particular structural form fits well into a historic region of a city, but transportation of special materials and storage on site is costly. Choice of a final design is not as straight-forward as it might seem and historic treatment of such by the DOT usually can give a good indication of what will finally be approved. A simple arrangement of bridge elements is shown in Fig. 5-16.

Plans and specifications issued by the engineer should state enough information to establish the basis for design, including standard used and publication date, type and magnitude of loading, material strengths, foundation bearing values, design method, expected flood elevation of waterways, tolerances for workmanship, quality control measures to be implemented during construction, and sometimes intended maintenance plans.

Data for a hydrology study are often gathered from as-built plans, site investigations, field surveys, bridge maintenance books, files from those who have studied

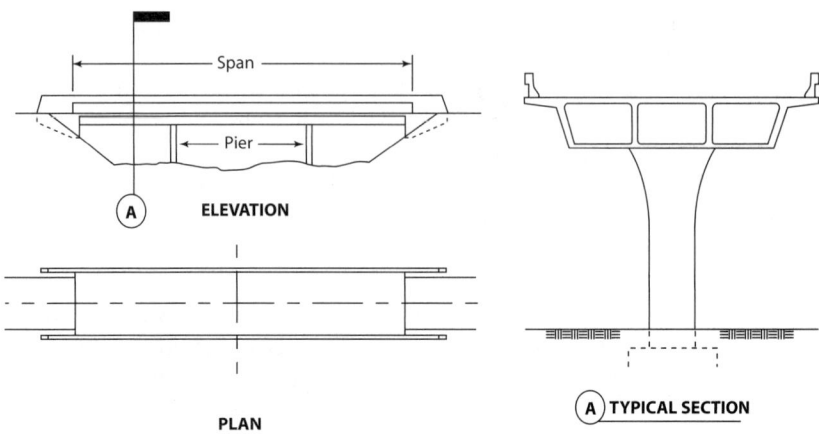

| **Figure 5-16**  Typical structural elements of a simple concrete bridge system.

the same region or from government agencies (U.S. Army Corps of Engineers, U.S. Geological Survey, U.S. Department of Agriculture, FEMA), meteorological data, stream and reservoir gauge data, and aerial photographs.

### 5.5.3 Construction Phase

Bridge construction is a formidable task, but an experienced contractor can ease pressure that an owner receives from the public and funding agencies. All members of the construction team must be coordinated and working together, since no one wants the project to fail. Other entities or persons who take part in construction include a resident project engineer to perform special inspections and monitor the quality of materials and labor. This engineer represents the owner and typically has a team of inspectors to help coordinate testing and observation at different phases of development. If the bridge is to serve highway traffic, the state's DOT is closely involved as well to make sure that its own requirements are met, regardless of whether the state actually owns the bridge. The designer is responsible for incorporating details and specifications into the project documents so the bid correctly reflects all applicable regulations, but the DOT has an important interest to observe that what is planned is, in fact, constructed.

It is also important to understand that each entity has its own reasons for desiring successful completion of a project that are at times mutually exclusive. The contractor wants to maximize profits, onsite inspectors want to maintain a solid relationship with the owner (usually the DOT), and the owner wants to minimize traffic disruption by getting the bridge erected and open as soon as possible. Everyone watches one another, trusts the experience and reputation that all parties bring to the table, but conflict is inevitable as the goal of completion is sometimes approached from different angles of individual importance. Conflict resolution techniques are important tools for a resident engineer in trying to maintain solid progress.

Costs associated with physical construction of large bridges are far greater than what is spent on materials, therefore methods used in this phase are critical to meeting budget. A launched construction approach was used for the Millau Bridge shown in Fig. 5-17, by which the deck is slowly moved out across permanent and temporary supports, thereby producing temporary stresses that may exceed the magnitudes of a structural analysis of the completed structure when subjected to its own weight and thermal or differential support movements. The mechanics of this approach are shown in Fig. 5-18 as related to the Millau project. All sections of the superstructure are cast individually at the starting point of a construction site under carefully controlled conditions, then pushed out along the intended alignment (led by a launching nose). Subsequent segments are coupled to the rear of the finished deck and pushed continuously. This method has been successfully applied to simple slab-type bridge

**Figure 5-17** Launched deck construction of Millau Viaduct cable-stayed bridge, France. (*Courtesy Enerpac, Milwaukee, Wisconsin*)

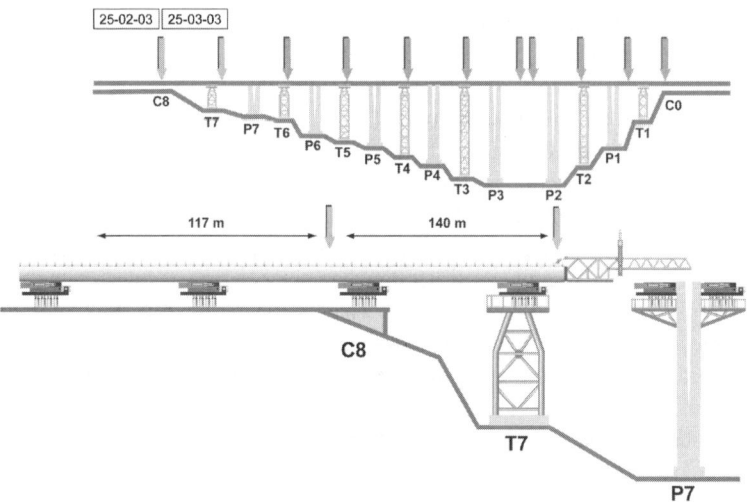

**Figure 5-18** Structural mechanics of bridge deck launching process. (*Courtesy Enerpac, Milwaukee, Wisconsin*)

decks, ribbed slabs, and box girders and long, varying spans, girder depths, and curvatures with other advances being made through research and technology. Cables are prestressed prior to launching and are augmented by additional cables added once the bridge has been launched into its final position. Downhill launching presents a special problem, as the system must both push the bridge to initiate movement and hold it back following release of initial friction at the saddles.

The cost effectiveness of incremental launching is due to repetitive operation and reduction of basic erection equipment requiring only formwork, a pushing device, and a launching nose. Every section of the deck experiences maximum moment and shear due to these loads, thereby resulting in greater material costs. However, if the launching method results in a savings of labor and construction time, any added material expense will most likely be offset. It is likely that construction loading due to launching will not exceed service load combinations, but these conditions must be compared nonetheless. Load transfer devices impose high concentrated stress in areas that are already carrying prestress forces, therefore each point of connection and support must be carefully designed and detailed to resist higher construction loads.

Balanced cantilever segmental construction of bridges has proven to be highly effective in situations where the use of scaffolding or other temporary supports would be difficult due to anticipated disruption of traffic, position of deep gorges, or hazardous waterways. This form of bridge begins with the construction of towers near either side of an obstacle to be crossed and the bridge supporting structure is built outward from the towers, being stabilized through cantilever action, and a simply supported deck segment is set between the larger arms. Possible out-of-balance forces include the presence of a stressing platform on one cantilever (precast method), relocation of the form traveler (cast-in-place method), live load imbalance across supporting piers, wind load prior to final tie-off of segments, and the possibility of one cantilever arm having a dead load that is 2.5% greater than the opposite cantilever (Chen and Duan 1999, p. 11–8).

Bridges are also erected by use of the progressive and span-by-span methods, where construction is performed primarily at the deck level, beginning at one end and proceeding to the opposite side. By the progressive method, segments are placed and secured together successively with support assistance of temporary bents or a tower-stay cable assembly. Span-by-span is often the chosen method for long bridges having fairly short spans, less than about 165 ft (50 m), where continuous forms for cast-in-place construction or continuous steel girders for precast segments are provided between piers and all elements are secured together by prestressing or other means prior to removal of temporary supports.

Bracing, shoring, and protection of surrounding roadways or structures are critical to the successful erection of a bridge. On the morning of June 18, 2000, the outer precast/pretensioned AASHTO Type V girders of the Souvenir Boulevard Bridge in Laval, Quebec, Canada collapsed prior to placement of the concrete slab diaphragm atop (Tremblay and Mitchell 2006). Four outer girders at the southerly bridge's interior spans slid off their bearings without prior warning, tilted over on the cap beams, and fell down onto Highway 15 below. Four interior girders did not collapse. This was not found to have been caused by an extreme loading condition, but rather was due to inadequate bracing of the girder system, as the four which collapsed sat atop pot bearings with sliding surfaces that offer very little resistance to transverse movement or rotation. The four inner girders, on the other hand, were placed atop guided bearing units. The slab and diaphragm formwork kept the girders separated, but did not have enough shear stiffness to prevent twisting and no other form of bracing was present during construction. In this condition, very little lateral load or disturbance is necessary to cause unwanted movement and may have resulted from direct wind loading, vibration and wind-induced pressure from passing traffic below, or thermal effects.

**Shop Drawings and Submittals**   Shop drawings do not ordinarily require a professional engineer's seal and signature unless the fabricator proposes significant changes to the original design. Any such changes require the approval of the owner, the contractor, and the original engineer. Value engineering proposals would also require a professional engineer's stamp.

A general notes sheet from the fabrication facility should contain the following information:

1. *Specifications*: Typically identifies AASHTO or individual state DOT and other standards with governing editions that were used in preparing the drawings,

including that which governs the quality of materials for main and secondary members, such as ASTM/AASHTO.
2. *Fabrication and workmanship*: Indicates requirements for making reentrant cuts in steel, reaming, or drilling procedures. Specifies method of curing concrete and gives instructions on cold- or hot-weather placement. Typical notes and diagrams may also be included in this section to avoid repeating procedures throughout the drawings.
3. *Shop welding and testing*: Describes welding processes to be used and specifications that control their use. Specifies the extent and type of testing to be performed. Fracture-critical welds should be clearly labeled.
4. *Shop cleaning and painting*: Lists blast cleaning requirements, including time limits between blasting and priming, and indicates the type of paint, system, and color, and whether the structure is to be fully or partially coated.
5. *Handling and storage on site*: Describes facility requirements for storage, such as temperature and humidity control. Introduces the need for lifting points and specifies the appropriate position to maintain in storage.

Although the engineer-of-record is not expected to review and confirm every dimension and fabrication condition shown on the shop drawings, certain critical areas need to be checked for conformance with the approved design plans. Essential controlling dimensions and material properties include length of span between bearings, dimensions and grade of steel plates in primary members and splices, primary weights (concrete elements) and dimensions of main girder elements, placement and profile of reinforcement and prestressing tendons, diameter and grade of fasteners, testing requirements for all materials, and sometimes elevation of seats or other supports (usually noted on abutment or pier plans for a given bearing height). Information that was not clear to the fabricator will be flagged as "engineer to verify" and special or unusual fabrication conditions always should be carefully checked.

**Field Work and Continuing Maintenance** The importance of preconstruction meetings cannot be overemphasized. The contractor should clearly explain the sequence of operations so field inspectors know what to expect and can prepare accordingly. If proper handling and bracing are not carried out, unexpected deformation after construction can occur in deck support elements, such as curved steel girders that have a natural tendency to twist and warp during erection. Field inspectors, though not specifying means and methods, act to remind the contractor of his obligations toward safety and of the plans carefully laid out in the preconstruction meetings.

Bridges are designed not only for ultimate load capacity, serviceability, and durability, but also for economical regular inspection and maintenance throughout the intended life of service. Periodic inspection of all bridge components plays an important part in preventing catastrophic failure as potential problems may be discovered locally in a somewhat timely manner depending on the schedule.

Loose and deteriorated deck material should be cleared away and the surface repaired in order to prevent further damage from the infusion of moisture or deicing chemicals. Periodic cleaning and flushing of deck drains, curb outlets, and downspouts is necessary to prevent ponding of excess water. Sand and other debris can accumulate around bearings and become frozen during winter months, transferring excessive tensile stress to the structure below the pads and forming cracks through the anchorage. This also causes unintended restraint of girders which can lead to failure or spalling. Cables should be readily accessible and, if necessary, individually exchangeable.

It is immediately apparent that one of the biggest concerns related to movable bridges is the longevity and maintenance of the intricate mechanical system necessary to operate them. These include motors, gears, bearings, wiring, controls, panels and switches that all can contribute to failure, especially since these elements are also subjected to a harsh environment of salt air (depending on site), humidity and extreme temperature variations. Movable bridges also impose greater load to piers and footings than a fixed bridge of similar construction and greater variability of load can accelerate deterioration of supporting elements.

The maintenance needs of any bridge must be determined by a qualified inspector who is knowledgeable in materials and means of construction, structural behavior, codes, and modern methods of bridge design. An inspector will need to aggressively seek out minor problems before they have a chance to turn into catastrophes, often with the aid of monitoring equipment, accurate tools, and a trained support staff. Some discovered components may only need repairs as opposed to being completely replaced, which should be adequately documented (along with everything else). Lead inspectors in the United States require certification through the National Bridge Inspection Standard, established with the Federal Highway Act of 1968, which sets federal policy regarding procedures, frequency of inspections, qualifications of personnel, format of documents, and procedures for rating inspected structures.

# 6 Building Your Own Competence

Colossal buildings and bridges, such as the suspension bridge in southern France shown in Fig. 6-1, inspire awe and bring to mind courage, determination, and astuteness. However, veterans who design such wonders still find themselves timid, hesitant and naïve at times, much like one who is only beginning their engineering adventure. New challenges inspire fantastic growth in character and competence, but the reality of risk should bring about sober consideration of how best to build and nurture that competence.

The whole process of learning and training is to encourage growth in technical as well as personal areas, so that an engineer will be equipped not only to be a better problem solver, but also a trustworthy colleague and consultant. To truly be successful in the profession, it is important to face every working day, and the good or bad that accompanies them, with patience and courage, then to teach others to do the same.

## 6.1 TECHNICAL GROWTH

Life as a professional structural engineer begins with curiosity, but it is maintained through technical competence. Courses at the high school level set the foundation into place, which is further strengthened by directed study on the university level. Upon graduation, an engineer-in-training chooses which basic path to follow, whether toward instruction and research or common practice, and the examination process for receiving recognition as a professional in the field is not far from consideration. Much remains to be learned prior to taking the necessary

**Figure 6-1** Millau Viaduct cable-stayed bridge, France. (*Courtesy Enerpac, Milwaukee, Wisconsin*)

licensing exams, but long-term survival in this profession also depends on technical growth.

### 6.1.1 Continuing Education Regulations for Licensure

The National Council of Examiners for Engineering and Surveying (NCEES) creates and scores the exams used by most states in the licensing process for design professionals. The NCEES has a long history, having been introduced in 1920 as the Council of State Boards of Engineering Examiners, and continues diligent work in establishing a uniform set of rules, laws, examinations, and record maintenance programs to assist in the licensure of professional engineers. One of the contingencies placed upon license renewal is continuing professional competency, stated in the NCEES Model Law as follows: "The board shall have the power and authority to require a demonstration of continuing professional competency of engineers and surveyors as a condition of renewal or relicensure." (NCEES 2004, p. 1) The *board* stated within is the jurisdictional (state, territory, or other possession of the United States) organization responsible for licensing professional engineers. Each board maintains its own autonomy, but is free to adopt the rules and laws created by the NCEES, or to modify them to suit regional needs. The NCEES Model Rules define a continuing education unit (CEU) to equal 10 professional development hours (PDH) of class in an approved education course. A base of 15 PDH per year for license renewal is recommended, but many boards add to that number. A qualifying course or activity must have a clearly defined purpose and objective which will maintain, improve, or expand the participant's skills and knowledge relevant to their field of practice. These activities would include attending courses, seminars, or in-house training programs; attending technical or professional society or committee meetings; teaching a course for the first time; attending video courses or webinars; authoring and publishing articles, papers, or books; completing courses that will improve one's business skills, such as learning a new language or completing an ethics or management course; and correspondence courses. Some jurisdictions also allow a certain amount of self-directed study to be included in the accumulation of PDHs. Each enforcement organization also establishes a minimum period of time that a licensee must maintain records of PDH completion, usually given in the form of a certificate by the administrators of the educational endeavor.

## 6.1.2 Advanced Educational Degrees

Most structural engineers enter practice with a college or university undergraduate degree in the field of civil engineering, which includes the necessary technical requirements for designing structures, though not to the level provided by an ABET-accredited degree in structural engineering, such as is offered at the University of California, San Diego. In order to expand competence and specialized knowledge, some choose to pursue a master's degree in civil engineering with emphasis on structures, or to continue on toward a doctorate if their goal is research and development or instruction. Advanced degree programs introduce concepts of a higher order, such as dynamic earthquake engineering and nonlinear methods of analysis, and some have argued that these subjects may not be absolutely necessary for practice in a particular region of the country. However, licensed professionals who are expected to carry out their duties with a standard of care should be able to recognize when a particular situation requires the use of advanced forms of analysis or design, and understand when they can complete the work themselves or need to assign it to someone better prepared or suited in a particular area. The American Society of Civil Engineers (ASCE), having been a leader in promotion of the industry since 1852, encourages university staff, governmental organizations, employers, licensed professional engineers, and others to endorse the attainment of higher levels of knowledge through education beyond a bachelor's degree (ASCE Policy Statement No. 465: October 19, 2004). It is further expected that this degree be obtained prior to entrance into the field of practice on a professional level in order to assure flexibility for a wide range of roles that a civil engineer assumes in the course of service to society. The position of ASCE is based on observations that the profession is experiencing rapid, significant changes related to globalization and expansion of the boundaries of practice, availability of information through new means, advancing technology, societal diversity, enhanced public awareness of broad-reaching influence and effects of engineering projects, and a change in focus of civil infrastructure toward maintenance, renewal, and improvement of existing systems. Again, some argue that these broad areas of knowledge can be learned through professional practice and continuing education as an alternative to a mandatory advanced degree. It seems that there could be some middle ground on the issue, but providing quality education and testing retention of that knowledge appears to be problematic apart from a university setting. Some respected institutions of higher leaning have established on-line computer courses pursuant to a master's degree in civil engineering, but the programs would require validation by a licensing board if used for those purposes.

### 6.1.3 Active Professional Involvement

Engineering is a vital part to the advancement of society and those within the profession are involved at different levels. Some are honored with the intricate work of space exploration, yet others are honored by simply bringing a homeowner's dreams to fruition. Whatever the charge, many lives are affected by the work of engineers. Professional involvement in a functioning society does not only include the course of one's work, but expands through all levels with the help of engineering affiliations and government promotional foundations. An engineer is already involved: The only question is to what level and of what effect?

The federal government not only encourages active participation in the affairs of society through paid positions within organizations such as the National Science Foundation or the Federal Emergency Management Agency, but also through many volunteering opportunities such as those pursuant to developing the National Earthquake Hazards Reduction Program. Though participation in some type of program or association serves to benefit society, each engineer benefits tremendously through their own involvement by learning new things, building friendships, and defining future direction. There is a committee or gathering of professionals for every notion related to the advancement of engineering, including code development and interpretation, advocacy, licensure and education, and many others directed on national, state, or local levels. Committee work can be as simple as offering ideas or as broad as writing new guidelines for practice, but every opportunity adds to one's positive technical growth and to name recognition among other professionals.

Active involvement means that engineers give of themselves to the advancement of others, either within or without the profession. Where would any career be if those who have learned the trade do not pass along their knowledge and experiences to others? Writing an opinion piece or an article for publication in a nationally distributed periodical takes discipline, courage, and vision, but the ability can be found within all who are passionate about their career. Passion drives motivation toward thorough research and training in communication, and genuine concern for others encourages the dissemination of truth appropriate to any subject.

### 6.1.4 Seminars, Conferences, and Personal Research

Active involvement in the profession may include writing articles or delivering presentations. Courses of university study usually assume an active/passive role in a person's development, where information is absorbed (passive) and recalled during a process of evaluation or through class projects (active) that turn out to be useable within society. Many engineers, however, choose a passive approach to technical growth offered by a variety of seminars and conferences where they can attend and blend in behind the shadows of others, unnoticed. These meetings present further opportunities for learning how other engineers perform their duties in other regions of the country, and possibly for picking up new clients. It is important to add subject diversity to engineering knowledge and one- or two-day seminars are great nonthreatening venues.

Personal research may be necessary to complete a specific project, such as for the design of a particular structural element that has not been attempted by the engineer before, or for the sake of expanding personal knowledge related to the profession. Some of the best learning can be had through personal effort as study is directed toward current needs, led by greater desire than if a seminar were selected from a catalog for the sole purpose of obtaining continuing education units. Even periodic review of courses taken at the university can bring fresh ideas and renewed understanding to matters which contribute to success in a particular field. Source material for study, however, must be chosen carefully as not all published works have the same authority and unpublished works may be of questionable value and accuracy.

### 6.1.5 Making Proper Use of Technical Research

Responsible use of technical information, whether obtained through the course of a seminar or from a published work, is the keystone which holds all knowledge together. All data has a purpose and an intended audience and this serves to identify the truth of a matter to the attentive user. The burden of using data responsibly lies not with the instructors or authors, but with the engineer who seeks to apply what has been presented. Most publishing companies, for example, include some warning verbiage at the front of technical literature:

> Reasonable efforts have been made to publish reliable data and information, but the author and the publisher cannot assume responsibility for the validity of all materials or for the consequences of their use. (CRC Press, LLC, 1999)

> This work is published with the understanding that McGraw-Hill and its authors are supplying information but are not attempting to render engineering or other professional services. (McGraw-Hill, 2003)

> Reasonable care was applied in the compilation and publication of Jefferson's Welding Encyclopedia to ensure authenticity of the contents. The American Welding Society assumes no responsibility for use of the information contained in this publication. An independent, substantiating investigation should be made prior to the reliance on or use of such information. (American Welding Society, 1997)

Additionally, authors and publishers recognize the importance of future editions of various works that are assembled based on feedback received from readers related to the existing resource. Because the field of structural engineering is a dynamic one, new discoveries continue to improve and enhance understanding of material and assembly behavior, and errors are sometimes identified in existing thought or approaches to an engineering problem. Does this mean that earlier editions of a publication are of no value? Certainly not! The point of the matter is one of due diligence in keeping oneself current with any technical changes in the field of practice and having a thorough understanding of fundamental engineering concepts that have been tested and proven over the course of hundreds or thousands of years. These fundamental engineering concepts serve as an anchor or a magnifying glass by which the merits of all other information may be judged. Consider the following excerpts from some later editions of popular university textbooks:

> The authors are greatly indebted to students, colleagues, and other users of the first three editions who have suggested improvements of wording, identified errors, and recommended items for inclusion or deletion. (Salmon and Johnson 1996, preface)

> The interpretations are those of the authors and are intended to reflect current structural design practice. The material presented is suggested as a guide only, and final design responsibility lies with the structural engineer. (Breyer et al. 2007, preface)

> While earlier texts have tended to feature the individual classes of materials—metallic, ceramic, polymeric, electronic, or other engineering materials, the Sixth Edition represents a definite advance in providing a fresh access to modern MSE (materials science and engineering), now portrayed as an integrated field instead of merely the sum of its parts. (Van Vlack 1989, foreword)

Technical authority in terms of trustworthiness or accuracy varies over the range of sources available to the field of engineering. One reason for this is based on the reputation or financial interest a publishing company may have vested in a particular work, to which such a company would strongly desire defensibility. Books written to be authoritative on a particular subject, published by a well-known house with many years of experience, should be afforded a higher status of recognition than relatively brief articles found within a technical periodical published by an engineering society. Some magazines are identified as trade periodicals and though they may be of high quality and circulation, accuracy of information is not afforded the same level of scrutiny as the aforementioned sources. Conferences and seminars provide information of limited exposure due to the constraints of time and purpose, and lecturers occasionally offer tidbits from the podium that have certainly not been verified by anyone, all of which leads to the conclusion that careful use of data is warranted.

**Experiments, Testing, and Engineering Reports** Experiments in structural engineering are generally performed to either discover properties of individual elements, such as beams or walls, or to determine behavior patterns of elements that are subjected to external forces. Tests are conducted in a controlled environment and may not adequately represent service conditions, but information recorded and published still gives a strong picture of what may be reasonably expected in terms of real behavior. Relationships between measured phenomena in laboratory tests and of what actually occurs in service are empirical by nature, but the coordination of knowledge within the industry continues to strengthen because of a large volume of data that continues to be gathered from tests and from direct measurement or observation during service.

It is important to read the qualifications included with an engineering report as to the conclusions offered relevant to what was being studied. Scientists strive for particular understanding of what is being reported and notation regarding project set-up, conditions, and testing methods used can be somewhat verbose. Consider the following warning offered by a team of scientists studying the cyclic behavior of steel double angle connections:

> "These conclusions only apply to the specimens tested and the reader is cautioned against generalizing and applying the conclusions to cases with different configurations and materials" (Astaneh and Nader, 1989).

Conclusions are usually specific in terms of offering a new approach or methodology for analysis and design, but all reports include observations of how a member or system behaved when subjected to the parameters of the test. This simple information gives valuable insight into a building's expected response under the conditions and set up described in the report which, when added to a collection of facts and figures in an engineer's toolbox, helps to remove some of the mystery of building response and create a more intimate understanding of such things.

Research is carried out by a wide variety of scientists, from those within academic institutions to consultants, on almost any topic of relevance to the engineering profession. Papers and reports are published every month in literally hundreds of periodicals and widely posted on the internet. One must always ask, "What exactly does this prove?" The authors typically explain abnormalities in the test cycles ("extensive vibration in steel beam supports of the testing apparatus") and identify the resulting questionable aspects of the test ("this vibration made it difficult to identify distinct natural frequencies").

Even results that seem obvious to an experienced engineer can be useful for the sake of scientifically justifying a hunch. Falk and Itani (1988) conducted tests on plywood and gypsumboard sheathed diaphragms and concluded that plywood walls were stiffer regardless of the presence of openings. Additionally, walls without openings were consistently stiffer than ones that contained openings (see Fig. 6-2), regardless of the sheathing material used. Taking this knowledge, the authors also add a conclusion that may not be so obvious: "The reduction in stiffness in a wall

**Figure 6-2** Concrete shear wall system with openings.

diaphragm due to the presence of openings is about equal to the proportion of the wall occupied by the openings." Therefore, conclusions are offered with a value that may be great to some and meaningless to others, but there is always something of use at any given time.

## 6.2 THE ART OF PROBLEM SOLVING

The process of solving problems can correctly be described as both art and science—seemingly contradictory, yet codependent allies. Artistic thought allows for consideration of a broad range of solutions, which can be properly narrowed down to the best choice with the help of established scientific fact and reason. Creative solutions are discovered by considering the value of presumption and courageously pursuing truth in a hunch. To bring these families of thought and action together, a strategy must be employed.

Such a strategy begins with an exploration into how the problem is defined: What is the issue at hand? What are the requirements and restrictions? What is the benefit or what is the end result supposed to accomplish? Sometimes the meaning and intent of a problem will only be apparent as its solution is attempted. View things from different perspectives and identify smaller parts of the whole that can be solved more simply. It almost goes without saying that the more familiar a person is with the ins and outs of a particular subject (structural engineering), the easier it is to solve related problems (design and detailing of elements and systems) in such a way that external parameters (economics, strength, serviceability) are adequately satisfied.

In a nutshell, this strategy continues with proper exploration and execution of the problem. As much information that is relevant to a given situation should be collected for the sake of broadening the scope of possible solutions. Research code requirements, find out what scientists have tested that may approximate the condition of loading and reaction, brainstorm with other professionals or code administrators, and review old calculations and drawings. When the problem is properly understood, and all relevant data has been logically arranged, the next step is to formulate solutions to the problem that will meet the needs of the greatest number of individuals (safety of society being the highest priority). All solutions involve some element of risk, some more than others, but properly channeled creativity and conviction found in the adequacy of engineering facts can help build comfort and confidence in those discoveries.

## 6.2.1 Critical Thinking

The word *critical* comes from a Greek word that means to question, to make sense of, or to be able to analyze. Critical thinking incorporates all of these definitions as a total approach to understanding how a person can make sense of a world that includes many disparate parts. It begins with a commitment to actively use one's wisdom, knowledge, and skills to solve a particular problem and being put into meaningful service through the process of making decisions when others remain silent, by following through on commitments even in the face of obstruction, and by taking responsibility for unintended consequences while working to set things right at the same time.

The process of critical thinking does not end with the solution of a particular problem. The right choice may have been made at a specific time, but a flexible thinker understands that a better solution may be discovered sometime later. A solution is continually being evaluated, either during passive service or in response to stimulus, such as seismic motion applied to a lateral force resisting system. At times, this process of evaluation involves an honest look at any bias that may have influenced the result. Others who have contributed to the process could have offered their opinions from the basis of a predetermined mindset, such as a misapplied proclivity toward an environmentally sensitive building material or function. This person's contribution to the solution may not be entirely genuine and the manager of the problem needs to filter through possible bias by asking questions, conducting thorough research, and making difficult decisions when necessary.

Critical thinking includes a willingness to consider other options, solutions, or suggestions and to examine them carefully without a predetermined judgment. This automatically requires flexibility in changing one's own ideas in light of different considerations and insight, provided there is adequate justification to do so. For example, suppose an architect recommends a roof framing system that is different than what the structural engineer had originally proposed. There are likely different reasons driving the change and facts of the matter may not all be shared, so in the spirit of cooperation, the engineer should consider the request as a problem solving opportunity. Benefits of the revised system are to be weighed in comparison with the offerings of the old; perhaps consultation with the owner or a trusted contractor can bring in other facts for consideration. An engineer cannot hold rigidly onto independent thought as though exercising monopoly powers, but must be willing to consider the creativity, and sometimes greater experience, of others.

Another important part of being able to critically analyze an engineering problem is to have some idea of what the end result is to look like, often through preliminary estimation or foresight. This ability comes with experience, but even those new to the profession have an ability to estimate the input, the manipulation of such, and the conclusion. There are general principles of estimation that can easily be applied to an engineering problem:

1. *Estimating mathematical results begins with a manipulation of numbers to a more manageable form through the process of rounding.* Structural engineering begins with an approximation of loading applied to the member or system. Once dead, live, and other forces have been determined through the use of charts, tables, or simplified formulas, tributary widths or areas of application can be approximated to create concentrated and uniformly distributed loading patterns. The order of magnitude of a trapezoidal load, for example, can easily be determined and applied to the system either as a concentrated force applied at the load shape's center of gravity or distributed across the load's length uniformly. Quantities themselves should be rounded up or down to the nearest 10 or 100, almost to the point of where the mathematics can be completed in one's head.

2. *Estimation also involves the manipulation of a problem's parts through simplification or rearrangement.* A structural engineering problem involves several basic steps: Determine loads, apply them to the member or system, analyze the element's reaction to the loading (internal moment, shear, deformation), design the components, and explain the construction to those who will build it. Internal bending moments and shears from multiple loads are determined by the process

of superposition, which will be managed quicker when fewer different load types have to be dealt with. Charts and simple equations are available in a variety of resources related to the solution of internal reactions for beams, columns, frames, and many other shapes and configurations, which can be easily worked through with the abbreviated loads and patterns described in the first principle above. Boundary conditions can be simplified for purposes of estimating member reaction or carrying load from one place to another in the model.

3. *To be good at estimating, a person must be able to recall basic facts and formulas quickly.* This includes mathematical shortcuts and memorization or speedy reference of engineering equations suited for a particular need.

### 6.2.2 Reaching a Conclusion

Chaffee (1991, pp. 453–455) proposes that our beliefs are the main tools used to make sense of the world around us and this can certainly be seen in technical professions. If the workings of things around us are based on verifiable investigation, we are said to be reporting factual information ("The concrete beam is cracked"). Other beliefs are offered through inference, which describes things in a way that begins with factual information, yet adds unconfirmed suppositions to known data ("It is cracked because there is not enough reinforcing present to absorb the carried load"). Beliefs are also expressed as judgments that are based on our evaluation of certain observable or testable criteria ("All beams without sufficient reinforcement will crack"). All modes of forming and expressing beliefs are important in helping engineers understand things that are seen every day, but in matters of a technical nature, beliefs must have a solid basis in science that is neutral in regards to bias or preconceived notions.

In the engineering profession, a conclusion or problem solution can exist in the form of a completed structural design, the summary of an investigative report, retrofit of a deficient member or system, strengthening measures employed for a historic structure, a maintenance program established for a bridge, or a detail needed to repair a mistake made in the field. With a basis in belief and experience, a conclusion often begins to take shape before all of the evidence for a matter has been collected. This is natural and not necessarily a bad thing, provided a final decision is not made until all planned and relevant facts have been considered.

## 6.3 IMPROVING YOUR PRODUCTIVITY

Organization helps to stimulate productivity and keep it fresh. Before working on any project, an engineer should evaluate the tools or decisions needed to maximize the potential for being productive. A good plan and a predictable, consistent list of procedures will help minimize error in completing any engineering project. The following list is suggested:

1. *Establish a dead line or final completion date of the work.* Clients do not always verbalize this and there are often misunderstandings as to when a project will be delivered. A written schedule that everyone involved can agree to is the best way to avoid problems, but some smaller work may not have a large paper trail to substantiate every nuance. In those cases, an engineer should still establish a time schedule that can be met and is approved by the client. Knowing the allotted time helps a person make informed decisions on the amount of work needed to finish everything and what resources will be required.
2. *Know the demands of the work and decide how best to delegate responsibility.* While it is important for an engineer's growth to work in unfamiliar territory, it is not the most productive way of completing a project. When time or finances are of the essence, choose the best people for the job who are either most familiar or better able to finish the task quickly. Delegation of certain portions of a larger project, or an entire design itself, is one of the greatest stress-relieving strategies available and often means the difference between making or breaking a schedule. Group segments of work to be completed, acknowledge the skills and abilities of available staff, and work together.
3. *Engineers who know their own limitations will be more productive than those who make things up along the way.* Understanding your own strong and weak points will help to organize the work schedule and leave little unattended. Call to mind what is most familiar about the subject of work and use that knowledge to guide the project through areas that may not be known so well. When limitations are known, it is easier (and quicker) to chart the right course of design and help can be sought early to fill in the gaps, making coordination easier to handle. Never be afraid to ask questions and to seek wise council.
4. *Familiarity with the tools of the trade will also speed along the process of design, analysis, and detail work.* Building codes, relevant standards, methods of bridge construction, and business skills need to be quickly consulted or brought to mind to keep work flowing smoothly. A familiar sight is one in which a professional engineer is surrounded by reference books that are opened to example problems or engineering formulas and descriptions in an effort to help remove the guesswork of uncharted territory . . . diligence displayed in the most productive way, recognizing that there is not a whole lot new under the sun in terms of engineering mechanics and many solutions that have gone before can also apply in some way to new work.
5. *Exercise the subtle art of accepting and rejecting work.* It is far too easy to be "yes" or "no" people, depending on our level of busyness or willingness to branch out into different areas of practice. A fine balance must be maintained that can only be truly learned through experience. A quick reply to a submitted request for work may lead to problems if the necessary schedule and budget is not given proper consideration with other members of the design team. Work load is certainly part of working productively, as too much work can lead to a

temptation to cut corners whereas too little work can quickly inspire laziness if "free time" is not managed properly.

### 6.3.1 How Quickly Can (or Should) You Design?

An engineer may commit errors of omission (a failure to perform a particular task), commission (mathematical or procedural error), or execution (incorrect assumption of model behavior or approach to a problem) and these can be difficult to detect and correct in the heat of the moment. Self-checking is carried out during design in an effort to reduce error, but it does constitute another step in the process, one of several control measures that often gets cut when pressed for time. Emile Troup is of the opinion that engineers cannot design structures at an extremely accelerated pace and meet the owner's vision and expectations of risk at the same time (Troup and Metzger 2001). The danger of reducing design time is higher risk and lower quality, often costing more money than if the design were done at a normal pace. It is important to be efficient, but not at the cost of cutting corners. Advancing technology has led to economies and better management of resources in the design office, but advancement also continues to bring about new, more complicated regulations and different ways of determining forces and building responses, the absorbing of which cannot be accelerated beyond the pace of the human brain to comprehend.

Engineering design involves many variables, all requiring careful attention. One study (Ellingwood 1987) of 23 structural failures revealed 8 basic causes: Inadequate consideration of design variable uncertainty; errors in methods of structural analysis, including a fundamental misunderstanding of how a structure is to behave; ignorance of extreme hazard loading; misunderstanding of relevant failure modes, such as those stemming from dynamic wind effects; improper treatment of failure modes that were recognized in the analysis; construction errors due to control and communication; financial, political, or other pressure applied to the participants; and misuse or willful abuse. Consideration of these events requires the limited attention span of a responsible human being, not the superhuman speed of an impersonal computer, therefore time is a precious resource that must be properly managed.

The ability to design members quickly often involves a simple process of organization. Flexural members, for example, are first reviewed for limit states related to global bending and shear, then checked for local buckling due to concentrated loads or other attachments (giving consideration to the effects of connections, such as a reduced cross-section from drilled holes). The combined effects of axial, flexural, and torsional loads would be reviewed next, giving consideration to methods of combination dictated by respective material standards. Finally, serviceability issues will be reviewed, by which the member's appearance, vibrational characteristics, deformation to load, and general integrity of function are preserved through conditions of normal service.

Proper use of simplified methods of design and analysis can speed along the process, but the justification for the use of such must be sound. It is not necessary to model everything in three dimensions and perform a complicated finite element analysis, provided simpler methods have been tested and confirmed to give acceptable results. For example, testing done by Nevling, Linzell, and Laman (2006) indicated that a coarsely modeled three-dimensional computer analysis of a curved horizontal bridge structure did not show a significant increase in accuracy over a two-dimensional grillage analysis of the individual girders.

## 6.3.2 Time Management

Engineers are never given an open-ended budget and deadlines are very real. There certainly are incentives to reduce the time spent in design, but how can this be done efficiently without cutting much needed corners? The answer can be found in the proper usage of technology and the wise management of time. Although technology has improved the speed with which tasks can be accomplished, it also opens the doors to further opportunities for wasting time. Technology encourages us to do more things than humanly possible and this expanding list of tasks should be managed by setting priorities for each day and sticking to them.

The internet and e-mail have greatly simplified business and communication, but they can be the two biggest distracters and time wasters. After researching a subject or a product, log off immediately. Set aside a particular time of day to answer e-mails and try to avoid the habit of checking your mailbox every time you approach the computer. Don't allow the electronic mailbox to hide your humanity, which so often happens. Speak with clients and colleagues over the phone from time to time to maintain that personal element of the business that causes others to feel appreciated and important. Remember that sending an e-mail message does not guarantee a two-way conversation and sometimes messages get lost, both in total or even just in meaning or attitude.

**Caffeine** The absorption of caffeine to keep alert cannot be relied on to reduce the amount of sleep a body needs to remain healthy, but it is often done to complete work with a fast-approaching deadline. It is important to plan a schedule around sleeping hours and to take caffeine in whatever form in such a way that the effects do not interrupt needed rest. According to the Drug & Alcohol Services Council of the Government of South Australia, the effects of an ordinary amount of caffeine found in a medium-sized drink can last up to 4 hours (Government of South Australia 2002), but the effect of multiple doses is unclear overall and would not be exactly additive.

Long-term ingestion of large quantities of caffeine on a regular basis can lead to serious health consequences. These can include the formation or agitation of stomach ulcers, a change in heartbeat pattern, muscle spasms and convulsions, delusions, breathing problems, and may also affect the health of an unborn baby. Many people who drink caffeinated beverages every day tend to become physically dependent on it and will experience strong headaches if consumption is halted for an extended period of time. Moderation is the key to taking advantage of the benefits of caffeine when alertness is desired. 500–700 mg/d seems to be the accepted maximum amount for healthy adults, while pregnant women need to limit it to only a third of that figure. When too much is ingested at one time, the human body responds nervously and somewhat rejects any control that may have been sought after: You chug a pot of coffee before taking an important exam only to find that your mind is racing in too many different directions and you can't concentrate on the matter at hand, such as is illustrated by the character shown in Fig. 6-3! For purposes of comparison, 8-ounces of drip coffee typically contain between 110 and 170 mg of caffeine and a 12-ounce can of a cola-type soda will contain between 34 and 45 mg. Caffeine has also been packed into powerful little pills that can be purchased without a prescription, but they can be dangerous if not taken according to the manufacturer's directions.

Some other steps to help reduce the chances of negative effects of caffeine consumption, as well as simply taken for the sake of not requiring caffeine for alertness, include the following:

**Figure 6-3** Too much caffeine in your system can certainly be counterproductive! (*Art by Ron Wheeler*)

1. *Get enough rest at a regular time period every day.* Current health recommendations are for 7 or 8 straight hours of sleep every night, which should be established as a pattern that is bodily recognized. A healthy person should be able to insert isolated periods of extended waking hours into their schedule to complete a project without seriously disrupting their body's need for rest. If daily sleeping periods are limited to only 5 hours or less, not only will the body accumulate something called a *sleep debt*, which is never fully repaid, but there can be other health concerns.
2. *Establish a regular exercise schedule.* If daily energy needs to be constantly derived from caffeine, then you aren't getting enough exercise. This is one of the most common traps an engineer falls into since a typical work day is spent behind a desk, sometimes long after the doors have closed, and exercise is not followed with any regularity. A common formula for aerobic-type exercise (swimming, running, brisk walking, bicycling, jumping rope) is three to four 30-minute periods every week. Even a 15-minute walk to get the mail is better than nothing at all. Regular exercise not only builds up energy throughout the day, but also lowers the risk of heart disease.
3. *Have a healthy diet.* Eating healthier reduces the occurrence of sickness and fatigue as the body uses and stores the best types of energy. Health experts believe that 6 to 9 servings of fruits and vegetables are required every day, each "serving size" being about as large as what can fit within the palm of an adult hand: A medium-sized piece of fruit, 6 ounces of 100% fruit or vegetable juice, $^1/_4$ cup of dried fruit, $^1/_2$ cup of cut-up fresh fruit or vegetables, or one cup of raw salad greens. In addition to fruits and vegetables, grains, nuts, and at least eight cups of water per day are important for a healthy diet. There are a lot of publications available to help you find the diet that best fits your lifestyle or needs.
4. *Exercise your brain by reading or learning something different.* The more active your brain stays, the better it will be able to use its energy for daily tasks. Engineers need to think in both linear and abstract terms, though favor typically leans toward the more rigid and absolute side of things. Our brains can be exercised by learning to appreciate and understand more fluid or changing-types of activities, such as art or music.
5. *Take regular breaks, or downtime, during the day.* It's important to have some periods of the day when you don't have to think very hard. Give your brain a break, possibly with a short nap, a video game or two, a short television show or movie, or whatever. Reruns of *The Three Stooges* can provide endless opportunity for irrational thought and mindless entertainment.

**Sleep** As has already been stated, a healthy body can handle intermittent periods of extended waking hours if it has been conditioned to a regular pattern of sleep. However, extended periods of restlessness can lead to negative job performance and some of the following patterns of behavior:

1. *More effort is needed to concentrate on a task.* Engineering is often about challenging thinking: Unconventional solutions to difficult problems are often necessary, being discovered using methods of analysis not fully understood, while being pressured to meet a deadline. When concentration is hampered, it takes longer to complete a design and it is more difficult to focus on the best solutions.
2. *It is more difficult to recall important facts and figures.* Engineers are expected to think on their feet, whether responding to an emergency phone call or reviewing a contractor's imaginative detail in the field. An answer or proposal should never be given in haste, but often the best solution depends on several factors that can be recalled from memory. The more knowledge that can be accurately recalled from memory, the better chances are for finding the best solution to an engineering matter.
3. *Irritability and stress is magnified in even small things.* A lack of sleep causes different phenomena to be amplified: Dogs tend to bark louder, trash trucks arrive during the worst time of day, kids choose to play with the most annoying toys, and tempers flare more quickly. The body is somewhat upset that it hasn't experienced much needed rest and all systems are working to make things miserable. Even with an appropriate amount of rest, it is difficult to set aside selfish needs and desires in return for a harmonious existence with those around us, and inadequate sleep only gives us a narrower field of vision.

Randy Gardner presents an interesting example of what can happen to a body that has been deprived of sleep for a great length of time. While still a highschool student in 1964, he set out to break the worlds' record for staying awake, and accomplished the feat in 264 hours (Coren 1998). There have been differing summaries circulating since then about his health and behavior during the period of the test coupled with a confusion of activities after he began to sleep again. Some reported no "obvious lasting physical or mental problems," but observations made by Lt. Commander John Ross of the U.S. Navy Medical Neuropsychiatric Research Unit in San Diego, at the request of Randy's parents, portrayed a disturbing picture. After Day 2, Randy had difficulty focusing his eyes. After Day 4, memory lapses, delusions, and irritability were reported. After Day 6, Randy's speech apparently had slowed and he had difficulty naming common objects. After Day 10, Gardner showed signs of paranoia and fragmented thinking. By the end of Day 11, Gardner exhibited an expressionless appearance, was slurring his speech, required encouragement to respond to any questions, showed a reduced attention span, and failed to complete some mental activities because he forgot what he was doing.

## 6.3.3 Developing Consistency and Clarity

Being based on historical activity, the phenomenon of consistency has some linkage to memory, both short- and long-term. Short-term memory refers to the quantity of things that a person can focus on at the same time and is often lost within 30 seconds. A common way of combating this rapid rate of forgetting is repetition, which keeps the information in short-term memory longer and sets up the transfer to the long-term. Long-term memory is often divided into three types: Procedural (remembering how to do something), semantic (recalling factual information), and episodic (personal events). It is relatively permanent and has a virtually unlimited capacity (Higbee 1996, p. 23). Basic principles of improving the memory, thus influencing consistency in accomplishing tasks, include the following:

1. Create meaning for that which is to be remembered and become familiar with it. The more a person knows about a particular subject, the easier it will be to remember associated principles or data.
2. Organize material while it is being learned rather than randomly picking and dumping.
3. Associate new material with something that already exists in long-term memory by use of analogy, metaphor, comparison, or contrast.
4. Create a mental picture in order to increase the possibility of remembrance, being both in textual and an equivalent pictorial form.
5. Pay attention to and carefully study that which is to be remembered, including small details.
6. Repetition is often the key to planting items in the memory.
7. Instead of only listening, write down what is to be remembered. Learn how to outline books of particular interest as they are read by deciphering the main point of each paragraph, section, and chapter. Notes can be brought out and reviewed at a later time, which strengthens the memory of these things.

A company needs to produce a consistent product in order to establish a solid reputation in professional circles, which means that all employees need to be on the same page. Calling to mind principles that are used over and over again, through the use of a sharp memory, leads also to clarity in the communication of a product to a client. Memorize company protocol and philosophy. Be thoroughly familiar with existing work and what the profession expects from the company.

## 6.4 BUILDING YOUR CONFIDENCE

Sometimes just the image of what can happen to a structure after a major event, such as the flood shown in Fig. 6-4, can drive an engineer of any experience level to desire confidence. It is this confidence that fosters encouragement to be the best that a person can be in a chosen profession. Responsibility to the public and to a large number of people who are involved in the building industry helps an engineer keep proper perspective: It is not a solo effort, as many people depend on the quality of the engineer's work.

Building Your Own Competence | 157

I **Figure 6-4** Ohio River flood in Paducah, Kentucky, 1937.

## 6.4.1 Working Within Your Means

Professional responsibility involves an ability to know what structures to avoid, or to deal with after a time of further training. A seemingly simple structure may have some elements that drive a need to think outside the realm of comfort, such as might occur with the building shown in Fig. 6-5. This structure is somewhat

I **Figure 6-5** Simple, yet interesting floor support challenge.

regular (for the most part), but there are some cantilevered floor elements present that support a wall, floors, and a roof above. It is different from what one expects, but it is an exciting aesthetic feature, one that highlights the importance of structural engineering. An acute understanding of one's own technical abilities is the best defense in these cases and asking questions during the process of training will be critical for building this knowledge.

## 6.4.2 Computer Usage

It is critical to remember that neither a computer nor its programming is intelligent. The user must be intimately familiar with the process of inputting data, interpreting the results, and verifying the conclusions. "Garbage in/garbage out" is a familiar warning, serving as a reminder that all aspects of an engineering problem (loads and combinations, stability against sway or buckling, boundary conditions in all principle directions) must be carefully accounted for during the creation of any computer model, both on local and global scales. All elements of a model experience local reactions according to their own boundary conditions, but these pieces work together in dictating behavior of the whole assembly. If this global picture becomes too difficult to visualize and understand, simplify it, change it, or seek assistance.

The internet can also be a powerful resource for dissemination or storage of project information. Some companies set up websites on which a variety of things can be posted, such as meeting minutes, project photographs, timelines and schedules, and other documentation that should be easy to access. It is important to remember, however, that it is relatively simple to post research material on the internet, even papers or treatises with a professional appearance but no financial backing and therefore little quality control, therefore caution in using data is warranted. Always know the source of information, conditions for which it was discovered, and reasons as to why it is being posted.

Many architects and engineers in supervisory roles have observed that increased reliance on computer programs for the sake of analysis and design has resulted in a decreasing understanding of fundamental design concepts and often hamper an ability to visualize the transfer of loads in three dimensions. Some of the most important skills in structural engineering include the ability to quickly define a load path, determine very basic patterns of support, and design simple members and assemblies using free-body diagrams and basic statics. When the computer is used to such an extent that even simple concepts are built into a fancy model, the human engineer loses touch with the essence and joy of the profession—to be able to operate intimately in the realm of design and determine solutions "on-the-fly." Intuition and judgment are developed as a result of hard work, repetition, and a solid understanding of basic building blocks founded in a good college degree's worth of study and research.

### 6.4.3 Defending Your Results

It is nearly impossible to predict how a regulatory agency will react to a set of submittal documents. Some staff engineers scrutinize every number and detail, whereas other groups may grant a building permit based solely on the presence of a professional engineer's stamp and signature. A designer may be required to explain, justify, or defend construction documents issued from the office, in writing, verbally, or face-to-face. Young engineers are also called upon to defend their analysis and design results to supervisors, but need not feel intimidated. The art of delivering a successful argument involves solid preparation, careful listening, and a keen sense of precognition. A winning argument gives the engineer something he or she wants: A building permit, acceptance of work completed, or possibly establishment of confidence in the results of an investigative report. The ends are achieved by careful and honest selection of substantiating data and a persuasive personality.

Credibility is an engineer's first line of defense, which is solidly founded upon truth and honesty, and others know when someone is not being genuine. One of the most important parts of delivering a successful argument is the ability to listen carefully, not only to the other party, but to the research materials as well. This practice makes the essence of credibility easier to bind as a character quality. It is easy to fall into the trap of reading without understanding, especially when it comes to building codes. Every regulation is based on an intention and it is that intent to which an engineer must listen. Once understood and carefully validated in light of sound engineering principles, the defender may proceed courageously in making a case to whoever challenges the results for whatever reason.

## 6.5 COMMUNICATION SKILLS

In speech or writing, there may be an emotive spin to the message that is either delivered or heard. Certain words may be used or strung together to cause another person to accept or reject a premise or conclusion to an argument on the basis of emotion. How can any reasonable person disagree with the declaration that a concrete beam "... has obviously failed because of the intentional lack of maintenance and blatant disregard of visible, severe cracking." While the elements of this conclusion may be true, there is an unfair emotional stain to the content that is difficult to see beyond. It encourages the listener to simply ignore any evidence or contrary viewpoints and to blindly accept only this discovery.

Communication of a technical nature, whether written or spoken, must be made in clarity and absolute truth, not hidden behind petty bias or pride. The very definition of a fallacy indicates how damaging an emotional spin can be to the truth of a matter. Fallacies are unsound arguments that are often persuasive because they appear to be logical, because they usually appeal to our emotions and prejudices, and because they often support conclusions that we want to believe are accurate. The listener should be able to distinguish factual data from inference or judgments. Linguistic clues are used to identify suggestions or guesswork, such as "seems," "appears," "likely," and "possibly," and these become important to the overall picture and clarity of a technical issue.

## 6.5.1 Philosophy of Good Communication

A message will never be understood if the speaker fails to communicate in a manner that impacts the audience. Listeners comprise a variety of different backgrounds, both technical and layperson, and each requires a slightly modified approach. Engineering jargon will be lost with the nonengineer and educated recipients require depth to a discussion to maintain interest. Good communication is not found in the message itself, nor even in the most eloquent of speakers, but rather in the attempt made by the speaker to be certain that the message is understood.

Communication between two or more parties not only involves speaking, but also listening. A good listener will concentrate carefully on what the speaker is saying, being certain to set aside distractions so that the message may be understood. If there is any doubt as to what is meant by a phrase or an expressed idea, questions must be asked to avoid confusion and head off possible conflict. Listening also becomes an exercise in patience, as it may take time for a speaker to get to the main point of discussion. Fight off the temptation to formulate a response while the other person is still talking. This can be difficult because it is also important in a fact gathering dialogue to anticipate what the opposing party will say in response to your own presentation, but in general it is best to wait until people are finished speaking before you gather your own response because only then will you have the completed point. Eye contact or some other cultural means of showing that you are recognizing another's control of the conversation at that moment helps the other to feel comfortable enough to share everything they can offer about a subject.

Interpersonal communication not only involves speech, but also body movement that can oftentimes scream out louder than the actual words themselves. Good communicators know how to control their own nuances and habits in order to avoid distraction from the truth of the message. Facial expressions and posture of the body can betray a sincere effort to consider what another person is saying because many times we are not even aware of our contortions, automatically embedded within our style of speaking and listening. They may not be intentionally rude or obtuse, but the perceptions of others often vary widely.

Good communication involves consistent discipline. Exercise calmness and self-control, not being combative. The human element also goes a long way toward establishing a comfortable sharing environment, where a speaker will not slip into sarcastic or arrogant tones in an effort to demean the listener or cause them to feel inferior. These things have nothing to do with professionalism. Engineers should be approachable and able to relate to their human clients and colleagues with realism, humor, patience, kindness, and genuine humility.

### 6.5.2 Verbal

A good command of the spoken language is not only important for clarifying any position on a matter, but it also displays a sense of competence. When words are carefully chosen to assure clear communication of an idea or subject matter that is relevant to the work of others, those who are listening to the speaker are grateful to get the story in a concise, yet accurate and thoughtful, manner. Good, hearty words that display the true power of linguistics should be learned and used appropriately to both define an idea as precisely as possible and to also inspire an atmosphere of seriousness.

One who leads a conversation is mindful of the words being used, rather than speaking just for the sake of being heard without making any real points or progress toward understanding or convincing. Words have an effect on the listener and a project manager chooses the appropriate words to define a particular subject in accurate enough detail, for clients and consultants alike, to leave as little room for misinterpretation as possible.

### 6.5.3 Writing

Engineers are often accused of being lifeless and uninspiring in presenting the evidence to support their conclusions on a matter. It is true that the energy of a scientific report must focus on the factual data and clearly explain how that data has been used to arrive at the author's conclusions, but many authors fail to realize that the clarity of that connection is also found in the author's own style that goes into writing. An author's style is what keeps the reader interested in what is being stated—all the facts in the world cannot hold attention as strongly as careful and stylistic writing.

An engineer is also an author, whether the writing is in the form of specifications, technical reports, structural calculations, or published articles, and the essence of their style in writing will stem from who that person is. Writing that flows naturally from an engineer's passion about a subject is far more enjoyable to read, and therefore easier to remember, than the robotic recitation of bland and boring facts. Even a natural ability to write must be founded on sound principles of style and grammar, some of which in the realm of science may include the following:

1. *The reader should be stimulated toward the essence of the writing, not to the author.* A writer who includes silly anecdotes in an effort to lighten the mood of the subject, or relates every "interesting" side story about his own life and

experience, does a disservice to the reader and will likely be remembered not as an important figure of the profession but as a pompous buffoon. Real-life examples play an important role in illustrating the facts of a case, but it is easy to overuse them and difficult to detect when the threshold between sensibility and arrogance has been crossed.

2. *Avoid the use of excessive wording that rambles along, dragging the flow of the writing and detracting from the important message.* Adjectives and adverbs are fun to insert haphazardly throughout a document, but in scientific writing they can easily lead a reader to lose interest or patience with a document, or even worse, to lose respect for the author. An observed crack, for instance, can simply be "ragged," it doesn't have to be "very ragged."

3. *Whenever possible, define subjective adjectives and adverbs in a way that makes them scientifically understandable.* A reported crack may be visually "ragged," but what does that mean in useable, unambiguous terms that everyone can agree to? Is there a material standard that defines the appearance of cracks in a measurable way (i.e., amplitude of deviation from an assumed centerline of crack propagation)? If so, what is the relevance of that measurement to the conclusions of the report?

4. *Clarity throughout the document is the key to persuading readers to accept the conclusions.* Factual data may be communicated in words or in pictures, depending on the subject and pattern of the output, and it is often best to use different means of presentation in the same document to keep the reader interested. Relationships between measured quantities can only truly be understood by a visual representation—a chart, graph, table, or photograph. Evidence used to support a particular claim may best be understood in the form of a written list, embedded and numbered within the text.

5. *If it takes a thousand words to describe something, use a picture.* Emphasize important points through the writing itself, but recognize when other tools may be called for. Do not overuse italics or underlines to highlight a word or phrase, but recognize their importance when needed.

6. *When using initials for brevity, be certain to define the full meaning of those initials earlier in the work.* Not everyone knows who the ASCE is, but if they are initially defined as the American Society of Civil Engineers (ASCE), further uses of the abbreviation will not interrupt the flow of the writing.

7. *Do not be tempted to use figures of speech, nonuniversal grammar, or obscure engineering jargon.* Write in a way that will be understood by the widest audience possible and do not assume that everyone knows the "lingo" of the profession. Slang terms will also cause havoc in a professional report, leaving the reader wondering about the credentials of the author. Use a dictionary frequently to be certain that words and phrases are not colloquial, but rather are accepted in the mainstream of the language. Keep the sense of the writing professional and don't give in to the laziness of casual writing, such as is common with e-mail.

8. *Something that is described in many different ways without the introduction of relevant, useful facts will often be dismissed by the reader as unimportant, contrary to what the author had hoped for.* Include fresh facts only to the extent that they are necessary to guide the reader in forming an opinion on the subject.
9. *There is often a temptation to give life to inanimate objects, but remember that computers do not think, consider, or philosophize.* Beams cannot "feel stressed out," but they can "be overstressed."
10. *Read reports or presentations out loud.* The human ear has a knack for detecting poor writing and sloppy grammar and the more it hears, the quicker it is able to suggest improvement.

Even completely original written works of engineering will rely to a certain extent on previously known facts, discoveries, or ideas and proper documentation is critical to lending validity to the original work and for the purpose of giving credit where it is due. Conclusions of an engineering report are easier to accept if similar findings have also been documented elsewhere. It is important that such documentation not turn into a plagiaristic effort on the part of the reporting engineer, whether that material is paraphrased (rewriting another person's idea in your own words), summarized (the main points of an idea are laid out), or directly quoted. Consistency and accurate note taking during research will be crucial to successful citations. Generally, it is not necessary to ask permission to use facts or fact-based theories, which include findings and conclusions of laboratory research, but a writer might have expressed those facts in a particularly artistic way that triggers a need to obtain permission to duplicate. Many times, however, facts that are used in an engineering report or publication can only be stated in a limited number of ways, where both the actual fact and the expression of that fact are somewhat inseparable. Successful engineering writing, therefore, should be clear, interesting, defensible, and relevant.

# 7 Communicating Your Designs

One of the most important skills an engineer must have is also the most difficult for many to learn to do well. The ability to clearly communicate a complicated design concept to people from all walks of life in writing, drawing, and verbal form takes time to develop and is not based on academic knowledge or social position. It is partially learned at college, but is easily lost within a sea of technical subjects. It could be enhanced through a greater selection of creative writing, drafting, and humanities courses, but a typical engineering curriculum does not offer many reasonable openings to increase the load.

Another skill that takes more time for some to develop than others, and some never really seem to quite get the hang of it, is the ability to visualize a structure before it is put onto paper. Consider the overhang shown in Fig. 7-1 of the Geisel Library on the campus of the University of California, San Diego. An engineer needs to see this picture in the mind before it can be adequately sketched in rough form on paper, after which structural design becomes an easier chore prior to generating a final set of plans. A rough sketch of the form an architect desires helps the structural engineer to see potential problems with the intended shape or layout, allows for appropriate coordination between the work of other trades (window anchorage, space needed for lighting or power conduit or other piping), and paves the way for completing final plans.

Although this chapter cannot cover all aspects of engineering communication, the intent is to highlight important points that the reader can explore further. The four focal areas an engineer will be concerned with include structural calculations, project specifications, project drawings, and engineering reports. Academic study

I **Figure 7-1**  Geisel Library at the University of California, San Diego.

provides simple enough information on each subject to send an engineer in the right direction, but most of the skills discussed in this chapter can only be learned in detail in a practical working environment, where they can be put to use under imposing consequences when appropriate seriousness is not given to a situation. Some consequences of poor communication that may be experienced in the "real world" include loss of job, loss of client, failure of a structure with subsequent loss of life and finances, loss of reputation, and certainly other unfortunate situations that cannot be taken lightly.

## 7.1  STRUCTURAL CALCULATIONS

A structural engineer's first line of defense for any project is rational judgment. Being able to choose the right solution to a problem and to justify that solution logically and scientifically inherently involves mathematical calculations. These are not to be considered the beginning and the end of all engineering practice, which is sometimes how they are looked upon. Engineering calculations are subject to sound judgment which is learned and strengthened through practical experience, adequate training, and a spirit that is willing to consider unfamiliar solutions to any problem. An engineer must strive for accuracy and a thorough understanding of practical limitations, including when to use a more sophisticated form (nonlinear, inelastic) or tool (finite element analysis computer program) and when to perform analysis and design computations with the simplest of tools (hand, nonprogrammable calculator) in order to maintain a strong sense of how these things are actually performed.

For structures so regulated, the 2006 International Building Code (2006 IBC) allows the building official to request calculations prepared by a registered design professional that demonstrate how compliance with code objectives are met (2006 IBC, Section 106.1). Section 1.7.6.3.1.1 of NFPA 5000 affords the building official the same leverage during the process of checking construction documents for compliance. At the same time, these codes also allow plans examiners to exercise a level of professional opinion and judgment based on the character of work submitted, for which it may be obvious to determine compliance with code requirements without the need for substantiating calculations. This can sometimes be a difficult call to make, however, and it is usually better for all parties to err on the side of caution in relation to justifying a design.

### 7.1.1 Analysis and Design

Once project requirements have been determined and relevant codes or standards are lined up, the next phase of an engineer's work is to break the larger whole into smaller, more manageable segments that will be easier to analyze and design. This process best involves an initial large split, perhaps a division of gravity-supporting elements and systems (post and beam) from the lateral force resisting system (diaphragms, frames), then a merging of the work of these first two groups to complete design of footings, connections, and working drawings. Analysis then proceeds with the creation of an appropriate mathematical model, determining loads and load combinations, applying them to the appropriate structural elements, and determining their effect on those members and overall structure reactions or supports. A successful solution to the model is judged based on expectation and careful checking, using simplified methods of analysis and possibly historic records of similar models. Results of structural analysis, which is simply the tracing of loads and their effects throughout a structure, are then used to design members, connections, and systems.

Until the 1970s, the slide rule was used as the basic tool for engineering computations, allowing for up to three or four significant figures of accuracy. These handy instruments seem meager when compared to even the simplest of pocket calculators, but consider the monumental structures that have been created with their use: The Empire State Building, Eiffel Tower, Brooklyn Bridge, and many more. In the 1980s, the convenience and greater accuracy of most early calculators led more engineers to move on from the slide rule, but it is likely that this move also caused the profession to lose sight of what precision is actually possible in the work of structural engineering. In terms of precision, it is commonly accepted that the result of a computation can never be more precise than the least precise element of that calculation.

Computer programs have become incredibly sophisticated with great potential for analysis, involving entire structures (with connections) that can be automatically integrated into a drafting program. While these tools are helpful in meeting deadlines, managing volumes of data, and improving structural analysis, they can also be a distraction from the exercise of true engineering judgment and logic. They are not intended to replace engineers, but only to enhance and improve the work of engineers. In fact, some who have been working in the field of developing sophisticated computer programs for analysis and design have offered a great respect for the simplest of structural elements that have been used for many years, including the nonprismatic frame element (Wilson 1998, p. 4-1), deriving further work with a solid basis on fundamental theories long in use. A structural engineer must develop a feel for what is right, which cannot be gained by serving the function of a technician blindly plugging numbers into a computer program. Checking analytical results becomes a critical part of using computers of any significance for projects of all sizes.

Whether by hand or computer, analysis of structural elements involves a discovery of six independent degrees of freedom, three serving translation, and three defining rotation about these translational axes by which every member follows. Refer to Fig. 7-2 for a three-dimensional frame structure that would require careful consideration of all dimensions. As load is applied, this frame will not only deflect but also twist, and forces must be resolved back to provide supports. Under some conditions, it may be necessary to look at other characteristics to serve as degrees of freedom, which can include beam deflection shapes or mode of vibration amplitudes in a dynamic analysis. All variables and constants used in structural analysis are subject to two basic physical laws: Equilibrium (static or dynamic) and compatibility (preservation of structural continuity). Coupling these physical truths with principles of virtual work provides a basis for analyzing support of forces by elastic and inelastic bodies or systems, since the concept is not restricted to any particular assumptions of material behavior.

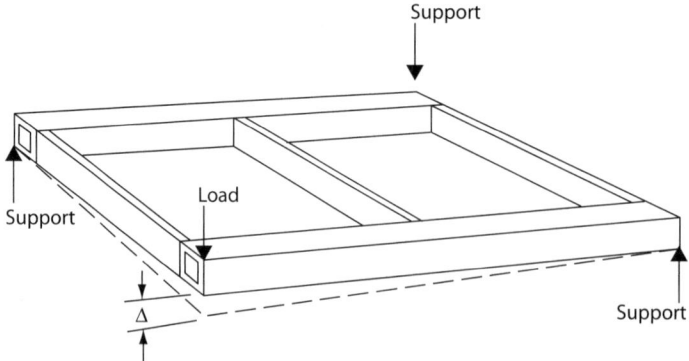

**Figure 7-2** Three-dimensional steel frames can be difficult to visualize regarding support of loads and analysis strategy.

Fortunately, for simple beams subjected to well-defined load patterns, internal moment, shear, and deflection (using material elastic properties) are given by many references in a tabulated format so that an engineer can quickly perform checks on the results of a more complicated analysis, or for the sake of performing hand operations for simpler elements and conditions in order not to lose touch. Results are summarized from multiple integrations of a linear equation that defines a loading pattern, along with constants based on boundary conditions. In terms of reality, simple equations and relationships help an engineer to remember that even the most sophisticated analytical results will never represent the "exact" behavior of a true structure, only as "precise" as reality allows. Unavoidable discrepancies between model and real structure are due to material property variability, inexactness of loading or support conditions, and limitations of analysis techniques.

While simplicity of analysis and visualization of a structural system are crucial to maturing as a competent engineer, there are most definitely cases and conditions where more complicated forms of analysis and design are warranted and should be carefully considered. It may not be possible to simplify a structure with a high level of complexity into smaller, more manageable and visible pieces due to conditions of support or material arrangements. The concept of structural integrity involves a complete structure, not merely a summation of response from individual parts, where all elements work together to form a redundant web of support (in many cases) with a global response to imposed load. It is this unity of action that helps a structure maintain overall integrity even when local failures occur and proper consideration can mean great economical difference between engineering solutions, or can play a pivotal role in maintaining true stability and strength against failure.

Contractors want to have an assurance that details proposed by a structural engineer are buildable and suitable to a particular project, which may require direct communication with field personnel in cases of repair or retrofit, and conditions may necessitate a change of direction that can only be accommodated by human flexibility in design—flexibility of decision making that could never be automated or computerized. A simple beam design is shown in Fig. 7-3, indicating the time savings available through charts and diagrams of common scenarios that have already been considered. There are certainly many more time-saving tools available, even for combined elements like rigid frames, and taking advantage of these is one step toward a quick yet competent design.

### 7.1.2 Presentation

Model building codes require that an engineer justifies compliance with the provisions through the use of engineering reports or computations, as may be applicable, necessary, or requested by the building official. It is easy to see that

170 | Chapter 7

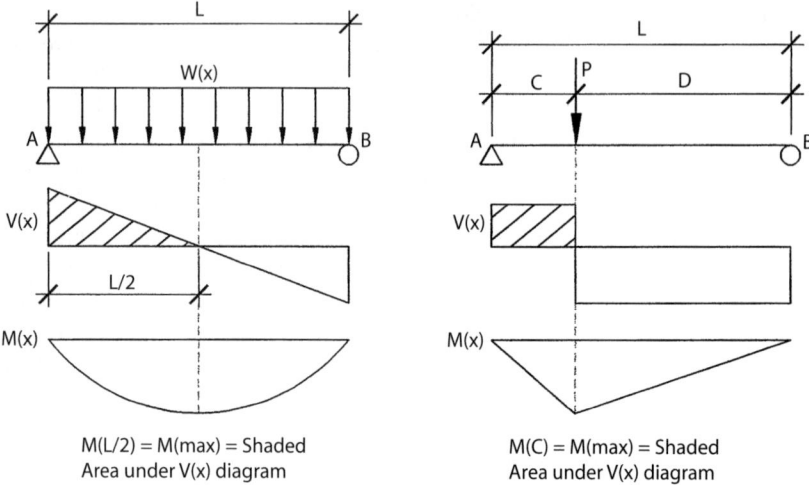

| **Figure 7-3** Typical simply supported beam loading and reaction diagrams.

compliance with seismic force minimum levels cannot be proven without calculations, as magnitude of load cannot be specified through intuition. A building official reserves the right, however, to approve submittal of drawings without extensive calculations if the engineer-of-record has a solid reputation for quality, if the work clearly demonstrates compliance without need of further documentation, or if the project is minor in nature. Because the purpose of structural calculations is to verify, confirm, or justify compliance with legal requirements, and that others will be making that determination, it is essential to submit (and keep a copy of) documents that can be clearly read and interpreted, even if set aside for a number of years and reviewed for some future work.

Some universal guidelines for presenting the results of a structural analysis and design are summarized as follows:

1. Methods of analysis and design used should be obviously stated in some form or another, either through basic documentation from a computer analysis program or citation of reference material.
2. Calculation pages should be neat and orderly, using a table of contents if necessary.
3. Number all sheets within a calculation submittal package in some logical order so that a reviewer can clearly identify an area needing correction or further work.
4. In the process of designing a member, lay out the most significant steps and highlight the final solution in some way (underline, box).
5. Provide some form of member identification to tie presented work to construction drawings, either by a direct copy of the plans or something that has been gleaned from architectural drawings and sketched atop to show framing

**Figure 7-4** Erection of structural steel-framed, antenna-supporting tower. (*Courtesy Conlan Engineering and Construction*)

members or other conditions, including shear wall or frame layout and identification of drag struts or chords. Having a physical picture of what the intended system is to look like (even in rudimentary form), in simple plan or stick-figure elevation, serves to keep the engineer on the right path as well.

6. Where the stability of multiple-level elements is based on layout of members below, sketch an elevation view to trace path of load from origin to foundation. This step can actually help speed the design process as a global picture of what is being modeled reveals short cuts and pitfalls to avoid or design for, even for a simple tower structure as shown in Fig. 7-4.

## 7.2  PROJECT SPECIFICATIONS

A specification is something that describes an element, either through performance requirements or identification of acceptable manufacturers, and methods of installation, execution, and inspection in extensive enough detail so that everyone with an interest understands what is expected. It is intended to describe the thoughts and intents of the design professionals and to establish a legal basis from which to develop contracts, secure appropriate bids, and guide the process of construction in a predictable way. Simple technical requirements can be listed directly on the drawings and extensive instructions may be reproduced in as many volumes as necessary.

It has often been found that a project's book of specifications is easier to interpret than the actual drawings, likely due to the fact that specifications contain extensive descriptions of the quality of work that is intended, from materials to manufacturers. Specifications also clearly identify the legal restrictions, bidding terms and submittal requirements that a contractor must agree to in order to successfully close out a project. They are intended to protect the owner against unscrupulous or incompetent contractors, but specifications should also be viewed as a protective mechanism for the contractor himself, as the scope of work and the expectations of the owner are clearly defined therein. All parties expect to be able to understand what is required to complete the work and problems most certainly arise when clarity is lost.

Specification writers must not be ignorant of real-world construction techniques, jobsite safety issues, sequences of operation, or limitations of building materials and systems. Most contractors are honest in their business dealings, but change orders are approved many times because specifications were interpreted to mean something other than what the designer intended. These types of problems can be managed through the use of time-tested specification sections carefully arranged in a manner that a contractor should reasonably expect. All material must be applicable to current codes and standards, using the correct designations and reference numbers, and extraneous text with no relevance to the project under consideration must be trimmed out.

Goldbloom (1992) explains that there are four main troubled conditions that writers produce in specifications or project notes: Sentences that are the wrong length, improper use of terms, ambiguous wording, and poor word choices. Use of too many words causes confusion and often will result in careless reading by a contractor who is pressed for time. Get to the point quickly in as few words as possible. Familiarity with the subject being described in a particular section, in addition to having an understanding of how the requirements will likely be met in the field, helps greatly in reducing the length to a manageable size.

In all sections, word usage is of utmost importance due to the fact that different readers have a variety of ideas of what something actually means. For example, the words *shall* and *will* convey different legal effects but are often used interchangeably. The first indicates an obligation of the subject whereas the latter conveys an intended course of action of the owner, engineer, or other party. The terms *working day*, *substantial completion*, and *to the satisfaction of the owner* can have ambiguous meanings related to time and completeness of work if they are not clearly defined. The term *calendar day* requires no definition and is a good starting point for defining a working day, which would exclude specific calendar days like weekends and legal holidays. Additional description is necessary to allow for unexpected contingencies or other discovered conditions, such as findings of an archaeological nature or to allow for reasonable release from restrictions imposed by severe weather.

Arora (1994) notes that writers often confuse the words *furnish, install,* and *provide* which require very different duties of the contractor. If a piece of equipment is to be furnished, it will be supplied, delivered, unpacked, and made ready for assembly or other use by the owner—it does not mean to install, test, start-up, or assemble. When a contractor is directed to install a particular element, it must be in a condition ready for use by the owner and may involve assembly, testing, anchoring, cleaning, or other operation. The term *provide* usually combines the definitions of both *furnish* and *install.*

### 7.2.1 General Organization

Completed specification books are organized into three divisions: Bidding documents, project conditions, and technical requirements, though one set of specifications may look quite different from another. Some projects require extensive paperwork for bidding and contractual definitions, but these are often given the term *boilerplate* because the content isn't expected to change from project to project. Larger corporations or government entities attempt to keep these front-end documents in as current a legal state as possible in order to avoid being railroaded by some technicality in the process of securing or administering the contract. There are certainly special projects that warrant changes, but in general common sections should be predictable to those in a position of approval.

### 7.2.2 Bidding Documents and General Project Conditions

The bidding division includes a comprehensive list of instructions for all bidders, including important dates (bid submittal, prebid job walks) and necessary forms that need to be completed. Project or contractual conditions include items such as insurance, document ownership, protection of the work, procedures for filing claims of extra work, payments, and termination clauses. General conditions are always the same from project to project, but a set of specifications may also include special work-specific requirements. The American Institute of Architects, National Society of Professional Engineers, American Consulting Engineers Council, Construction Specifications Institute, and others have written forms of general contractual conditions that have been tested over time on a variety of different projects. Some of the subjects addressed within the general and bidding section of common specifications include provisions for alternates or contract modification, unit prices and payment, administrative requirements for meetings and submittals, quality assurance, temporary facilities, demolition, and contractual paperwork such as the bond.

It is important to include provisions for quickly resolving disputes as to the meaning or intent of the project documents in order to keep the schedule moving. A contractor should have some idea of what to expect in a set of documents, but special procedures for a particular element of work may not be immediately understood and the engineer must make careful effort to clearly define the scope of work. If something appears to qualify as an extra expense, the contractor will submit a request for a change order to obtain additional money over the bid amount and the contractual procedures for this (sometimes inevitable) activity must also be clearly explained. Typically, the project engineer or architect must review and approve all change orders and she should have some idea of what costs are reasonable. Justification can be requested in the form of bills, receipts, or proposals.

### 7.2.3 Technical Section

Most technical specifications are assembled according to a predefined numbering and organizational system known as *Masterformat*, created by the Construction Specifications Institute (CSI) and recently updated in a combined effort with Construction Specifications Canada (CSC), and is available in different software packages tailored to the needs of a particular group, such as *Masterspec* which was developed by the American Institute of Architects (AIA). Each division is further subdivided into three sections. A *General* section describes administrative or procedural requirement that are specific to that division including an outline of definitions, a listing of related sections elsewhere in the project manual, submittals, quality assurance, delivery and storage, sequencing (with caution), owner's instructions, maintenance, and others. A section devoted to *Products* will define various materials, equipment, systems, mixes, fabrications, or assemblies that are to be incorporated into the project as related to a specific division. This section identifies allowed manufacturers, finishes, and specific definition or description of what is to be supplied by the contractor. A final section concerns *Execution* and describes installation methods for previously defined products, including measures for cleaning and protection against damage. Installers must meet certain qualifications, and those will be further defined within this section.

The writer must be consistent in wording and organization throughout all of the technical sections, regardless of whether they are related to the structural or mechanical systems. This is often difficult to achieve because each consultant assembles sections pertinent to their own work and forwards these to a project manager, who often does not have time enough to review and return for grammatical or consistent-wording forms of correction. Following or prescribing a common master set of specifications helps achieve consistency, but some clarity is inevitably lost due in part to a mixture of writing styles and different ideas of what sounds best. The technical section of specifications has been in consistent form for years and product manufacturers work hard to keep their paperwork up-to-date, otherwise they get phased out due to noncompliance or irrelevance. Common divisions include site construction (clearing, excavation), concrete (cast-in-place, architectural), masonry (unit assemblies, stone), metals (structural steel, fabrications, gratings), wood and plastics (rough carpentry, decking, sheathing), and much more.

Opening paragraphs define the quality of an upcoming section, so it is important to make a good first impression. It is not necessary to explain that "the contractor shall complete all of the work described in this section," partially because statements to this effect already appear in the general conditions of the contract and also due to the fact that everyone understands the contractor is supposed to complete everything described in the project documents. A better opening to each technical section clarifies what portions of the work identified on the drawings will be hereafter defined, such as "the work specified in this section includes . . . as shown on the project drawings."

It is also important to describe events in the order they will be performed to reduce the possibility of missing steps. For example, the cast-in-place concrete section defines material and mechanical properties required from a mix design, construction and maintenance of forms (unless described elsewhere), placement, finishing procedures, curing, stripping of forms, and final treatments.

### 7.2.4 Special Sections or Conditions

General contractual and project conditions suffice for most work, but there are many reasons for the inclusion of a section of special conditions that general and subcontractors must be aware of and agree to abide by. In case of disagreement between the general and special conditions sections, the special conditions should govern since they are written for the particular work at hand. There may be site restrictions to be aware of, such as limitations of noise and dust, and the contractor's response to these issues are to be clearly defined so as to leave no room for misinterpretation. A section of general conditions identifies the importance of time to a project's success, but the special conditions section will denote actual dates for completion and associated bonuses or penalties.

It may be necessary to restrict the times and days that work can be performed, such as after a specific hour and identification of legal holidays or those specific to the proper function of the community, such as disallowing church-associated work on Sundays.

## 7.3 PROJECT DRAWINGS

Section 1603 of the 2006 IBC requires specific design information to be included on project drawings, including identification of structural member sizes, locations relative to floor level and columns, and design load data. Identifying dead and live loads used is fairly straightforward, being selected from a model code and summarized from material estimates, but the reporting of snow, wind, and seismic loading used in structural design can be cumbersome.

When ground snow loads are reported to be greater than 10 psf (0.479 kN/m), project drawings must also identify the flat-roof snow load, snow exposure coefficient, importance factor, and the thermal factor determined from Chapter 7 of ASCE/SEI 7-05. Whether wind pressures govern the design of a structure or not, drawings should indicate the basic design 3-second gust wind speed, importance factor, exposure coefficient, and other coefficients found in Chapter 6 of ASCE/SEI 7-05 that may be needed by consultants to design premanufactured (or other) elements. Factors used to determine seismic loads that are to be recorded on the plans include the importance factor, mapped spectral response coefficients for both short and long periods, site class, seismic design category, and any determined seismic response coefficients or factors that are plugged into code equations as found in Chapter 12 of ASCE/SEI 7-05.

Drawings must be as clear to the intended result as possible, including sufficient detailing and material specifications because if they are not, the contractor will likely find something to hold up as an extra cost once bids are accepted. The contractor's scope of work must be absolutely clear; it is not enough to simply add a note to the plans stating that the contractor shall be responsible for "reviewing the contract documents" and "obtaining clarification for items not immediately clear," however it is not possible to state every model code requirement on a set of plans. There are many conditions to which an ordinary contractor should already have a working familiarity with. Some common problems associated with project drawings have included the omission of certain materials, architectural plans that were not coordinated with structural drawings, structural drawings that were not coordinated with mechanical plans, and dimensional errors. In fact, Section 106.1.1 of the 2006 IBC and Section 1.7.6.3.1.4 of NFPA 5000 require that plans be submitted with sufficient enough detail to demonstrate compliance with the code, to clearly show the nature, character, extent, and location of work.

In order to properly define the structural system, both to the building official and to the contractor who will be putting things together, a complete set of plans is necessary. These plans will include the following at a minimum:

1. *Structural Notes*: Define the standards of quality expected in materials used and construction measures employed, including identification of relevant codes.
2. *Foundation Plan*: Complete dimensions must match the floor plan, while sizes and detail references must be accurate and complete enough so that a bidder can put together a reasonable cost for completion of the work.
3. *Framing Plans*: Must also include information on strap ties, drag struts, or other special elements necessary to tie the building together. It may be necessary

to divide different portions onto different sheets for clarity, such as a *Low Roof Framing Plan* and a *High Roof Framing Plan*.

4. *Framing Elevations or Sections*: A completed assembly is best shown in smaller segments, usually through whole building cross-sectional views or individual wall sections. A stacked lateral force resisting system can quickly be interpreted through the inclusion of proper elevations and schedules.

5. *Typical Details*: Commonly expected in the building industry, or as pictures of important code requirements that may be too difficult to explain in the structural notes or a set of project specifications.

6. *Special Details*: Needed to define the unique properties of that specific project. Liberally apply detail cuts at any location on a plan or section where it is difficult to see what is to be framed.

### 7.3.1 Goals and Methods

In order to reduce errors, omissions, or ambiguities, all drafters within a company should perform their duties in much the same way, using the same checklist to produce a unified set of construction drawings regardless of project type or size. Common procedures for all drafters include project setup within the computer system; adherence to company protocol regarding the use of colors, line weights, text size and type; drawing assembly in relation to numbering and order of appearance, following a logical sequence of construction (foundation information comes before roof framing); use of the most recent version of standard notes or details found in a company database; striving for a uniform appearance of all details and sections, including alignment of text and consistent terminology; and communication with supervisors or other project managers regarding schedule, design or scope changes.

In the process of assembling structural drawings, an engineer needs to be aware of possible limitations of site information provided by another party. Errors in an original survey can result in significant costs if they are not discovered before the construction phase. Some data points may be difficult to obtain, such as those which are at a significant difference in elevation as the others, and they should at least be used in design with caution. Additionally, the actual position of relevant existing structures may not match what is shown on record drawings and critical circumstances should be defined by a new survey. One of the difficulties of laying out building plans is determining where to begin. Some buildings are regular for the most part, but contain certain areas that present special challenges, such as shown in Fig. 7-5, indicating a structural column near an elevator pit, set at an angle from a curved glass block wall, adjacent to a series of concrete shear walls that are placed in some relation to a domed entry structure.

In terms of computer-aided drafting, successful companies have their own set of standards to which all engineers and draftspersons are expected to adhere. Items for standardization include layers (names, colors, linetypes, and usage), plotted

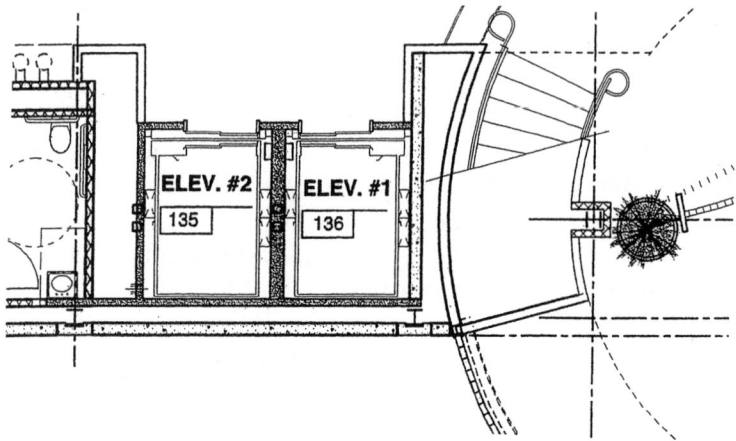

**Figure 7-5** Partial floor plan in region with different shapes and special challenges.

lineweights for line work and text, font size and style, dimension styles (arrows, ticks, placement of text), as well as usage of blocks and externally referenced objects or drawings. Office standards should be incorporated into drafting protocols in the form of templates, manuals of practice, a database of typical details and structural notes, or by the use of some form of third party software that manages data and recognizes the real power of a good drafting program.

The most important goal of structural details is to clearly identify the work. Clarity involves a number of important considerations, including the following:

1. Show everything that will be seen in the field, but dash or hide elements that are set into the background.
2. Confirm that all special, and many typical, details are cut from or identified on a plan, section, or another detail.
3. Keep repetition of information to a minimum, if any at all.
4. Carefully check dimensions, conditions, and elevations.
5. Construction details should properly match the mathematical model used in analysis.
6. Pay attention to sequence of framing the detail.
7. The content may affect the work of other trades, therefore confirm that someone else hasn't already taken care of a specific condition.
8. Construction tolerances are a reality of life that can throw a huge monkey wrench into the works, therefore allow for them in all configurations.
9. Confirm that deflections of structural elements do not interfere with other pieces, such as crossing members or glazing.
10. Identify special testing or inspection requirements.

## 7.3.2 Presentation

All drawings used in the construction of any structure must be reliable, readable, and present information that is constructible. These things are so fundamental, but the art of drafting is close to being lost in the push to increase an engineering graduate's technical knowledge. In fact, an engineer's problem solving prowess is closely linked to his ability to perceive a structure in both two- and three-dimensions (2-D and 3-D) and to accurately describe what is imagined to a person of limited experience. Some engineers in larger companies do little of their own drafting and the duty is delegated to an engineer-in-training or to a nonlicensed draftsperson, which may work reasonably well provided the company has some form of quality control and management of every single document produced such that there is adequate technical review and consistency in the presentation.

Consistency implies that the following will be true:

1. *Anyone who is hired to produce drawings must have this same ability to perceive and communicate in 2-D and 3-D.* Recreating ideas onto paper is so much more than simply understanding how to manipulate a computer drafting program: It is the ability of a draftsperson to raise their awareness to a higher level of perception and thought that will make a project (and a career) successful. A second set of eyes that are trained to properly build a detail as it is drawn, not to mindlessly recreate lines or a rough conceptual sketch, are invaluable to a company's risk management philosophy. A successful draftsperson is also curious about things that are observed in the real world, where there exists a curiosity of not only how things come together, but also how to recreate those things onto paper in a way that others can understand. Training programs must focus on encouraging, building, and strengthening this curiosity not only in engineers but also in the drafting team.

2. *All line work and organization of information (i.e., aligning text, producing schedules) is to fit within company protocols.* A good way to teach an appreciation for line weights and organization is to have a draftsperson complete sketches of structural details or building and bridge sections by hand. This will be discussed further within this chapter.

3. *Though different persons are involved from project to project, the same chain of command will be respected for every design team.* Communication begins among team members before the first line is drawn. A draftsperson needs to understand the goals and intended end result of a project just as much as the engineer-of-record, even though they will not typically be involved in meetings with the client. Sometimes it may be extremely helpful to include a lead draftsperson in meetings with a client in order that they can hear first-hand accounts of how the project is being visualized and directed. The stream of communication between all persons involved in a project's execution cannot be allowed to break down into ambiguity or confusion and the engineer-of-record is directly responsible for disseminating and clearly explaining information and instructions in a timely manner, regardless of how the work is ultimately delegated.

4. *The technical requirements of a project will be adequately checked for correctness and completeness prior to formal issuance of construction documents.* This involves a sequence of steps: Initial production of drawings, check-set, and back-check. The first generation of project drawings should be of sufficient clarity to describe what is to be built.

**Teaching Presentation Skills and Visualization**  It is always an amazing experience to watch a skilled artist recreate something from the real world onto a canvas or piece of paper. Famous oil painters like Bob Ross demonstrate an intimate appreciation for spatial relationships, the role of color and blending, and realism that serves to inspire those who stand by as witnesses.

There is still great value in the manipulation of simple drafting tools like the triangle, T-square, and scale. Because the perception of quality or cleanliness is often subjective, a company needs to have a distinct set of instructions in place that are quickly appreciated the more a draftsperson has to write, draw, correct, erase, and scrutinize in comparison with an example of professional, acceptable work. Pencil lead is commonly available in the following degrees (from hardest lead/lightest lines to softest lead/darkest lines): 6H, 4H, 2H, H, F, HB, B, 2B, 3B, 4B, 6B, and 8B. Linework created by softer lead has a greater tendency to smudge, so a thin metal erasing shield often helps guide the writing hand over paper without damaging completed work. Of course, these skills of identifying quality and readability can also be taught using computer-generated details, line types, colors, plotting standards, and skills of organization (lining up text, consistency in wording, and the like).

All persons who are responsible for completing drawings should be taken on multiple tours of different types of facilities, especially those in different stages of construction. The guide, preferably one in the company who serves as engineer-of-record, points out connections and asks probing questions: "What is the basic function?" "What are some other ways to achieve the same support condition?" "What design standards are necessary to get the dimensions and conditions correct?" "How could things have been made simpler for the builder?"

### 7.3.3  Reviewing the Work of Other Consultants or Clients

There is often a temptation to merely refer to the work of others when it comes to dimensions and the layout of different elements, but this can be a troublesome approach. A foundation subcontractor needs sufficient information directly on the structural plans related to that work in order to make sure that there are no conflicts with the superstructure or subgrade components. For example, layout and geometry of raised concrete curbs in slab-on-grade construction are frequently determined at a later time in the sequence of an architect's design and may be subject to change, but it is important to identify their placement in the most logical fashion directly on the foundation plan to allow construction to proceed quickly and smoothly. Minimize references to the work of others and carefully coordinate, with sufficient documentation, critical dimensions or conditions. The key to successful use and review of the work of others is consistent, clear communication. A structural engineer needs to be familiar with the duties of other trades (refer to Chapters 4 and 5) in order to anticipate when conflicts may arise or when something needs to be reviewed or modified to accommodate a sensitive or physically restrictive situation.

Building Information Modeling (BIM) is a system of document control being developed by design/drafting software providers in an attempt to streamline the availability and dissemination of information relative to a structure's complete design, including elements of all trades. It is a completely different approach to project completion that requires a team of experts for successful and meaningful transition to the practice. Due to the complexity and intended availability of modeling information, engineers and architects who are using or transitioning to this technology must carefully assess the relevance of data to update, the extent and involvement of sharing that is to take place between consultants, security of electronic data, and coordination of budget and responsibility.

### 7.3.4 Responsibility

Teamwork and solid communication are keys to a successful project. No one wants to meet or cause failure and everyone is usually ready to fix problems should they arise. Unintended problems may require a buildable detail from the contractor that is checked for soundness by the engineer and dressed up by the architect to fit within the form of a structure. Everyone must work together in order to keep costs down and to meet a project's schedule. The best policy for resolving problems that arise in the field is to call a meeting of all associated parties, where solutions may be offered without fear of ridicule and with the expectation of being considered. All options which meet the intended engineering function should be assigned some form of value and there should then be no reason to reject the lowest-cost solution. Only after a problem has been resolved should the issue of payment responsibility be considered.

The issue of project responsibility begins within the engineer's office and is directed by company leadership and office standards. All construction documents must be subjected to a system of quality control that includes oversight and established protocol, such as *A Guideline Addressing Coordination and Completeness of Structural Construction Documents* published by the Council of American Structural Engineers (CASE 962-D, 2003). Larger companies may have an engineer on board whose sole function is mentoring new graduates, setting up and enforcing office standards and procedures, reviewing all contract documents for completeness, and checking to see that clients and contractors have a level of confidence with the work that is received.

A licensed professional is expected to follow the rules established by state licensing boards and national engineering societies, but sometimes there is confusion when companies of different disciplines and professional registration work together to meet a client's needs. An engineer needs to be careful not to *aid and abet* another person in the commission of a crime, such as practicing engineering without a license or outside their area of expertise. Precedent set by the State of Maryland [*Anello v. State*, 201 Md. 164, 168 (1952)] defined the terms *aider* and *abettor*, which is similarly defined by other states, as follows: "The legal definition of the word 'aider' is not different from its meaning in common parlance. It means one who assists, supports, or supplements the efforts of another. The word 'abettor' means in law one who instigates, advises, or encourages the commission of a crime. To be an aider or abettor, it is not eseential that there be prearranged concert of action, although, in the absence of such action, it is essential that one should in some way advocate or encourage the commission of the crime."

The responsible engineer affixing stamp and signature on a set of plans is asserting that he has complied with the state regulatory agency's directives on who is allowed to use such a seal and what that seal implies. *Direct supervisory control* and *direct professional knowledge* are phrases in common appearance within state engineers acts, but there is considerable debate on how they are to be applied to practical situations. For example, if an unlicensed individual produces a complete set of construction drawings for a structure requiring a building permit and delivers those plans to a licensed engineer for a stamp and signature, can the professional truly exert direct supervisory control or direct professional knowledge at this stage of development? The answer strongly depends on how a state engineer's board, as well as the court system, has interpreted such cases in the past. Many documents produced within an engineer's own office are completed by unlicensed staff members, so what would the difference be here, provided the licensed individual carefully reviews the documents?

Supervisory control, it seems, can still be managed after the completion of documents provided the engineer who bears responsibility retains the right to review and change any or all engineering decisions that may have been made by an unlicensed designer, that all code requirements are thoroughly reviewed with the addition of supplemental calculations when necessary, that all dimensions and conditions are reviewed for constructability, and the need for other professionals (architect, mechanical engineer) is established and arranged.

## 7.4 ENGINEERING REPORTS

Engineering failures, such as the one shown in Fig. 7-6, require written reports to summarize an engineer's conclusions on what caused the damage to occur, and possibly to identify further retrofit measures or repairs depending on the scope of

**Figure 7-6** Wood truss bottom chord tension failure. (*Courtesy Lane Engineers, Inc.*)

work agreed upon with the client. Many times, a person who reads a report with some knowledge of the event which precipitated the investigation has already formed his or her own opinion on the matter and may only be looking through the report to justify that position. This cannot be helped, but it highlights the fact that an engineering report must state facts in a truthful manner, to the best of the writer's ability, without intent to malign someone else or to boost one's credibility in one way or another.

A good report is well organized according to a logical sequence of events. Namely, a request for services was initially made by the client, a site investigation took place by the engineer where things related to the call were documented, research occurred back at the office to pursue a decision regarding the purpose of the investigation, and conclusions were considered and finalized.

Consider the following report subdivisions that seem to work fairly well:

1. *Introduction*: Describe specific reasons the report was requested and identify the property.
2. *Investigation*: Describe any special procedures followed or equipment used during the course of the site visit to help readers see the activity. If any tests were performed or materials sampled, those would also be described here.
3. *Observations and conclusions*: Depending on the scope of work, it is helpful to list problems or issues observed as related to the initial project request with relevant conclusions in the same breath, in order to begin connecting the dots for the reader.

4. *Research and further investigation*: Any other points that might help solidify conclusions of the investigation would be presented here, along with recommendations for further investigation, presented with specific directions and whether testing is necessary or if specialists need to be called in to investigate or confirm some observation or condition.

Photographs are an important part of any engineering report, since the pattern of a developed crack or the physical location of damage can speak volumes in terms of identifying cause. These can be placed in an appendix, along with structural calculations that may be necessary to justify a hunch or show conclusively that some condition does not meet absolute principles found in a model code or building material design standard.

# 8 Engineering Mechanics

The practice of structural engineering hinges on principles of material behavior and engineering mechanics, which identify the response of elements and completed structures that are subjected to load. Figure 8-1 shows a residence in Louisiana constructed in such a way that the living quarters are raised above an anticipated flood plain. This is a desirable solution for management of flood damage risk, provided the effect of a soft lower story is taken into account in the design of a suitable lateral force resisting structural system from wind, earthquake, or flood forces. Therefore, a structural engineer is concerned with the behavior of structural systems and must understand the basics of loading and rational engineering mechanics to arrive at any defensible solution.

Because of space limitations, it is not the intention of this chapter to provide complete guidance on determining or distributing loads that apply to a structure, as there are already many fine references available for doing just that. Rather, background and general information is summarized from a variety of sources in such a way that readers of all experience levels learn how and why these principles serve the purpose of discovering better solutions to structural engineering problems.

## 8.1 STATIC LOADS

Statics is a branch of engineering mechanics dealing with solid bodies at rest. These bodies are assumed to be perfectly rigid in order to simplify methods of analysis. It is an assumption that is justified more often than not because local deformations under load are usually small enough so as not to appreciably affect

**Figure 8-1** Residence in Louisiana with first floor raised for flood protection.

equilibrium or structural motion. In order to fully understand this principle (and others), an engineer must first discover the types of loads that can be imposed upon a structure. Static loads are considered to be those which are stationary by nature and continuously exert influence on the supporting structure unless they are relocated by separate phenomena.

## 8.1.1 Dead

Dead loads include the weight of all elements considered as permanent fixtures, including structural members themselves, roofing or flooring materials, insulation or fire-suppression systems, ceilings, finishing materials, equipment, and the like. These are considered to be fixed in magnitude and position and are relatively predictable based on manufacturer's data or tabulated amounts, such as is identified in Chapter C3 of ASCE/SEI 7-05. Dead loads are applied to a structural member along its length in the direction of gravity, regardless of whether the member is sloped or not, and may be present in a uniformly or trapezoidal distributed pattern, as concentrated forces, or as surface pressures depending on the arrangement of structural members.

## 8.1.2 Live

Live loads applied to the roof of a structure are commonly specified by a state's adopted building code and relate to construction-type loading in the form of material storage (balanced or unbalanced), safety tethering, or some other. During occupancy of the building, roof live loading exists in the form of reroofing personnel and materials or fire-fighting equipment during an emergency. Section 4.9 of ASCE/SEI 7-05 allows roof live loading to be reduced based on pitch, curvature, and area tributary to an element being designed.

Floor-supported live loads, on the other hand, are defined as those which a structure carries in addition to its own weight, commonly attributed to the usage or occupancy of that structure. For example, a building must be able to support occupants, floor finishing materials, furniture, movable equipment, and storage. A bridge, on the other hand, is to carry traffic in the form of vehicles, trains, or pedestrians. Live load applies along a member's horizontal projection, regardless of whether it is sloped. Table 4-1 of ASCE/SEI 7-05 lists historically accepted live load values for different floor uses (catwalks, libraries, offices, storage, and the like.) and Section 4.8 of ASCE/SEI 7-05 presents equations for reducing these base values according to supported area.

It has been debated as to whether vehicle live loading should be combined with seismic forces in bridge design. Opponents suggest that tires and suspension systems actually serve as damping devices and may actually reduce seismic response (ACI 341.2R-97, Section 4.1.3). Special bridge configurations, however, sometimes demonstrate the importance of including at least a percentage of live loading in seismic load combination equations that are presented in bridge codes for purposes of advance planning for emergencies.

### 8.1.3 Snow

Snow loading is defined regionally as *ground snow* and factors found in Chapter 7 of ASCE/SEI 7-05 are multiplied into simple equations for translating those established values into *roof snow* loads for structural design purposes. A building's proximity to large trees, hills, or other buildings can affect the amount of snow accumulation on the roof and special consideration is warranted in these cases. Drifts, sudden violent storms such as is shown in Fig. 8-2, and other factors affect the design of framing members in certain areas and the local building official has file records to provide direction and some consistency in expectations. Appropriate design values are obtained for roof elements by considering slope and the covering material also (how well does it allow snow to slide off?); but floor balconies or other flat projections will not be suited for special reductions offered with these considerations. As with roof live loads, snow loads are typically applied to framing elements along a horizontally projected plane.

Values of ground snow load are maintained by building departments, based on statistical analyses of meteorological data and historic measurements. Depending on the geographic elevation of a constructed facility, required loading can be quite high, therefore the influence of adjustment factors becomes a crucial part of structural

188 | Chapter 8

**Figure 8-2** Blizzard in Jamestown, North Dakota, 1966.

design. Roofs at different elevations present danger of accumulation from sliding snow (in absence of dams or guards), increased drifts, or dynamic effects of falling snow.

The presence of snow on a bridge does not usually affect design to a great degree, since it is small in comparison with other applied load, but ice accumulation is cause for concern. Icing not only adds weight, but builds up in such a way that increases the exposed area of structural members to wind pressure based on the amount built up.

## 8.1.4 Soil Pressure

The magnitude and character of soil pressure experienced against a fully or partially buried structure is based on physical properties of the soil (bulk density, coefficients of friction and cohesion, profile of soil types, moisture content), interaction between soil and structure (frictional binding or slippage of materials), presence of overburden weight (adjacent foundations or vehicles), slope of retained material, and relative displacements. Pressure is distributed normal to a vertical surface in a triangular shape with the greatest intensity at the bottom and must include effects of water pressure when the structure extends below the water table. This will be discussed further in Chapter 9.

## 8.1.5 Others

A flood delivers both static and dynamic load to structures and may include additional forces from tidal surge in coastal high hazard regions identified by the Federal Emergency Management Agency (FEMA), through the National Flood Insurance Program. Structures built in flood zones must be protected from flotation, lateral movement, and collapse, as well as maintain a solid building envelope in order to keep wind and water from entering. When flooding occurs in response to another natural disaster, such as a hurricane, it can be difficult to identify the primary source of damage. Hurricane Andrew, for example, was a Category 4 storm according to the Saffir-Simpson hurricane scale when it came ashore in Dade County, Florida on August 24, 1992. An assessment team assembled by FEMA observed that flooding damage was comparatively minimal, though damage from wind pressure and associated debris was extensive (FEMA 1992, p. 5). The majority of observed flood damage occurred along the immediate coastline due to hydrostatic pressure and inundation by storm surge. Typical static flood loading to a basement wall is shown in Fig. 8-3.

Rain loading can be significant on flat roofs which deflect to such a degree that water ponds during a heavy storm faster than it is allowed to run off, usually requiring a close review for roof systems with a drop of less than $1/4$ in over a run of 12 in (1.19°) (ASCE/SEI 7-05, Section C8.4; NFPA 5000, Section 35.1.2.8.8). Water is often channeled off the roof through a gutter and downspout combination

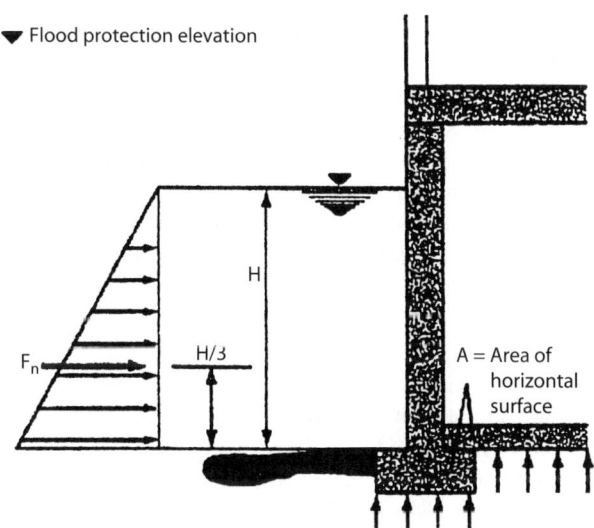

**Figure 8-3** Static flood loading on basement wall.
(*Courtesy Federal Emergency Management Agency*)

or an internal plumbing system with inclusion of overflow drains and scuppers in case primary drains are clogged.

Other phenomena covered by load combinations in model building codes include hydrostatic or bulk material pressure (similar to soil loading), miscellaneous fluids with well-defined pressure based on density, or self-straining loads due to temperature or moisture changes, material creep, or differential settlement. Impact or other extreme loads also can play a role in structural design, as a significant force is applied over a relatively short duration of time, playing the role of a static-type force yet setting up significant dynamic effects in both buildings and bridges.

## 8.2 DYNAMIC-TYPE LOADS

Wind and earthquake forces are both dynamic in nature, though there are fundamental differences in how they are applied to a structure. The phenomenon of wind is caused by variable solar heating of the atmosphere, initiated by differences in pressure, and affects a structure through direct pressure in proportion to exposed surface area. Seismic loading, on the other hand, occurs as a sudden release of energy from the earth's crust and upper mantle causing ground movement of different intensities. This arises from built-up stress and is imparted to a structure as a function of its mass.

### 8.2.1 Understanding Structural Dynamics

Under dynamic loading, a structure vibrates in at least one mode for a period of time. The first mode is identified by the entire structure moving in a particular direction, whereas higher modes are noted as situations where one part (or parts) of the structure moves in one direction as another part moves in the opposite direction. The most basic dynamic property of any structure is the fundamental period of vibration, T, which is defined as the length of time (in seconds) it takes for the structure to cycle through the first mode of vibration freely and return to its original undeflected shape. It is calculated using theoretical equations or simple approximations that are presented in ASCE/SEI 7-05, based on height and type of bracing system present.

Damping of a structure also affects dynamic performance and can be thought of as resistance of the construction materials to imposed motion, though it is a difficult measurement to pin down in precise terms. It exists in the form of frictional resistance, yielding of steel elements, crushing of wood framing, and crack formation in concrete and masonry. Special damping devices are sometimes installed during construction to improve the dynamic performance of a structure and control damage. They are classified as either passive (base isolation or supplemental energy dissipation devices) or active (mechanical devices whose

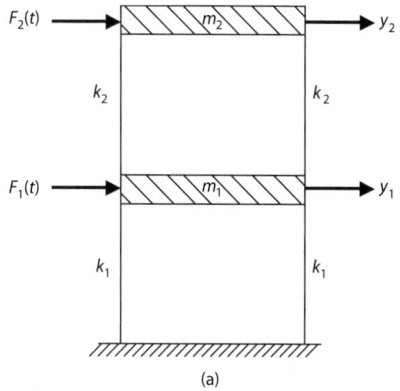

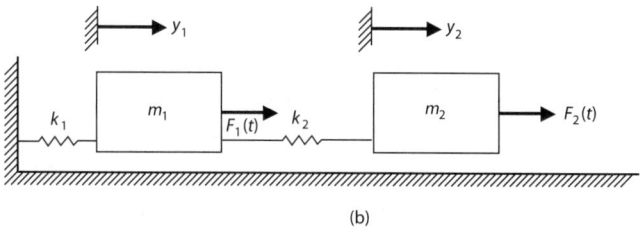

**Figure 8-4** Mathematical model of multiple degree-of-freedom dynamic system.

characteristics change during building response based on measurements of motion). Even in buildings without special devices some damping, or energy dissipation, has been recognized for many years with an accepted value of 2 to 5% of critical viscous damping in a linear response analysis (Hanson 1993).

Typical equations of dynamic motion use a lumped-mass approach, where the mass attributed to each level is lumped at respective floor and roof center of gravity elevations that are subjected to lateral displacement. The inertial force that corresponds to a particular degree-of-freedom is dependent only on the acceleration in that degree-of-freedom. A multiple degree-of-freedom structure, including mass, damping, and stiffness can be visualized as in Fig. 8-4.

### 8.2.2 Wind

Although wind loading is not truly considered "dynamic" in the purest sense of the word, it does have certain dynamic characteristics and will affect a structure in a different manner than those that are static in nature. Wind conditions that are important in engineering applications are measured by the National Weather Service with consideration given to averaging time of wind gusts through a sensor

(3 seconds), height of instrument placement above ground [typically at 33 ft (10 m)], and roughness of surrounding terrain (assuming a reasonably uniform roughness over a distance of about 100 times the instrument height). The 3-second gust speed of a wind event occurs as a considerable change of speed and direction during a storm and is the constant value used for structural design. As wind travels along, the surface of the earth restricts flow with a frictional force based on topographic features, having less effect farther above the ground.

In the United States, hurricanes are a reality of life that affect many people every year. A structure that has been properly constructed using principles outlined in current model building codes based on region (i.e., high-wind load provisions where applicable) can survive an extreme wind without collapse, but may escape with extensive damage. Hurricanes are typically classified by category according to the Saffir/Simpson Scale, identified by number, maximum wind speed, and expected level of damage to common structures:

No. 1: Winds to 95 mph (42 m/s); no real damage to anchored structures.

No. 2: Winds to 110 mph (49 m/s); some damage to roofing materials, windows and doors; no real damage to anchored structures; minor flooding near coastal areas.

No. 3: Winds to 130 mph (58 m/s); some damage to roofing materials, windows and doors; some structural damage to buildings; serious flooding in coastal areas, smaller coastal structures destroyed.

No. 4: Winds to 155 mph (69 m/s); extensive damage to roofing materials, windows and doors; complete roof failure in small residences; major flooding and damage to lower floors of coastal buildings.

No. 5: Winds greater than 155 mph (69 m/s); considerable damage to complete failure of roof structures; very severe and extensive damage to windows and doors; some complete building failures; severe flooding and damage to lower floors of coastal buildings.

Due to their rare occurrence, tornadoes are not included in development of basic wind speed maps published in design standards, but some jurisdictions require special structures to resist the effects of these phenomena. Tornado Alley is described roughly as an area that runs north from the middle of Texas through South Dakota and northeast to Ohio, being known historically for some of the largest tornadoes ever experienced. Tornadoes classified by Fujita (Taly 2003, pp. 370–374) as F-4 with wind speeds between 261 and 318 mph (117 and 142 m/s) can lift strong-framed homes off their foundations and carry them a considerable distance to disintegration, send automobile-sized missiles a distance of 100 yards (91 m) or more, and can completely debark trees. Tornadoes with an F-rating between F-6 and F-12 are theoretically possible, but believed to be nonexistent. An F-5 category tornado swept across Oklahoma on May 3, 1999 and left a trail of tremendous damage to residential structures. Observations and experience led to the conclusion that structural damage will be inevitable within the path of any tornado classified as F-2 or stronger, typically falling apart at the load path created to resist wind uplift forces.

**Characteristics and Derivation**  Failure of a building opening during a violent wind storm can generate large internal pressures that, when coupled with exterior pressures acting in the same direction, often govern the design of roof and wall element anchorages. Flexibility of the building envelope lends to fluctuation of these internal pressures, as does suction produced along the exterior surface, whose combined effect is greatest along the windward roofline. Pressures on a building due to wind occur as streams of air move against or past it. Exterior walls provide primary resistance to these forces, but as air moves over the top of the roof, suction translates into an uplift load applied to members and connections that can be quite severe in a high wind event. It is modeled in testing as a monotonic load, which applies as pressure without reversal.

Basic wind service-load pressure in pounds per square foot, P, corresponding to the 3-second gust wind speed at a height of about 33 ft (10 m) above the level of the ground, is determined by Equation 8-1, where V is the basic wind speed in miles per hour:

$$P = 0.00256 * V^2 \qquad (8\text{-}1)$$

[In S.I.: $P = 0.613 * V^2$ (V in m/s, P in N/m$^2$)]

This basic equation is a starting point and it must be multiplied by different coefficients that account for exposure, proper gradient height, topographic effects, and building type to arrive at a useable design pressure that can be applied to a building model. Chapter 6 of ASCE/SEI 7-05 includes maps, tables, charts, and equations for determining design pressures using peak gust wind speeds based on hurricane predictions of 50- and 100-year return periods, including studies along the coast and inland areas.

**Application**  Regardless of the direction in which they act, wind pressures always apply in a normal (perpendicular) direction to the impacted plane in proportion to the width of area that is tributary to the element or system under consideration. Building walls resist horizontally applied forces, both into and away from the surface, whereas sloped-roof assemblies are impacted in both horizontal and vertical directions from suction or direct pressure. For design purposes, the main windforce resisting system (MWFRS) is that which provides resistance to wind pressure for the entire structure, including horizontal diaphragms with connections to vertical lateral force resisting elements (shear walls, frames). These are the same systems that resist seismic and other loads which set up racking, sliding, and overturning forces on a structure, described further in Chapter 4 and also found in bridge construction described in Chapter 5. ASCE/SEI 7-05 describes 3 different methods for deriving wind pressures: A simplified procedure

(Section 6.4), an analytical procedure (Section 6.5), and by wind tunnel testing (Section 6.6).

Individual elements that make up the building envelope, which transfer wind pressure to the MWFRS, are given special consideration as components and cladding, possibly located in regions considered to be discontinuous, which give rise to dynamic wind effects such as vortex shedding. Forces applied to components and their connections within these discontinuous regions (wall corners, roof eaves, and the like) can be quite high, even greater than what would be determined for the MWFRS, therefore, multiple load combinations will need to be reviewed.

### 8.2.3 Seismic

Earthquakes have shaken the world for many years and valuable lessons are learned about human nature and our state of preparedness (see Fig. 8-5). In the United States, for example, the U.S. Geological Survey has indicated that seismic activity can occur in at least 39 of the 50 states, but many people still mistakenly

| Earthquake | $M_w$ | $M_o$ (dyne-cm) |
|---|---|---|
| 1960 Chile Earthquake | 9.6 | $2.5 \times 10^{30}$ |
| 1964 Alaska Earthquake | 9.2 | $7.5 \times 10^{29}$ |
| 1906 San Francisco, CA Earthquake | 7.9 | $9.3 \times 10^{27}$ |
| 1971 San Fernando, CA Earthquake | 6.6 | $1.0 \times 10^{26}$ |
| 1976 Tangshan, China Earthquake | 7.5 | $1.8 \times 10^{27}$ |
| 1989 Loma Prieta, CA Earthquake | 6.9 | $2.7 \times 10^{26}$ |
| 1992 Cape Medocino, CA Earthquake | 7.0 | $4.2 \times 10^{26}$ |
| 1994 Northridge, CA Earthquake | 6.7 | $1.3 \times 10^{26}$ |
| 1995 Kobe, Japan Earthquake | 6.9 | $2.5 \times 10^{26}$ |

1 dyne-cm = $7.4 \times 10^{-8}$ foot-lbs

**Figure 8-5** Sampling of recent earthquakes measured on the Richter scale. (*Courtesy U.S. Army Corps of Engineers*)

associate shaking potential with only the western region. Scientists can only predict the occurrence and magnitude of future events with limited accuracy and cannot say with absolute certainty how the earth's crust will behave. The 1994 Northridge Earthquake in California, for example, occurred on a previously unknown fault (Todd et al. 1994, p. 5) and held some surprises in terms of motion and character. The event was recorded at a greater depth than any previous large earthquake in the region and it did not match with any surface geological structures in an easily recognizable way (Hall 1995, p. 10).

Earthquake magnitudes are commonly reported in terms of two major scales: Richter and Modified Mercalli (MM). The Richter scale is based on direct measurement of seismic wave amplitude recorded on a seismograph and results are reported in scales of surface wave, body wave, or moment magnitude. The latter most accurately represents an earthquake's energy (Olshansky 1998, p. 8). The Richter scale is logarithmic: Each whole number represents an approximate 31.5-fold increase in seismic energy released. The MM scale reports relative values of intensity as a measurement of how people feel and react to the shaking, being in variance from place to place for the same quake. It is expressed in terms of specific events, such as movement of furniture, difficulty in walking, damage to structures, and others. The two scales of measurement compare as follows:

Richter 2.5 (MM I and II): Generally not felt, but recorded on instruments

Richter 3.5 (MM III–V): Felt indoors and outdoors by many, swaying trees

Richter 4.5 (MM VI and VII): Everyone feels movement and runs outdoors, some local damage occurs, and poorly built structures may be considerably damaged

Richter 6.0 (MM VIII and IX): Destructive event, many buildings shift off their foundation, noticeable ground cracks appear

Richter 7.0 (MM X): Major event, ground is badly cracked, many structures destroyed

Richter 8.0 (MM XI and XII): Great event, bridges wrecked and many structures destroyed, waves seen on ground surface, very wide ground cracks

Major earthquakes typically only last for seconds at a time, very rarely longer than a minute, but they can cause much devastation including direct damage to structures (see Fig. 8-6), ground surface ruptures, tidal waves resulting in flooding, and breakage of utility lines which in the past has led to raging infernos.

**Characteristics**  Seismic forces are cyclic in nature, evidenced by subsequent pushing and pulling, and can vary in direction, magnitude, and duration. Because of this, a different approach to detailing lateral force resisting connections than simply assuming a monotonic application of force (one direction) has long been recognized. Vertical accelerations have also been observed, which can have a pronounced effect on structures, such as increased axial force and bending moment in the ribs of a concrete arch bridge.

**Figure 8-6**  Damage to building sustained in the Northridge Earthquake, California 1994. (*Courtesy UCLA Department of Earth and Space Sciences*)

Scientists discovered this dynamic reaction to ground motion during their studies of the San Francisco Earthquake of 1906 and surmised that if accurate ground accelerations were measured, buildings could be engineered with far greater reliability. With the invention of the strong-motion accelerograph in the United States in the 1920s, sometimes also called an accelerometer such as the one shown in Fig. 8-7 (though there are subtle differences), ground motion is measured on a continual basis in more and more buildings throughout the world, as well as increasing numbers of large bridges, providing data that helps code officials update seismic design provisions and measurements of motion in a more predictable fashion.

The motion of a flexible building helps to visualize the effect of soil and period. When founded on a soft soil, the system responds to ground acceleration with a longer period (it takes the building a longer time to return to its original position after being displaced), resulting in higher seismic forces to be resisted by the building. When founded on a stiff, bedrock-type soil, the building is not subjected to seismic forces as high due to the difference of periods between the foundation and the building.

**Derivation**  Determining seismic forces for buildings and bridges requires knowledge of ground motion characteristics of the construction site, soil classification and dynamic behavior, building occupancy or traffic index, the fundamental period of vibration, lateral force resisting system and response values. A 2500-year mean return (2% probability of exceedance in 50 years) has been typically used for collapse avoidance in most modern building codes. Severe structural damage

**Figure 8-7** An early strong-motion accelerometer built first in 1935. (*Courtesy U.S. Geological Survey*)

may occur, but the building should not collapse. This is also qualified through Section 1.4 of ASCE/SEI 7-05, which states that the intent of the standard is for structural elements to be designed and arranged in such a way as to redistribute loads away from locally damaged regions to adjacent areas capable of resisting the effects without collapse, though not technically being a "progressive collapse" prevention standard in the strictest sense of the term.

To derive seismic forces for building design using ASCE/SEI 7-05, an engineer uses several different chapters, which can at first appear quite complicated, but the material is actually laid out in a logical, understandable manner. Chapter 11 introduces the reader to basic design criteria and construction requirements for all structures regulated by the standard, including determination of Seismic Design Category (SDC). Chapter 12 summarizes analysis and design procedures applicable to buildings and their components, whereas Chapter 13 addresses nonstructural components that are permanently attached to structures. Chapter 14 lists material-specific seismic requirements and modifications to reference design documents (AISC 360, ACI 318, ACI 530, AF&PA NDS). Chapter 15 addresses normal, self-supporting nonbuilding structures that may either be situated within or isolated from a building structure in some fashion. Chapter 17 defines applicable requirements to seismically isolated structures and Chapter 18 follows those with active or passive damping systems. Chapters 19 through 23 include miscellaneous procedures, references, and design provisions for different purposes, almost a potpourri of leftovers from earlier chapters.

The basis of seismic design is a strength limit state beyond first yield of a structure, though equations may also be used with service-level load combinations in Section 2.4 by which magnitude is adjusted to service conditions (ASCE/SEI 7-05, p. 352). An engineer, therefore, must pay attention to design methodology that will be used for each material and levels at which applicable forces are determined for all load combinations. Earthquake forces are determined as a function of both horizontal and vertical components of ground motion and are multiplied by a structural redundancy factor for each orthogonal direction. These forces, as well as their vertical distribution throughout a structure, are permitted to be determined through an equivalent static lateral force procedure (standard or simplified), a modal response spectrum analysis, or through a linear or nonlinear seismic response history procedure, though there are limitations to each method based on structure type, regularity of the lateral force resisting system, and importance of use.

Defining an earthquake's response at a particular site involves determination of several factors, including mapped short-period (0.2 second) and long-period (1 second) spectral accelerations. Values that have been mapped through the National Seismic Hazard Mapping Project are based on a maximum considered earthquake for use in linear static and dynamic methods of analysis. These accelerations, modified according to site soil class, and the importance category of a structure are used to classify relative seismic risk by assigning an SDC to the project, ranging in associated risk from SDC A (lowest) to SDC F (highest). Greater detailing requirements apply to the higher categories in order to achieve a ductile response in materials and connections to seismic loading. The response modification factor, R, presented in Table 12.2-1 of ASCE/SEI 7-05, is used to reduce design seismic forces for ductility (the ability of elements to withstand stress in the inelastic range) and overstrength characteristics (reserve structural system strength) of the chosen lateral force resisting system. The element overstrength factor, $\Omega_0$, presented in Section 12.4.3.1 of ASCE/SEI 7-05, is defined as the ratio of idealized yield strength of a member or system to its maximum obtainable strength and is used to develop certain design criteria in specific locations of the standard, usually applied to collectors and regions where structural behavior will be classified as irregular.

**Application** Wind forces apply directly to a structure, whereas seismic forces are imparted to a structure by inertia of the ground beneath it. Weight at each level of construction are displaced by this shaking action and shear forces are developed throughout the structure as a lateral force resisting system works to hold the building together. Building behavior during an earthquake is a function of three contributing factors: The fundamental period of vibration (based on stiffness, size, and configuration), magnitude and distribution of carried weight, and energy absorption capabilities of the lateral force resisting system. Tall flexible structures will set up a type of whiplash effect under ground shaking, moving in different directions at the same time (higher modes of vibration), and require the use of advanced methods of structural analysis for design. Basic seismic design can be viewed as the process of defining the characteristics of a design earthquake, dictated by the code for a particular area, and translating the effects of that event into forces that building systems need to be capable of resisting.

The idealization of seismic motion for the design of structures can be force-based, where elastic models are used to predict the strength of a system according to its measure of ductility; displacement-based, which seeks to determine the effect of seismic displacement on structural components; or energy-based, where the dissipation of a ductile structure's energy is measured. It is generally recognized as uneconomical to design buildings that remain within a purely elastic range of response during a major earthquake. Energy absorption, which is the product of force and displacement, is key to structural resistance against damage or collapse during moments of inelastic behavior (Kuwamura and Galambos 1989). Therefore, when a structure is capable of absorbing more energy than an earthquake puts out, it can survive the event.

A response spectrum is often used to develop a building's elastic response to earthquake motion as though the problem were a simple one of statics. The response of multiple single degree-of-freedom systems is idealized as a graphical plot of relative displacement, velocity, or peak ground acceleration versus period of vibration. Figure 8-8 shows the basic shape of a design base shear response spectrum, which is constructed from measured site-specific, short- and long-period accelerations that have been truncated in recognition of physical limitations to the maximum considered earthquake within a region. Time histories of ground motion recorded on accelerographs during a seismic event can also be used as direct input for a computer model. The base shear coefficient, which is used in common strength-based design procedures, is determined from the spectra as a function of the fundamental period.

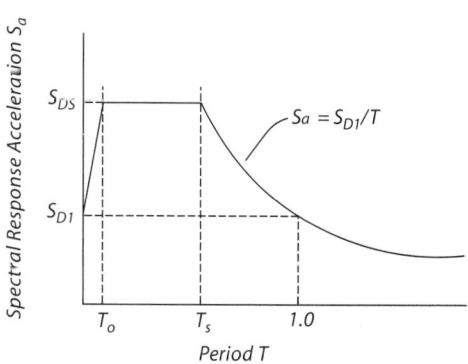

**Figure 8-8** Response spectra diagram (*Courtesy U.S. Army Corps of Engineers*)

Section 12.8.3 of ASCE/SEI 7-05 prescribes the vertical distribution of seismic force applied to a structure in terms of mass tributary to a certain level and the height of the centroid above grade, resulting in a distribution that effectively models an inverted triangle when lumped masses are similar in magnitude. Some studies, however, have shown that there is very little increase in upper story acceleration of woodframe residential buildings (Cobeen, Russell, and Dolan 2004, W-30a, p. 41), therefore code procedures may overestimate movement for certain structures. Seismic design story shear is distributed over the horizontal plane to elements of a vertical lateral force resisting system according to relative positions of mass centroid and center of rigidity of the lateral force resisting elements, illustrated in Fig. 8-9. Horizontal distribution of seismic forces in a single plane is done by considering the stiffness of the diaphragm, which is a phenomenon that will be discussed later in this chapter.

Current methodologies do a fine job of combining the art and science of earthquake design, but an engineer must not lose sight of the fact that a structure's true dynamic response to seismic excitation is quite complicated. Behavior of components located along the lateral force path often changes during a strong event and it is therefore necessary to use worst-case scenarios in the determination of structural response. Simplified models and assumptions are important, but an engineer can explore greater freedom in design and different performance objectives by consideration of inelastic system relationships and site or structure-specific parameters.

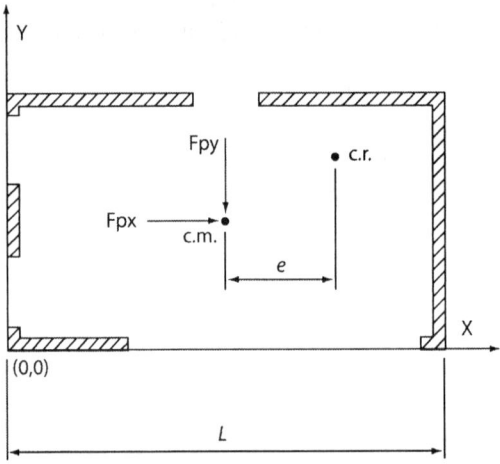

**Figure 8-9** Concept of lateral force distribution on a horizontal plane using rigid diaphragm theory. (*Courtesy U.S. Army Corps of Engineers*)

## 8.2.4 Blast, Impact, and Extreme Loads

Buildings are commonly designed to resist the effects of gravity, earth retention, fire, wind, and seismic activity with good predictability and a high level of confidence. The effects of fire are well mitigated through codes by restricting buildable area for different construction types requiring adequate space for the arrival of firefighting equipment and personnel; defining fire, smoke, and heat separations; imposing sprinklers, standpipes, and other elements to assist suppression efforts; and dictating means of egress for all occupants. Wind and seismic forces have been studied for many years and are defined with a reliable basis in probability. Performance objectives for these types of common hazards have long been recognized and lead to confident design work.

However, structural engineers cannot possibly design for every conceivable hazard, though agencies that regulate the construction and design of sensitive buildings and bridges, such as nuclear power facilities or major arterial pathways, often impose special conditions through specific codes or design standards that must be reviewed. All loads that are to be considered in design should be confirmed with government requirements at the beginning of the design phase and clearly identified within the project documents. Because extreme or unusual load cases and combinations are difficult to predict and quantify, design of a structure to resist them is based on historical performance, probability, and innovative philosophy (progressive collapse, combined hinging mechanisms). All design efforts will have a common repetition of steps: Define the hazards a structure is required to resist, determine performance objectives required by code or special consultation with the owner, and decide which strategies for meeting the decided performance expectations are to be pursued.

Jon Magnusson (2004) discusses some of the philosophies of designing commercial buildings to resist extreme loads and gives the reader insight as to the difficulty involved. For example, structural engineers are well-acquainted with the need for redundancy and a continuous load path throughout a building, but these desired qualities have been responsible for both restricting and encouraging progressive collapse from an unexpected hazard. Documents released in conjunction with the assessment of building performance during the World Trade Center terrorist attacks in New York in 2001 indicated by interview that structures which seem to be most susceptible to progressive collapse are those that are well-tied together and possess a high level of continuity. On the other hand, this continuity has also been responsible for the physical reality of progressive collapse design strategies, where strength levels of unaffected members and connections work together to share load when something has been exploded out of the assembly. It is truly a curious phenomenon.

**Fatigue** The fatigue life of a structural element is defined in terms of N-cycles of a repeating high load source, which has the effect of reducing the ultimate strength as N increases. For most of this period, however, a member will be stressed to a value that is much lower than full its capacity, therefore fatigue-type loading is not as severe as it may first appear (Blodgett 1966, p. 2.9–10). Fatigue failure is progressive over a period of time, beginning with localized plastic deformation where fiber stress exceeds the yield point (although the average distribution of stress over the entire cross-section is below yield). A crack forms in response to this phenomenon. Designing members to resist the effects of fatigue loading requires careful attention to detail, including provision of smooth corners or openings (no sharp edges); avoiding placement of welds, openings, or sudden geometrical changes at points of high stress; and orientation of the longitudinal axis of an anisotropic material in the direction of applied force.

## 8.3 COMBINING LOADS AND FORCES

During its lifetime, a structure will be subjected to a variety of different loads, some having a more severe influence than others. ASCE/SEI 7-05 includes a series of load combination equations that have been assembled based on historic usage and statistical analysis that apply to both service-level (allowable stress design) and strength-level (strength design) forces. These equations include multiplication factors to account for the likelihood and expected magnitude of a certain load type (dead, live, wind, and so on) in comparison to others within that equation and the probability that they will occur at the same time, selected in an effort to assure that each combination represents the same probability of exceedance (Chen 1997, pp. 26–33, 34). Low-probability events (sabotage, explosions, tornadoes) are given consideration through miscellaneous standards and design manuals.

Based on experience with certain types of structures (woodframe versus concrete) and region of consideration (low versus high seismic activity), engineers commonly preselect load combinations that have historically been found to govern a particular scenario, thus saving some time and effort in design. For example, design of the lateral force resisting system of a concrete shear wall building that is located in a region of strong seismic activity will nearly always be governed by seismic load combinations as opposed to those which include wind. However, the design and anchorage of individual components located in an area of structural discontinuity may still be governed by combinations that include wind, so care in determining which equations require mathematical review is important.

### 8.3.1 Design Methods

Loads that a structure resists are combined according to principles of statistics and are based on the design methodology chosen. Allowable Stress Design (ASD) is based on the premise that actual service load stress built up within a member due to its attempt to resist applied forces must not exceed specified allowable stress, having been determined through testing and experience, modified by an appropriate safety factor, and published by different material organizations.

The size of steel members may also be determined through the use of Plastic Design (PD), which recognizes the reserve strength of steel beyond initial yielding and defines a plastic collapse mechanism as the point when enough different cross-sections of a member have yielded, causing instability.

Load and Resistance Factor Design (LRFD) is a reliability-based method using strength-level forces applied to steel and wood structures, operating on the premise that the effects of factored load combinations must be less than or equal to factored predefined resistance limit states. Ultimate Strength Design (USD or SD) applies to concrete and masonry structures by the same principles described above for LRFD.

$$\varphi_i R_n \geq \Sigma \gamma_i Q_i \qquad (8\text{-}2)$$

The LRFD method is identified in Equation 8-2, where a member's nominal resistance characteristics for a specific limit state is identified as $R_n$ and the physical effects of applied forces is shown as $Q_i$. Service-level forces are multiplied by predefined factors, represented as $\gamma_i$, to raise them to strength-level and are based on different uncertainties within a particular combination of loads, defining the probable demand of a structure. Strength reduction factors are also applied to the resistance side of the equation, represented by $\varphi_i$, to account for variation in material properties, statistical uncertainties, and significance of failure mode.

## 8.4  INTRODUCTION TO BUILDING MATERIALS

Once the character of external loading is understood, an engineer is ready to investigate the behavior of materials and systems when subjected to combined phenomena. Model codes, published standards, and other regulations give instruction on how to create an appropriate mathematical model, to which an engineer applies sound mechanical principles and adequate knowledge of structural material behavior. Though terminology among each industry may differ, basic properties of interest for any material include ultimate tensile strength, yield stress, elongation, modulus of elasticity, compressive strength, shear strength, and fatigue strength.

### 8.4.1  Common Construction Materials

The most common structural elements can be divided into three general categories: Metals, ceramics, and molecular materials. Metals include structural steel, aluminum, and iron, which exhibit ductile behavior when subjected to load and have a high elastic modulus. Ceramics include concrete and masonry, which generally fail in a brittle fashion (little plastic deformation), and therefore do not have a well-defined yield point. Molecular materials include all species of wood, which are made up of longitudinal fibers and strings of molecular chambers, and are characterized by large deformation under small elastic stress. Each of these more common materials will be covered in later chapters.

## 8.4.2 Environmentally Sensitive Materials

Presidential Executive Order (EO) 13101 was issued on September 14, 1998 to advocate the use of environmentally preferred products in all aspects of federal government work and are defined as those having a reduced effect on human health and the environment when compared to products that might otherwise serve the same purpose. The order also defines recovered materials as waste materials and by-products that have been recovered or diverted from solid waste and recycling means a series of activities by which products are recovered from the solid waste stream for use in the form of raw materials in the manufacture of new products other than fuel.

Straw is considered to be a molecular material and has been used for construction in the form of bales (compressed blocks bound with steel wire or polypropylene twine) or panels (compressed straw sheets), formed out of the unused product of farm grains such as wheat, rice, wild grasses, bamboos, sugar cane, corn husks, or hemp. Structural design using straw products is not an exact science, though it has some basis in testing and rational mechanics. Elements of construction are commonly fastened together using dowels and are protected from moisture by polyethylene barriers and air passages. Because bales will experience nonuniform deformation or compression under load, it is common to introduce posts and beams as the primary gravity load resisting elements with bales providing insulation and resistance of nominal lateral forces by racking stiffness.

Rammed earth is similar to compressed adobe or unfired clay masonry, without mortar joints (King 1996, p. 55). The material is compacted into place in a series of lifts of varying densities and may be enhanced with aggregate, portland cement, lime, fly ash, or other agent. After compressing the material, wall forms can be removed in short order to allow surfaces to dry, harden, and strengthen, a process that can take up to two years for complete curing. Walls can also be strengthened with steel or bamboo reinforcement, though construction obviously becomes more difficult in this manner. A version of rammed earth construction called Pneumatically Impacted Stabilized Earth (PISE) involves spraying soil that has been prepared for use in construction under high pressure against a form, a process that is much faster than ordinary placing and compacting.

## 8.5 GENERAL BEHAVIOR OF STRUCTURAL ELEMENTS

No matter how small, an applied force will always cause deformation in a structural member and resulting stress can be calculated fairly easily within the elastic range of movement. This truth sets up the beginning of engineering mechanical principles. If deformation disappears upon removal of the imposed force, the material is said to behave elastically. Beyond this point, brittle materials will rupture before any noticeable change in length of a specimen subjected to load, whereas ductile materials exhibit large deformation prior to failure.

### 8.5.1 Solid Body Mechanics

An engineering student is introduced to basic principles of solid body mechanics in early physics courses and learns to appreciate the work of so many scientists that have gone before. Some of the more important theories relevant to structural engineering practice, which have been in relatively consistent use (little change to the original theory) for hundreds of years, include the following:

1. Newton's First Law of Motion proposes that if a solid body is not in motion, the summation of all moments of force about any point (or axis) must also equal zero.
2. Newton's Third Law of Motion states that materials will equally resist applied force in the opposite direction by a system of internal reactions.
3. Navier's Hypothesis states that the stress or strain at any point along the length of a bending member is proportional to its distance from the member's neutral axis.
4. Hooke's Law describes the relationship between stress and strain, where deformation of an elastic body is proportional to applied force or, in other words, intensity of stress is proportional to rate of strain. The constant of multiplication between these two values is the modulus of elasticity (Young's Modulus), E.
5. Bernoulli's Assumption states that planar sections of a beam prior to bending remain planar during and after bending.

Behavior of a solid body subjected to load is based on the nature, magnitude, and direction of that load relative to the member's principal axes. The principle axes of any cross-section are defined as three rectangular axes through any point, about one of which the moment of inertia is a maximum, about another a minimum, and some intermediate value about the third axis. The shape of a member's cross-section is to be selected in such a way that mechanical properties will best be put to use: Area is chosen to reduce shear, tensile, and compressive stress; geometry is arranged in such a way as to create a moment of inertia that efficiently resists the tendency towards rotation about a particular axis. Resistance to applied load requires consideration of the neutral axis, which represents zero strain and stress in any member and passes through a body's center of gravity.

Applied forces to a body are often resolved into components applied in the direction of respective axes. By this, the effect of each can be accounted for in materials that have different physical properties along each axis (anisotropic), such as for wood members with different strength parallel than perpendicular-to-grain. This is also an important practice for following a load from origin to final support location.

Equilibrium of a static member is determined through the use of a free-body diagram, by which all applied loads are resolved into components and summed in such a way that the result must be equal to all reactions in any direction.

An equally important physical law is that of compatibility, where member elongations (strains) and displacement of a corresponding joint of that member are required to be geometrically compatible. All elements that are connected to a single point must have the same absolute displacement at that location, even if the members are all of different materials. Compatibility is sometimes confused through long-term creep, connector slippage, or friction, but such conditions do not silence the immediate principle.

**Stress and Strain** The most basic diagram found in the study of elasticity is the stress-strain curve, such as that shown in Fig. 8-10. Strain is defined as change in a member's length, or total deformation, divided by the original length and is a dimensionless quantity. Stress is defined as applied force per unit area or moment per section modulus and exists in the form of tension, compression, or shear. Bending stress is simply an image of a tension-compression couple applied about a member's neutral axis, where stress has zero value along this line and varies linearly to a maximum value at the top and bottom (extreme) fibers of any cross-section. It is this understanding of how moments are resisted that allows an engineer to develop unique solutions to design problems, where an applied eccentric force creating moment in the carrying member may actually be transferred there by two simply supported elements setting up the necessary couple, facilitating less-expensive connection requirements.

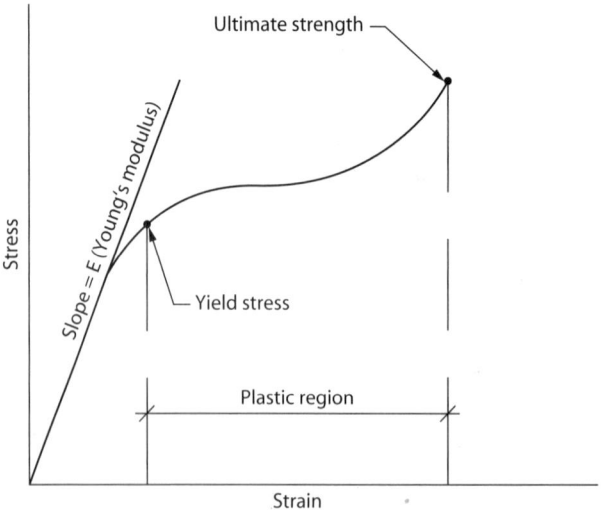

**I Figure 8-10**  General material stress-strain curve.

In addition to pure bending stress, a member experiences horizontal shearing stress if resisting moment varies along the beam's length. Horizontal shear stress has a parabolic distribution that is maximum at the neutral axis and zero at the outer fibers. Torsional loading may also be applied to a member, which is when a force occurs at a particular distance from the neutral axis or center of gravity and causes a twisting motion about this reference axis. Though action is principally established through transverse and longitudinal shear stresses, it is essentially a moment-type of load and resistance may be pictured in the form of an equivalent force-couple in the opposing direction.

The stress-strain diagram identifies a material's yield point and how much deformation occurs from an applied load before it is reached, following a slope equal to the modulus of elasticity. A material behaves elastically before reaching this point, which means that all strain experienced will disappear once the load is removed. Once the yield point is exceeded on the curve, a material retains some deformation when the load is removed: Plastic deformation, or permanent set, has occurred. For most materials, plastic deformation depends not only on the maximum value of stress reached, but also on the time elapsed before removal of the load (Beer and Johnson 1981, p. 39). As a general rule, most solids are linear-elastic at strains of less than 0.1% (Ashby and Jones 1980, pp. 71–72).

**Hysteresis**   Hysteresis is a phenomenon that occurs in all materials where a history of applied loading is identified in a diagram (see Fig. 8-11) as a lag between application and removal of load and is fundamental to the understanding of material behavior under a cyclic-type force. This lag is identified as a shift in the looping pattern that is plotted on a diagram as magnitude of force versus system displacement. The relationship can be developed from a single framing element or an entire assembly, such as a frame or shear wall.

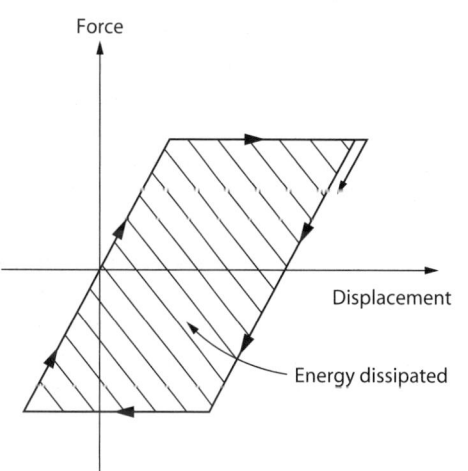

**Figure 8-11**   General structural system hysteresis diagram.

Quality of the hysteretic loop explains much about a system's character, such as the good ductile behavior of a semirigid steel connection being shown by a gradual increasing of loop area at larger amplitudes of displacement (Azizinamini and Radziminski 1989). Astaneh and Nader (1989) performed cyclic testing of double angle steel semirigid connections and observed pinched loops, indicating a reduction in connection stiffness by brittle effects. Loop pinching in most cases is due to material discontinuity from cracking or loss of bond in concrete members, or the creation of a gap or fracture in steel elements. In their experiments with double angle connections, this gap was created by permanent angle deformation and bolt elongation.

## 8.5.2 Serviceability

Deflection of a member due to flexural stress will be resisted in accordance with the modulus of elasticity, whereas the modulus of rigidity governs resistance to deflection caused by shear stress. One of the most significant things about deflection of floor-framing elements also happens to be one of the most elusive. Human beings have different levels of tolerance to perceived movement and it is impossible to design a system to please everyone. Model building codes stipulate restrictions, component manufacturers take those limits a step further, but all the while human perceptions will still vary broadly. Responsibility for sound performance begins and stops at the state-adopted building construction code in combination with commonly accepted industry standards. For floor framing systems that include a variety of spans and loading conditions, it is helpful to keep deflections relatively consistent throughout the entire floor in order to minimize vibration or perceived soft spots.

Any mechanical system that contains rotating parts can set up vibration in the structure which supports it. Coordinated physical fitness activities, such as aerobics and dancing, have been shown to induce vibration in floor systems and pedestrians marching in unison can set up undesirable effects over the length of a bridge. Effects of global vibration due to applied wind or seismic forces can be managed through increasing a structure's mass, reducing its height, changing the shape, or installing special components such as a tuned mass damper or a tuned liquid column damper to dissipate dynamic energies.

In an office environment, damping of vibrations is provided by nonstructural elements that are often subject to removal at a future time, including a hung ceiling, filing cabinets or other furniture, movable partitions, and ductwork. Reducing the potential for floor vibration during the design stage involves careful consideration of long span framing members, positions of fixed partition walls or ceilings, the source of vibration, and future arrangement or existence of elements installed prior to occupancy.

## 8.6 GENERAL BEHAVIOR OF STRUCTURAL SYSTEMS

A completed structural system is assembled of smaller building blocks: Beams, columns, beam-columns, vertical diaphragms, horizontal diaphragms, arches, plates, shells, cylinders, domes, joints, connectors, and connections (Adams 2006). Although a structure may need to be analyzed with consideration of all dimensions in a complete mathematical model, it is helpful to describe a more complicated assembly in terms of individual pieces, and combinations of pieces, to observe how a complete load path from roof, or top of a bridge tower to the foundation, can be achieved.

### 8.6.1 Horizontal Systems

The most basic horizontal structural element is the beam. Bending stress is distributed across the neutral axis, or axis of curvature, whereas an applied tensile or compressive stress is idealized as uniform over the cross-sectional area. The principle of superposition allows an engineer to simply add the effects of both types of stress when they act in conjunction, as seen in Fig. 8-12. When load is applied to an arched beam in the plane of its curvature, radial stress acting perpendicular to the span will combine with flexural stress in either a tensile or compressive manner. If applied loads tend to straighten the beam, radial stress is tensile, whereas if the beam's curvature is increased, radial stress is compressive. This stress can have a significant effect on curved wood members because of that material's weakness in a direction perpendicular to grain.

Analysis of beams that are continuous over multiple supports is often simplified by placement of live load in different spans to create worst-case effects. For example, the maximum positive moment within a span is determined by placing live load on that span as well as on all other alternate spans. Maximum negative moment within a span is determined by placing live load on both adjacent spans and all other alternate spans. At an interior support point, maximum positive moment occurs in the beam when live load is placed on the two spans immediately beyond the two spans adjacent to that support (one on each side), whereas

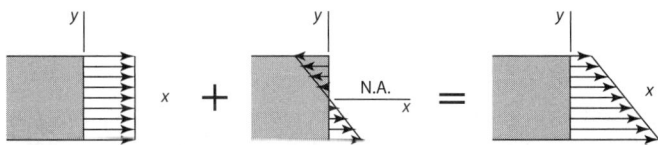

**Figure 8-12** Concept of superposition, combining the effects of stress from axial load plus bending moment.

maximum negative moment occurs over the support when live load is placed on each adjacent span and all other alternate spans.

Diaphragms, plates, or built-up frame assemblies intended to resist forces perpendicular to the plane of their major axes may be subjected not only to bending, but also to torsion depending on support conditions and placement of load. A three-dimensional case of torsion can be simplified by remembering that just as the torsional resistance of a single element is equal to the total resistance of its individual parts (flanges, webs), so that of a built-up frame assembly is approximately equal to the total resistance of its individual members (Blodgett 1966, p. 2.10–11).

In resisting laterally applied forces (those in the plane), roof and floor diaphragms are typically idealized as something like a wide-flange (I-shaped) beam, where the sheathing or deck itself transfers shear forces (web) and bond beams or other perimeter framing elements, such as the top plates of a wood-frame wall, resist moment from horizontally applied forces as chord members through a tension and compression couple (flanges). The vertical lateral force resisting system (shear walls, frames) functions as reaction points for the horizontal diaphragm. Boundary elements that run parallel with applied lateral forces must collect the reaction load and distribute it to various walls or frames within the lateral force resisting system below. Figure 8-13 shows a selection of horizontal diaphragms commonly used in building systems, indicating these functional web and flange elements.

Horizontal diaphragms are classified according to behavior as rigid or flexible. The former assumes distribution of seismic force based on vertical lateral force resisting system rigidity, whereas the latter is used to justify a simpler tributary approach (applied lateral load is distributed based on width tributary to a line of force resistance). A flexible diaphragm is defined in ASCE/SEI 7-05 is presented in Fig. 13-9 as related to a wood diaphragm, but any material that meets the comparison would also classify accordingly. Section 1613.6.1 of the 2006 IBC allows a designer to assume flexible behavior for horizontal diaphragms constructed of wood structural panels (plywood, oriented strand board, or composite panel) or untopped steel decking, provided: nonstructural cementitious topping over wood diaphragms do not exceed $1^1/_2$ in (38 mm), each line of lateral force resistance meets allowable story drift limitations, vertical lateral force resisting elements are light-framed walls sheathed with wood structural panels or steel sheets, and cantilevered diaphragm portions are designed within specified code restrictions.

### 8.6.2 Vertical Systems

The most basic vertical structural element is a column. Deflected shapes when bucking occurs will differ depending on support conditions at top and bottom, as shown in Fig. 8-14, and the effective column length is defined as actual height multiplied by a factor, Ke. For columns that are assumed to have pinned-type support at the top and bottom, for which sidesway is prevented, using a Ke factor

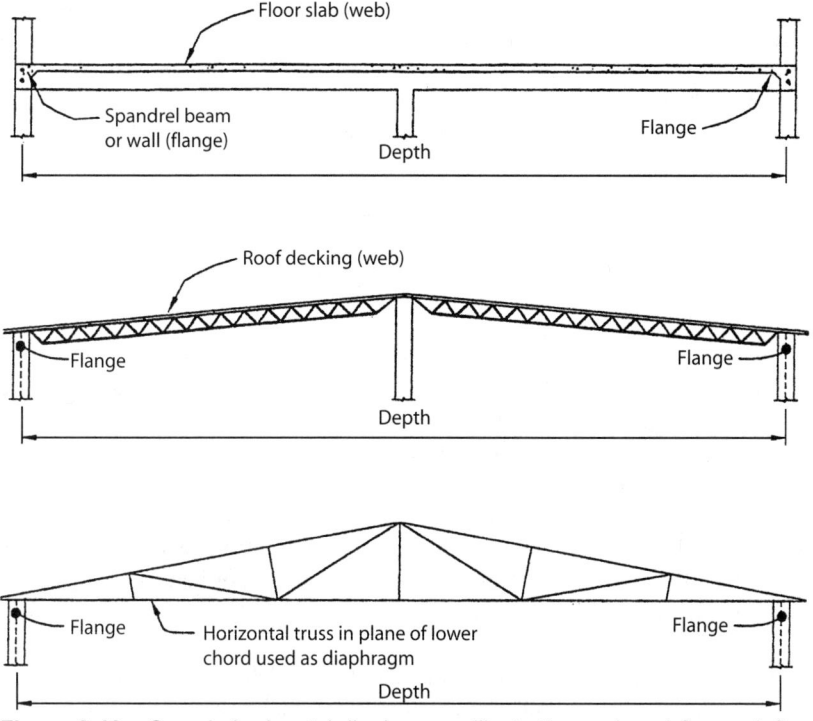

**Figure 8-13** Sample horizontal diaphragms, illustrating *web* and *flange* definitions. (*Courtesy U.S. Army Corps of Engineers*)

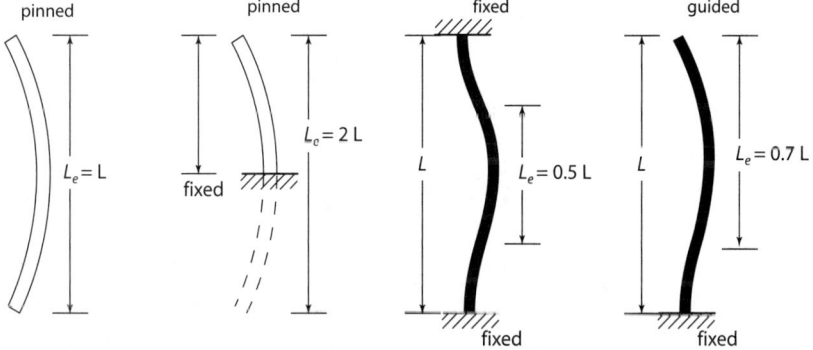

**Figure 8-14** Buckling stiffness comparison of columns with different boundary conditions.

of 1.0 in design equations will be somewhat conservative. A column's slenderness ratio, defined as effective length divided by radius of gyration, is foundational to column buckling theory. Swiss mathematician Leonard Euler published his findings on critical column loading in 1744, which is expressed in Equation 8-3 as the axial load at which a compressed element collapses in service or buckles during a loading test, where L = overall column height, I = least moment of inertia available in a cross-section (determined about potential buckling axes) and E = Young's Modulus.

$$P = \frac{\pi^2 EI}{(K_e L)^2} \tag{8-3}$$

It is important to remember that column equations for all materials are based on the concept of isolation, whereas a more complex situation exists in nearly all practical conditions. A column usually forms part of a complete frame system with varying end restraints and the stability of each piece is dependent to a certain degree on all of the others. A column may carry a single axial load, but other connected elements impose some unintended flexural and torsional forces resulting from the column's initial crookedness or inaccurate placement of carried load.

Shear walls act as cantilevered beams that transfer lateral forces acting parallel to their length to the foundation or other horizontal system of resistance. The rigidity of a wall segment, which is the reciprocal of its total deflection due to flexure and shear, is defined in terms of dimensions, modulus of elasticity, modulus of rigidity or shear, and conditions of support at top and bottom. Equation 8-4 identifies general wall deflection at the top with a cantilevered boundary condition, and Equation 8-5 indicates deflection with fixity at the top and bottom, such as for a pier that occurs between windows or other openings. P = applied force, E = Young's Modulus of the wall's mass material (i.e., concrete or masonry, not reinforcing steel within), h = wall height, A = wall cross-sectional area, G = wall mass material shear modulus, commonly taken as 0.3E. These relations hold for all materials, though adjustments are added for different phenomena in wood and steel assemblies (fastener slip relations, hold-down anchor elongation). Refer to Fig. 8-15.

$$\Delta_C = \frac{Ph^3}{3EI} + \frac{1.2Ph}{AG} \tag{8-4}$$

$$\Delta_F = \frac{Ph^3}{12EI} + \frac{1.2Ph}{AG} \tag{8-5}$$

High rise walls are considered as cantilevered for the full height of the building and individual rigidities are calculated for each floor level based on properties of the wall element below that level. It is also acceptable to consider wall elements

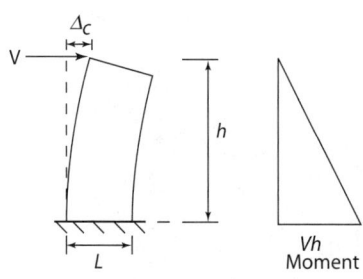

(a) Cantilever Pier

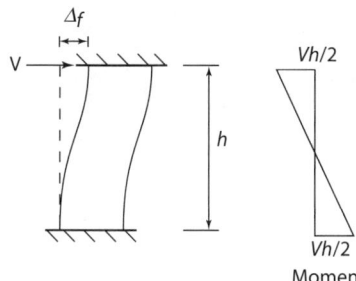

(b) Fixed Pier

**Figure 8-15** Behavior of cantilever-type and fixed-type shear wall piers.

as fixed between floors and each segment around openings calculated accordingly. The combined rigidity of piers in series is determined simply by adding individual rigidities, whereas piers that are stacked require a different consideration. Individual shear wall segments separated by large openings can be considered as stiffened through proper design and connection of intermediate beams or panel segments which resist axial, shear, and flexural loads. Shear wall segments that have been properly coupled together will then act as frame-type elements instead of pure cantilevers.

Frames are assemblies of elements that resist force through a stiffness of the beam-to-column joints (rigid frames) or which contain additional diagonal elements to brace motion against collapse (braced frames) as shown in Fig. 8-16. Instability of frames or their individual elements is defined as a condition reached during buckling under increasing load at which the capacity for carrying further load is exhausted and deformation continues. Frame systems described in Table 12.2-1 of ASCE/SEI 7-05 include concentrically or eccentrically braced; ordinary, intermediate, or special moment-resisting; composite or ordinary partially restrained moment-resisting; and buckling-restrained braced. These frame types will be discussed further in other chapters according to the material of which they are created.

**I Figure 8-16** Steel concentrically braced frame.

## 8.6.3 Redundancy and Reliability

Redundancy focuses on the ability of a structure to tolerate some level of overload without collapse, such as when loads are redistributed following the failure of a single member in order to maintain stability. For a structure comprised primarily of ductile framing elements, the limit state does not depend so much on order of member failure. However, the performance of an assembly of primarily brittle-type elements may depend greatly on this order of failure with respect to position from the initial failure point (Chen 1997, p. 26–29). Studies conducted in 1979 of damaged and collapsed bridges concluded that many bridges have alternative load paths that are not easily identifiable. For example, the LaFayette Street Bridge in Minneapolis-St. Paul, a steel bridge made of welded plate continuous over three spans, survived collapse because the superstructure redistributed loads after the failure of one girder in the central span (Frangopol and Curley 1987).

Structural reliability and redundancy is difficult to calculate precisely. When defining the reliability of a structure, there are two basic approaches. The series method defines a structure's limit state as the point when a single member fails, also known as the weakest link method. The parallel method, on the other hand, proposes that all of the members in a structure must fail to achieve its limit state. More complicated systems are modeled using a combination of series and parallel-type procedures, still being subject to inherent limitations for a true model, such as quantifying the ability of members to share load and time dependent effects on member capacity.

## 8.7 GENERAL BEHAVIOR OF COMPLETED STRUCTURES

The behavior of a completed structure when subjected to load is based on several factors. Important factors that generally contribute to good building performance during a severe loading event (seismic, hurricane, limited impact or blast) include a stable foundation, repetitive conditions of gravity-member support, redundancy in the lateral force resisting system, a somewhat regular building configuration, a continuous load path from top to bottom, ductility, strength, and stiffness.

### 8.7.1 Buildings

The effects of wind flowing past a tall structure are usually determined to some extent by a level of wind tunnel testing, though useful information can already be found for buildings that are regular in shape and not subject to aerodynamic disruptions in flow from adjacent structures. Though commonly associated with seismic loading, an offset distance between center of mass and application of the wind resultant force will also set up a torsional response in the building, which will always be present to a certain degree due to randomness of wind force fluctuations. Internal wind pressure that builds up due to building envelope permeability also causes random aerodynamic activity in many cases. Direct wind pressure is known as drag, but transverse pressure (that which is perpendicular to the drag force) is also set up that needs to be accounted for in building design. This combined response is complicated by the interaction of turbulence, motion of the building, and dynamics of wake formation, with effects being more pronounced in tall structures.

Rigid frame structures resist cyclic loads by absorbing and dissipating strain energies through ductile rotation of plastic hinges formed away from member connections, therefore the mechanics of a structure's collapse as a result of lateral forces may be studied by increasing applied forces incrementally until a sufficient number of plastic hinges develops, creating global instability. A building with semirigid connections, however, will experience relatively large cyclic rotations in the connections themselves (Astaneh and Nader 1989), resulting in pronounced effects on the building's overall dynamic response that needs to be accounted for in the study of its behavior.

Study of the effects of cyclic (seismic) forces imposed on a structure reveal a great deal about the benefits of continuity, redundancy, and proper detailing.

Damage observed in the cities of Armenia and Pereira, Columbia as a result of an earthquake with a magnitude of 6.2 on the Richter scale on January 25, 1999 confirmed the validity of concrete detailing regulations in modern codes. Columns in low-rise structures with very small amounts of transverse reinforcement experienced severe damage at locations where large inelastic deformation occurred (Pujol, Ramirez, and Sarria 2000). Full-scale seismic tests of a six story steel concentrically braced frame building, as part of a joint U.S.-Japan research program conducted in Tsukuba, Japan in the later 1980s, displayed extensive brace buckling and yielding under severe shaking (Roeder 1989). Beams and columns also yielded throughout the bottom three stories, while a composite slab cracked extensively near the K-braces of the frame. Once these braces failed, a significant reduction in strength and stiffness occurred, though the structure did not collapse.

Buildings with upper levels set back a distance from the main lateral force resisting system complicate the distribution of seismic forces because of a sudden change in stiffness or a torsional response from eccentricity. In concept, such a building can be represented by the taller wing (includes the upper set back portion) attached to any shorter wings at all floor levels where equilibrium and compatibility conditions are to be satisfied (Cheung and Tso 1987). Symmetrical plans provide the best resistance to seismic forces because the center of rigidity of the vertical lateral force resisting system is very close to the center of mass of the structure, therefore little torsion will occur. It is not always possible to divide a complicated structure into well-defined boxes using separation joints, therefore the behavior of irregular projections and shapes must be considered in a building system model. The basic approach is to add continuous ties, drag struts and increased fastening to account for odd behavior due to loss of symmetry.

### 8.7.2 Bridges

A bridge's suspension system is subjected to a symmetric mode of vibration due to dynamic wind effects, identified by the movement of towers toward each other as the center deck span deflects downward and the outer spans translate upward, or an asymmetric mode of vibration by which the towers move in the same direction, causing antisymmetrical motion of the deck spans. In order to counteract fluttering action of antisymmetric movement, which is agreed to have been responsible for destruction of the Tacoma Narrows Bridge in Washington in November 1940, a relatively high torsional and bending restraint system is required for the deck.

A cable-stayed bridge system, however, is not as sensitive to wind oscillations as the suspension-type bridge, due in part to inherent stiffness and damping characteristics afforded by different cable lengths and frequencies that tend to disturb formation of the first or second mode of oscillation (Podolny 1974). Proper design of many bridge structures requires an aerodynamic analysis of wind effects on the flow of traffic itself will also need to be studied using wind tunnel testing or computational fluid dynamic methods. Such studies are highly dependent on turbulent properties of terrain, proximity of large structures, and wind climate.

In cable-stayed bridges, pylons above the deck are either pinned or rigidly connected to piers below, which in turn are commonly supported by piles or caissons. Behavior of the entire bridge system is influenced by displacement of the piers under seismic or wind loading, therefore it is important to account for pier flexibility and soil-structure interaction in the analysis.

Testing has shown that slab-on-girder bridges are usually stiffer in flexure than predicted by analysis, mainly due to horizontal restraint provided by girder bearings. Analytical comparisons revealed that this restraint, even with use of new neoprene bearing pads, can reduce total moment from applied loading by up to 9% (Bakht and Jaeger 1988).

### 8.7.3 Progressive Collapse

The study of progressive collapse of structures has been approached in different ways as scientists not only try to understand the phenomenon, but attempt to define a building's resistance capabilities to such an event in a way that can be quickly applied by practicing engineers. Progressive collapse involves spreading of an initial local failure from element to element, eventually resulting in collapse of a large portion of the structure, if not all of it. It is common to think of moment resistance as primarily responsible for the redistribution of forces in resisting progressive collapse, but Hamburger and Whittaker (2004) remind engineers that this assumption is actually conservative due to the additional presence of compressive arching in a concrete floor slab and catenary behavior of steel framing. To justify this conclusion, a study is cited by the authors which demonstrated that a structure with 30 ft (9.1 m) bay spacing and W36 (ASTM A992) horizontal framing elements could support the weight of nearly 20 stories after column removal, though with significant deflection.

A building's behavior and resistance to progressive collapse can only be determined through an iterative process, as described in different standards such as the United Facilities Criteria document UFC 4-023-03, Design of Buildings to Resist Progressive Collapse (2003). One such approach involves the systematic removal of a heavily loaded column or bearing wall at the building's exterior and design of the remainder of the building in such a way that the loss of such an element will not cause collapse. An example of this would be the elimination of a column or two from a building line which has been constructed with additional vertical members, strengthened elements and connections in such a way that the building's remaining support is changed into a Vierendeel-type truss as shown in Fig. 8-17.

**I Figure 8-17**  Vierendeel truss configuration.

The doctrines of progressive collapse work under the same banner as studies in redundancy and reliability, though with a more complicated dynamic (typically nonlinear) effect. Most buildings are vulnerable to some degree upon local failure, but the most important contributing factor is a lack of continuity within the structural system and a lack of ductility in the members, connections, and materials themselves. Systems that are particularly redundant, though being somewhat of a low suitability for dissipating mechanical energy, can sometimes compensate for a reduced ductility. All of these issues lead an engineer to discover a proper understanding of structural mechanics and behavior when subjected to load in order to plan for all reasonable conditions of loading, and perhaps some that are not so intuitive.

# 9 Soil Mechanics

Geotechnical engineering involves the study of soil behavior when subjected to external forces or phenomena, where the principles of soil and engineering mechanics are applied to the evaluation of foundations, dams (see Fig. 9-1), retaining structures, tunnels, and the needs of site development. It is important for practitioners to have adequate knowledge of laboratory techniques used in evaluation and prediction, subsurface exploration tools and methods, as well as a good understanding of geological and geophysical character and their effects on engineered structures. Much of the necessary information is learned in university soils courses, but it can only be applied to real engineering problems through a broad base of practical experience.

The importance of a sound foundation support system should be obvious to any engineer, not only related to the structure itself, but also to costs associated with site preparation. It is beneficial to require as little site work as possible for placement of a foundation system, though excavation, reconsolidation, and compaction are essential operations in the whole preparation process. Whatever the case may be, understanding of soil types and site preparation measures contribute to the design of a sound, economical structural support system.

Due to space limitations and the existence of many fine references on the subject, complete design methods for foundation or soil-retaining systems are not intended to permeate this chapter. Rather, background information on soil types, foundation elements, and anticipated patterns of behavior is given to help engineers, both new to the field and close to retirement, develop good solutions to soil-related problems.

**Figure 9-1** Hoover Dam. (*Courtesy U.S. Bureau of Reclamation*)

## 9.1 CHARACTER OF DIFFERENT SOIL TYPES

Engineering properties of soil vary in all directions, even within the same strata, and sound judgment plays a significant role in the design of foundation systems. Some of the physical properties of interest to an engineer include unit weight, water content, angle of internal friction, compressibility, and permeability. Studies and tests performed on these properties allow estimations to be made of soil bearing capacity, likely settlement, lateral soil pressure, and frictional resistance to sliding.

In order to properly define soil properties, a geotechnical engineer classifies discovered strata according to the Unified Soil Classification System (USCS) per American Society of Testing and Materials (ASTM) Standard D2487 and will accompany these labels with further description, such as degree of in-situ consolidation (soft, loose, medium, stiff) and measure of plasticity. This system grew out of one originally created by Arthur Cassagrande in 1948 and has been used in all regions of the United States with a high level of confidence. Other systems of soil classification that may be employed, depending on the intended use of a site, include that generated by the American Association of State Highway and

Transportation Officials (AASHTO) based on performance of the subgrade below highway pavements and bridge abutments or approaches.

## 9.1.1 Rock/Granite

Large rock formations provide a solid base for foundation support, but they are often fractured, weathered, and faulted, making a determination of bearing capacity and stability a difficult task. Bedrock formations can extend to great distances along a horizontal plane and are generally defined into three classes: *Igneous*, which is formed either underground from trapped magma or above-ground from cooled lava, and includes granite and pumice; *sedimentary*, formed as a result of chemical reactions and pressure from overlying soil layers which have eroded from their original position over many years by wind and water, including sandstone, limestone, shale, and gypsum; or *metamorphic*, existing at one time as sedimentary or igneous rock but has changed due to the application of pressure and great heat, including schist and gneiss. Boulders or cobbles, pieces of rock that have sheared off during some type of event or weathering, can measure between 3 in (76 mm) to greater than 12 in (305 mm) in dimension.

Voids of different sizes will always be present over the expanse of a rock mass and there typically exists a layer of soil at varying depths. The most difficult types of rocky soils to design foundations for are those that are intermixed with large soil deposits. It is difficult to determine the actual depth of soil in every conceivable location during the design stage, and when it comes time to excavate and place footings, they will likely be dug to various depths in a nonuniform manner. In addition to this variability based on the presence of softer soil regions, the bearing capacity of a rock formation will be further affected by weathering of mineral constituents, size and frequency of fractures and other discontinuities, and resistance to further deterioration when receiving load.

Foundations that rest partially on rock and partially on softer soil have a high probability of experiencing differential settlement, as the portion sitting atop rock will not move at all in relation to settlement of the softer sand or gravel mixture.

## 9.1.2 Gravel

Gravels are pieces of rock that measure about $1/4$ in (6 mm) to not more than 3 in (76 mm) in size, also being known as crushed stone or pea gravel, and are frequently used as a base material for concrete slabs-on-grade. A well-graded sample of gravel is one in which the size of grains is fairly evenly distributed throughout, whereas a poorly graded sample contains a majority of particles of one particular size. Though cohesionless, gravels can be graded and compacted to achieve a suitable and stable bearing capacity through internal friction between grains. Anticipated settlement occurs within a short period after load is applied, therefore stability of a supported structure is not materially affected by long-term consolidation of the soil.

### 9.1.3 Sand

Sand is found loose and single-grained with little cohesion, yet simple to compact for use in a foundation system, with a particle size less that $1/16$ in (1.6 mm) in dimension and a gritty feel when dry. Confined sand makes an excellent bearing material, so long as a water source will not be flowing through it to cause erosion. It is further subdivided as coarse material, retained on a No. 10 (2 mm) U.S. standard sieve; medium, retained on a No. 40 (425 μm) sieve; and fine, passing through all sieves larger than No. 200 (75 μm).

Frictional resistance of individual particles in a sample of sand will be affected by presence of voids, shape and size distribution of grains, roughness of particle surfaces, and the presence of moisture. This in turn affects the angle of internal friction, which is used to calculate coefficients of lateral earth pressure in retaining or buried structures. In general, higher shear strengths may be expected in samples with a lower void ratio and moisture content, increasing angularity or roughness of particle surfaces, and more uniformity in grading.

### 9.1.4 Silt and Clay

Clays are produced from weathered rock and composed of hydrated oxides of silicon, aluminum, iron, potassium, sodium, calcium, and magnesium. These minerals are arranged in a specific crystalline pattern and define a clay's classification and physical properties. The loose crystalline structure permits moisture absorption, which leads to expansion and potential damage to supported structures. Depending on different conditions, clay is further classified as sandy or silty.

Silt and clay are both known as *fine-grained* soil types, though silt may also be classified among the sandy-type soils. Clay is defined by ASTM D2487 as passing a No. 200 (75 μm) sieve and can be molded with putty-like properties over a particular range of moisture contents, yet exhibits surprising strength when dry. Clay that has a low potential for expansion does not necessarily require special measures to support a structure. Silt is a material that also passes the No. 200 (75 μm) sieve, yet has a nonplastic consistency and little strength when air dried, making it a poor foundation material when considered in isolation (Bowles 1982, p. 28).

The *expansion index* (EI) of a soil is measured according to ASTM D4829 to provide a basic, unitless index to the expansion potential of a compacted soil when infused with distilled water. Soil with an EI of less than or equal to 20 has *very low* potential for expansion, whereas it will have *very high* potential if EI is measured above 130. ASTM D4318 describes methods for determining the liquid limit (LL), plastic limit (PL), and plasticity index (PI) of soils tested in a laboratory in an effort to define the fine-grained fraction within samples, define shrink-swell properties, compactibility, compressibility, shear strength, and permeability. Albert Atterberg defined six limits of fine-grained soil consistency as the upper limit of viscous flow, liquid limit (defined as the water content, in percent, of a soil at an arbitrary boundary point between semi-liquid and plastic states), sticky limit, cohesion limit, plastic limit (defined as the water content, in percent, at the boundary between plastic and semisolid states), and shrinkage limit. The PI is numerically equivalent to LL − PL. Values obtained for LL and PI are used to assist in determining a soil's family and characteristics.

Commonly occurring clays include montmorillonite (most expansive lattice-type structure), illite, attapulgite, chlorite, and kaolinite (least expansive, fixed crystal lattice-type structure) and are characterized by expansion with the entrance of water and shrinkage as water escapes from a soil mass. Risks involved in foundation design for expansive clay are far more significant than for nonexpansive soils and an engineer needs to choose the appropriate foundation type based on experience, sound judgment, and a proper understanding of the consequences of poor footing performance. Eric Green (2005) lists foundation types in order of increasing possibility of failure when used for expansive soils as follows: An elevated structural slab supported on piers (lowest possibility of failure); a stiffened slab-on-fill with piers; a stiffened slab-on-fill, possibly with chemical treatment or presaturation of the soil; and a stiffened slab-on-grade (highest possibility of failure).

### 9.1.5 Other Soil Types

*Loam* is an interesting material that has a relatively even mixture of different grades of sand (less than 52%), silt (28 to 50%), and clay (7 to 27%). It has a gritty feel, but is slightly smooth and sticky under certain conditions and retains moisture well. Depending on its texture and make up, loam is further classified as sandy or clayey. Most soils of agricultural quality are some type of loam with organic particulates.

Organic soils consist predominantly of vegetation in various stages of decomposition, classified as *muck* in higher degrees or as *peat* in relatively well-preserved conditions. Subterranean portions of a forested or heavily vegetated region seem to contribute more to the formation of muck and peat than surface matter decomposition. Organic soil is recognized by category according to the types of components present, including woody or nonwoody fibers, and wood particles such as roots and chunks. This material is commonly removed from a building site prior to construction, as further decomposition can lead to damaging differential settlement.

Loose sediment on the floor of oceans, streams, rivers, or creeks is specially classified as *marine soil* which contains both a terrestrial sediment (erosion from the shore or from suspension in the water body) and marine constituents (organic and inorganic remains of dead marine life). Coral and coralline limestone is present in more tropical regions, where effects of high humidity, frequent rainfall, and a warm climate influence erosion and subsequent marine soil accumulation than in more temperate areas.

Though not technically a soil type, *permafrost* areas are those which contain perennially frozen ground at a depth which is determined by the air thawing index, thermal surface radiation, moisture content, and unit weight of the soil. Behavior of foundations will vary according to freezing or thawing related to temperature changes and redistribution of subsurface water. In fact, the very process of construction alters ground surface temperature to the effect of changing thickness of the frost zone and reducing depth to top of permafrost. Seasonal heave or settlement occurs in a surface layer subject to cyclic freezing and thawing, a process by which moisture is drawn up through the soil from a greater depth and converts to ice crystals or lenses that are subject to thawing. Generally, finer-grained soils are more susceptible to frost effects because they have a higher propensity for this capillary action.

## 9.2 PREPARING A SITE FOR CONSTRUCTION

For the successful erection of any structure, surface and subsurface conditions of a construction site must be given careful attention. Evaluation includes determining the existence and placement of underground utilities or structures, selecting an appropriate foundation system based on soil preparation recommendations (often determined by a consultant geotechnical engineer), expected groundwater conditions, difficulty in excavation or stock piling soil, amount of grading required for positive drainage of surface water away from the structure, adjacent construction or traffic pathways, stability of natural landscape that may be used to support structural elements (abutments, outcroppings), and depth of frost penetration. Most of the measures taken in preparing a site for construction involve earthwork, which collectively refers to the process of moving and sculpting site soils to create different shapes and to modify its physical condition.

The precision of a survey crew in setting elevations and alignments according to design drawings is important to successful placement of structures, to defining the right amount of excavation for heavy equipment, and for achieving adequate drainage of storm water from the ground surface.

### 9.2.1 Geotechnical Reports

The best way to define behavior of site soils is to have a professional geotechnical report produced after a series of tests from samples and borings. In order to define soil characteristics, and thus anticipated behavior, such a report typically includes information on the following:

1. Type of foundation required, or best suited for the site, based on structural load
2. Allowable vertical and lateral bearing values
3. Settlement predictions and necessary mitigating measures needed to reduce the potential effects
4. Groundwater level
5. Seismic hazards relative to the site and soils, such as danger of liquefaction, slope instability, surface rupture, or lateral spreading and proximity of known faults
6. Potential environmental issues or proximity to sensitive regions
7. Possible construction "red flags," such as unstable excavation potential
8. Differences in application of net soil pressure (new stress applied to soil grains after excavating) versus gross soil pressure (includes original stress which was imposed by the weight of excavated soil)

A site-specific geotechnical report is always required where classification, strength, or compressibility of the soil is questionable, where expansive soil is likely to be present, when the ground water table is expected to be high, for pile and pier foundations, where characteristics of rock strata are doubtful or variant, or if the structure is located in Seismic Design Categories (SDC) C, D, E, or F (2006 IBC, Section 1802).

Where expansive soils exist, a geotechnical report provides the designer with different options for foundation type, usable bearing values, and detailed instructions on how to prepare the site soils to receive the footings. For example, a report might suggest the use of internal stiffening beams with slab-on-grade construction because they have been shown to reduce distortion or doming effects from soil heave, as well as reducing the magnitude of differential settlements. A geotechnical engineer can also require that more extreme measures be taken, including soil replacement and/or deep pier placement, for very poor soils, depending on associated risk and company philosophy for managing that risk.

Other avenues for obtaining soils information are available to a limited extent for certain projects. One of these includes the local building department itself, which maintains records of permitted construction documents for a certain period of time (generally a minimum of 180 days from the time a project is completed) that can be viewed by any person, though most often only over-the-counter. Copyright and other obvious restrictions to the official use of a soils report that has been developed for a different site warrant very careful consideration, but at the very least, record documents can be used to give an engineer an idea of what might be expected from site soils in the region. Since the early 1900s, the United States

Department of Agriculture (USDA) has published a variety of soil surveys that can also be used to determine general soil descriptions and characteristics in a region, though not all information has been updated or restored (see Fig. 9-2).

## 9.2.2 Clearing and Excavation

Clearing and grubbing occurs early in the development of bare land, by which tree stumps, shrubs, roots, and other vegetation are removed to an adequate depth. This layer of earth is often stockpiled for later use as landscaping base or placed to help control erosion. Work is commonly done by a tractor equipped with a bulldozer on the front, such as is shown on one of the earthwork machines in Fig. 9-3.

|  |
|---|

**Figure 9-2** Helpful soil survey documents may be obtained from the U.S. Department of Agriculture, Natural Resources Conservation Service.

I **Figure 9-3** A variety of equipment is available for grading purposes.

Earth excavation is easily done with scrapers, dozers, or trenching machinery for all particular needs of a construction project. Type of earth to be moved, distance it must be transported, and load support capabilities of the soil and movement pathway are the biggest factors in choosing the right equipment for a job. Front-end loaders generally work in tandem with hauling equipment if travel distance is greater than about 100 ft (30.5 m), whereas tractor-drawn scrapers are considered to be cost effective for self-hauling distances to about 1000 ft (304.8 m).

Excavation of rock usually involves some level of drilling and blasting, generally including boulders that are about 0.50 yd³ (0.38 m³) in volume. When rock is adequately broken apart, shovels, bulldozers, front-end loaders, backhoes, scrapers, scoopers, and clamshells have all been used successfully in removing fractured material. Use of equipment is dictated by roughness of terrain to the tires, workable space, and height of hauling equipment. After initial blasting and clearing, additional charges are strategically placed throughout the site in order to continue final shaping of the rock surface according to design depth and layout.

## 9.2.3 Grading

For all soil types, it is important to direct rain and other surface water away from any structure through the use of proper grading, swales, gutters, or drop inlets and piping. Section 1803.3 of the 2006 IBC requires the final grade to slope a minimum 5% over the first 10 ft (3 m) around the structure, though impervious surfaces (concrete, asphalt paving) only require a minimum 2% slope. Climatic or special soil conditions may allow a reduction in minimum slope of ground and a building official charged with enforcement of the code can assist in determining whether this is feasible.

Where sites are not level, the existing ground will either have to be cut to a certain extent and removed or voided areas require compacted fill soil prior to placement of a structure. The amount of cutting necessary is determined by intended placement of a building's lowest level, which may be nestled directly into a hillside, and by the most natural course for draining away rain water from the site. Figure C11-1 of ASCE/SEI 7-05, reproduced here as Fig. 9-4, depicts how a building's first level can be defined as either a basement or a story depending on how much it is cut into the surrounding soil. Engineered fill is placed in lifts, each receiving a specified degree of compaction, until the design height is reached.

Final grading of a site may be significant enough to dictate top of footing elevations and the structural engineer needs to coordinate work with that of the civil engineer to avoid conflicts. It is important to remember that adjacent asphalt or other paving is placed with a base of specified thickness and slope for drainage may actually be provided in a direction parallel to a building's foundation, which is to be placed below all surface features.

## 9.2.4 Compaction

Compaction of supporting soil layers is intended to create a somewhat uniform bearing surface and reduce the incidence of settlement, where soil volume is

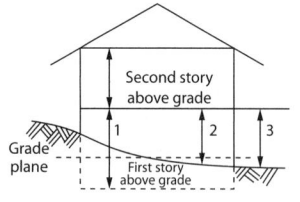

The lower floor level is classified as the first story if the finished floor surface of the floor level above is:

1) More than 6 ft (1829 mm) above the grade plane;
2) More than 6 ft (1829 mm) above grade for more than 50% of building perimeter; or
3) More than 12 ft (3658 mm) above grade at any point.

**Two-story above grade building**

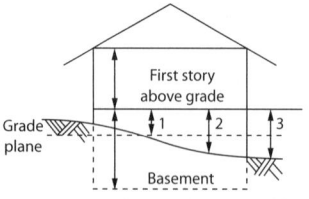

The upper floor level is classified as the first story if the finished floor surface of the floor level is:

1) Not more than 6 ft (1829 mm) above the grade plane;
2) Not more than 6 ft (1829 mm) above grade for more than 50% of the building perimeter; and
3) Not more than 12 ft (3658 mm) above grade at any point.

**One-story above grade and basement building**

**Figure 9-4** Definition of a building story from ASCE/SEI 7-05, "*Minimum design loads for buildings and other structures.*"

reduced by a momentary application of load through tamping, rolling, or vibration. The process generally increases shear strength of site soils while decreasing its compressibility and permeability. Cohesive soils (silt, clay) require special attention to moisture content during compaction, as those which have been placed too dry are likely to collapse when saturated under imposed loading. Cohesionless soils (sand, gravel) remain somewhat pervious during compactive efforts and are therefore not affected to a great degree by moisture content during these operations. For all soil types, compaction is accomplished by the amount and type of energy imparted by equipment appropriate for the job, which include sheepsfoot rollers, rubber tire or smooth wheel rollers, vibrating or tamping plate compactors, or crawlers.

Figure 9-5 shows a foundation that has been placed on a poorly graded site, where large boulders were simply overlain with a thinner section of concrete for footings rather than removed: Adverse settlement is extremely likely. Proper compaction begins with the removal of larger solids and organic material, followed by excavation of site soils to a predetermined depth. Engineered fill may also be specified if native soils are not particularly desirable in their natural condition for supporting a structure. This fill is created using reconditioned site soils or imported material, placed and compacted in lifts of thickness ranging between 6 and 12 in (152 and 305 mm).

Laboratory testing per ASTM D1557 provides the basis for determining percentage of compaction and water content necessary for a particular soil to achieve desired engineering properties. This standard describes the Modified Proctor

**Figure 9-5** Example of a poorly graded project site with standing boulders and loose soil.

Compaction Test, which has a history with original measures introduced by the U.S. Corps of Engineers in 1945. Field compactive efforts are to bring consolidation to a percentage of this optimal laboratory value, commonly set at 90% for building pads (2006 IBC, Section 1803.5; NFPA 5000, Section 36.2.4.3.2).

## 9.3 BEHAVIOR OF FOUNDATION TYPES

The behavior of footings in supporting load is affected by many different factors. Consideration is typically given to the effects of a footing's shape, placement of carried loads (centered or eccentric), ground surface slope, influence of soil compressibility or expansive potential, impact on soil pressure due to proximity of other structures, position of groundwater table, influence of different vertical soil layer types, and safety factors (Winterkorn and Fang 1975, pp. 128–145). The physical characteristics of soil directly below a footing also influences general behavior, but long-term consolidation of soil layers far below a structure is an equally important consideration, especially for combined or mat-type footings.

Courses in geotechnical engineering define mechanical properties of soil as the interaction between individual particles and moisture that is present. An approximate representation of the stress-strain properties of a soil mass is shown in Fig. 9-6. Because of its low tensile strength, soil almost always fails in shear (Marxhausen and Bagley 2006), which occurs when particles begin to slide past one another and is dependent on the frictional and cohesive strength present in that interaction. Modes of soil shear failure below foundations include general shear failure, local shear failure, and punching shear failure. General shear failure is identified as slippage of one footing edge in relation to the ground surface, where tilting of the foundation and bulging of the adjacent soil may also occur. Local shear failure

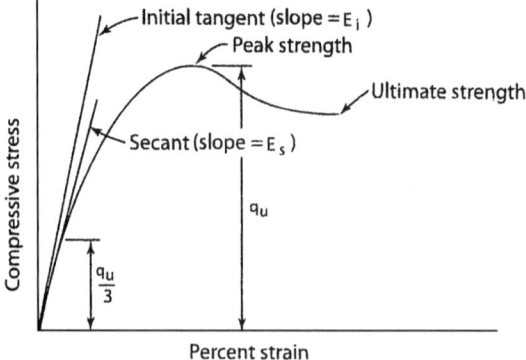

I **Figure 9-6** General soil stress-strain curve.

causes significant vertical compression of soil directly below the foundation and may also show signs of surrounding soil bulging. Punching shear failure occurs with minimal movement of surrounding soil, but considerable penetration of the footing into the soil.

There are a few things that must be accomplished by an adequate design. The depth of footings needs to be such that bearing occurs on firm, stable soil that will not be influenced by a freeze-thaw cycle in cold climates or by swelling/shrinking effects in other areas. Additionally, width of continuous footings and square bearing area of isolated pad-type footings must be large enough to keep vertical bearing pressures below either code-prescribed amounts or as specified in a geotechnical report, as well as to prevent excessive settlement.

## 9.3.1 Spread Footings

A spread footing is known to be *shallow* when its depth below the surrounding grade line is less than the footing's least width, as seen in Fig. 9-7, otherwise it carries the title *deep*. Simple spread footing models, as well as strip footings to an approximate degree, have shown that stress carried by the soil diminishes with increasing depth from the bottom of a footing (Brown 1997, pp. 50–51; Winterkorn and Fang 1975, pp. 164–166). Position of the groundwater table complicates this distribution, as bearing capacity is weakened below that level due to

| **Figure 9-7** Pad footing awaiting placement of concrete.

loss of grain cohesion and a reduced effective unit weight of submerged soils. Assuming a somewhat uniform and homogeneous subsurface above a very low water table, the following relationships have been shown to be sound approximations of bearing pressure below a foundation:

Say a square spread footing has a side dimension $W$ and supports a vertical load $P$.

1. Bearing pressure, $q$, directly below the footing equals $P/(2 * W)$.
2. At a depth of $W/2$, the pressure influence on the soil will be approximately $0.8q$.
3. At a depth of $W$, the pressure influence on the soil will be approximately $0.5q$.
4. At a depth of $z$, the pressure realized by the soil may be assumed spread out laterally from the centerline over a total distance of $W + z$ along each axis.

The influence of soil bearing pressures at some depth below grade from footings that are located close to one another is shown in Fig. 9-8, where overlapping pressures become additive, although the effects of such can vary considerably with angle of shearing resistance of the soil. On the other hand, the shear strength of overburdening soil contributes to a greater bearing capacity at these deeper regions. Soils are always nonhomogeneous to a certain degree, therefore modified soil bearing capacities should be determined with care.

During a seismic event, isolated footings behave in an elastic fashion when a plastic hinge develops at the base of a lateral force resisting column above. Other means of load transfer to supporting soil occur through tipping or rocking, still considered to be supported linearly, or through inelastic deformation of the footing itself.

### 9.3.2 Continuous (Strip) Footings

The geometrical condition of adjacent parallel strip footings (length-to-width >1) causes the effect of overlapping bearing pressures to be minimal when compared

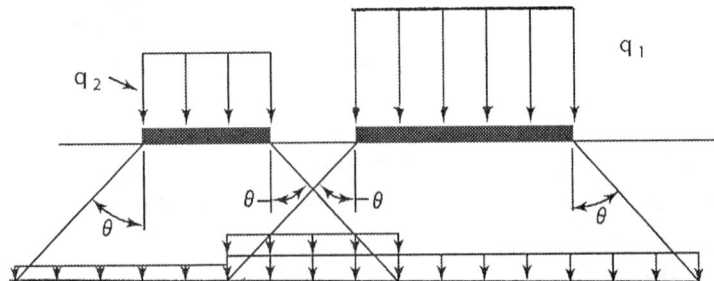

**Figure 9-8** Influence of overlapping soil bearing pressure from adjacent pad footings.

**Figure 9-9** Continuous footing with multiple pipes that may call for additional steel reinforcement.

with that of pad-type footings. Refer to Fig. 9-9 for an example of a continuous footing that has been fitted with piping and other embedments, the effect of which may require special consideration as voids are placed within the concrete mass.

Continuous footings commonly support a variety of loads of limited magnitude, including uniformly distributed and concentrated loads that are not large enough to require their own isolated pads. In the vicinity of columns, soil pressure distribution is higher than at other regions due to a dishing of the soil, but positive support can still be evaluated based on loaded area tributary to each column location.

### 9.3.3 Combined or Mat-Type Footings

When two or more isolated columns occur in close enough proximity, a combined rectangular-shaped footing is often chosen to assist in withstanding rotation, differential settlement, and in meeting the allowable bearing capacity. The shape of a footing can be created in such a way that uniform or linear distribution of soil pressure below results, thereby setting up an intuitive and straight-forward method of rigid body analysis, such as might occur with a trapezoidal-shape used to support two columns of unequal magnitude. Columns also transmit loads by rotation or shear, adding further considerations to the model. For analysis and design of a rigid body combined footing that carries any number of concentrated and distributed loads (uniform, triangular, trapezoidal, and the like.), the following procedure, which has a long history of success gives an engineer an idea of how combined footings behave rigidly under specified dimensions and conditions:

1. Determine the resultant of all loads and its point of application along the bottom of the footing. Including the weight of the footing itself will help keep eccentricity down, but will lead to somewhat erroneous results in determining reinforcing steel to resist flexure during a later step.
2. Determine minimum footing area such that allowable soil bearing pressure is not exceeded at any location, taking eccentricity of the resultant from the geometric centerline into account.
3. Define a linear equation for soil pressure distribution across the footing length based on resultant and eccentricity.
4. Construct shear and moment diagrams over the footing length using column loads as concentrated forces for simplification.
5. Converting resultants to ultimate levels, determine the required footing thickness based on a condition of either wide-beam shear or diagonal tension (punching) shear. It is usually beneficial from an economic standpoint to use footing proportions that will not rely on stirrups to carry shearing forces.
6. Determine the area of longitudinal steel required to resist calculated bending moments within the footing and compare with minimum values found in ACI 318-05 for flexural members or footings.
7. Select reinforcing steel in the short direction as directed by ACI 318-05 to resist the effects of shrinkage or temperature fluctuations.

Larger footings that do not qualify as rigid are designed as beams on an elastic foundation and in order to solve the appropriate equations, the *coefficient of subgrade reaction* must be determined. It is a quantity determined by experiment as the ratio between the vertical subgrade reaction and the deflection at a point on the surface of contact. Soil response, therefore, is approximated as coupled or uncoupled springs whose properties are calculated based on the modulus of subgrade reaction, area of the model that is attributed to individual elements, effective depth of reaction, and changes in the modulus that might occur over this range. It was discovered that a single, static value of this modulus might prove to be misleading, as consolidation settlement can vary the influence of reaction to carried loads (ACI 336.2R-88, Section 1.7), and it may be necessary to consider the results of different forms of analysis using a range of values for the subgrade modulus. Mathematically, the problem of a beam supported by an elastic foundation is a linear fourth-order differential equation and is solved in terms of footing width, constants of integration relative to boundary conditions, and elastic moduli.

Deformation of the footing itself will also affect the distribution of load to the soil. A flexible footing that supports a uniformly distributed load over granular material will cause an equally uniform transference of load, resulting in greater pressure and settlement along the edges since resistance is smallest there. A rigid body, on the other hand, settles uniformly over the same soil, resulting in greater pressure toward the center. Settlement at the center of a flexible footing on nongranular soil, such as clay, will be greatest even though pressure distribution is still uniform, whereas unit soil pressure will be greatest along the edges of a rigid body footing though experiencing uniform settlement. Foundation behavior will also be affected by the stiffness of the supported structure, as reactions are redistributed and stress may be increased in the footings and elements of the structure.

Mat foundations are typically used in situations where the subsurface is somewhat weak or nonuniform and a large number of closely spaced pad-type footings would prove unreasonable. It is also common to use mat foundations in the support of basements or pits located below the groundwater table in order to provide a watertight compartment. The total weight and volume of a mat can be economized by adding strategically placed ribs or cells, which are also effective in controlling settlement. Piles may be used to support a mat, considering each to have an elastic spring constant of *EA/L*, where $E$ is the modulus of elasticity, $A$ is the average cross-sectional area, and $L$ is the length of the pile. Conventional theories for design of a flat plate on elastic foundation also apply to mat design, though caution is warranted because the subgrade response is somewhat difficult to predict, soil properties actually vary in both horizontal and vertical directions, and there is typically a great variation in the type, placement, and magnitude of supported loads.

### 9.3.4 Deep Foundations

To a certain extent, *drilled piers* are deep cast-in-place spread footings, as depth below grade is larger than width or diameter. They can be excavated with a bell at the bottom to increase bearing area, but are only possible if an adequate layer of cohesive soil exists directly above the proposed pier bottom and does not have a tendency to slough into the fresh excavation. Deep foundations are special in that applied axial and lateral forces, torsional loads, and bending moments are resisted not only through bottom bearing, but also by the shaft within surrounding soil. Lateral or bending-type forces may also be present as a result of superstructure thermal movement, wind or seismic loading, unbalanced earth pressures, or axial forces applied at an eccentricity from the shaft centerline. Drilled piers placed in groups spaced not farther apart than three times the diameter of a single pier can be assumed to act independent of one another when load is applied perpendicular to pile spacing or eight times the diameter when load is applied parallel to pile spacing (ACI 336.3R-98, Section 3.5.4.9). Figure 9-10 shows a cage of steel reinforcing being lowered into an excavated pier.

*Pile* foundations serve the same function and are typically of very similar geometry as drilled piers, yet are driven into the soil by specialized equipment. They

**Figure 9-10** Placement of reinforcing bar cage into a drilled pier excavation. (*Courtesy Conlan Engineering and Construction Corporation*)

have been successfully driven as timber, iron or steel pipes, steel H-shapes, and reinforced or prestressed concrete. Many times, a driven pile will improve the properties of surrounding soil due to a consolidating effect.

Design and construction of these foundation types are influenced by soil type, stratification, permeability, stability, and depth of groundwater. If a drilled pier extends below the groundwater table, for example, dewatering and the use of casing to temporarily support the sides of an excavation are often called for, thus greatly influencing construction cost and design effort. Construction activity itself may also be responsible for changing soil properties, including shear strength along the side of the shaft, to such an extent that precaution in design is warranted. Different support configurations are shown in Fig. 9-11 for both types of deep foundations discussed.

Area that is available for construction and access to the site, as well as the proximity of existing structures, play a significant role in the selection of foundation type, especially when drilled piers or piles are given consideration. Pile drivers, for example, are machines with a tall vertical framework that house a heavy mass

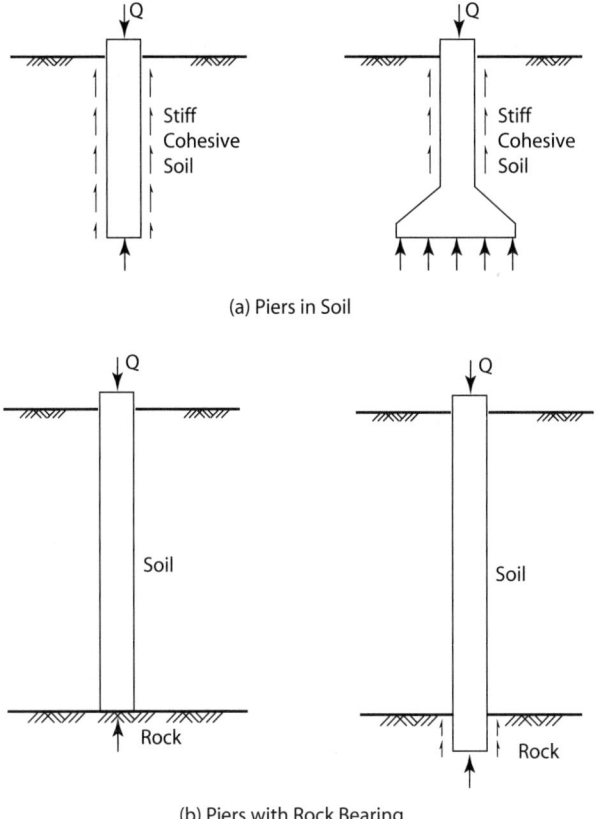

**Figure 9-11** Different load support mechanisms for deep foundations.

of iron, lifted by machinery and either pushed or allowed to drop down onto the head of a pile to drive it into the ground, therefore appropriate space and protection of surrounding structures from excessive vibration, settlement, or contamination is critical to successful installation.

Pile ductility is influenced by geometry and soil strata and is a desirable property when supporting axial, shear, and bending moments relative to seismic forces. Lateral deflection of a pier that is short and stiff relative to surrounding soil will occur about a point somewhere along its length and resistance to this movement is dictated primarily by restraint provided by the structure itself and load deformation characteristics of the adjacent soil. Lateral movement of a deep pier with sufficient length and slenderness, on the other hand, is exhibited by flexure and considered in relation to flexural rigidity, EI. In these cases, numerical relationships between pier deflection and soil reaction are considered in constructing a deflected shape. Whether flexible or rigid, design of a substantially surrounded deep foundation against the effects of bucking from compressive forces may be considered using an effective unbraced length of zero (Baker 2003, p. P–81).

Failure of pile foundations typically occurs either by slip, which is when the soil-pile interface fails and the pile slips down while remaining essentially undeformed itself; or excessive lateral movement, which occurs when the pile deflects laterally due to a combination of geometric instability and plasticity. Because the geometry of both drilled piers and piles is of critical importance, continuous special inspection during drilling or driving operations is required for verification of dimensions, plumbness, type and size of equipment, and materials supplied (2006 IBC, Sections 1704.8 and 1704.9). The geotechnical engineer who supervised testing and prepared design documents should also be present during excavation and placement to verify that anticipated conditions are, in fact, present.

### 9.3.5 Other Types or Systems

*Pier and beam* foundation types are constructed in such a way that all imposed loading is carried by continuous beams supported atop drilled piers that may or may not be tied together by pier caps for stability. Bearing capacity of the soil at the base of these piers is the primary agent for supporting load. Structural elements are usually made of concrete, but any material that is protected from the deleterious effects of surrounding soil is suitable. Structural behavior is closely linked to the distance that piers extend above the soil surface, though this measurement is usually very small for practical reasons.

Where expansive soil or permafrost exists, conventional deformed-bar or post-tensioned concrete *slab-on-grade* systems have been suitable means of supporting structural loading. These are uniform in thickness with stiffening edge beams and/or interior deepened ribs running in one or both directions, providing a high enough strength to resist expansive soil forces and sufficient rigidity to limit foundation deformation. Two primary soil forces for design are center lift (heave), which occurs as a result of progressive swelling beneath the center of the slab or due to shrinking of the perimeter soil regions; and edge lift (dishing), which is a cyclic heaving around the perimeter of the foundation system.

## 9.4 BURIED OR RETAINING STRUCTURES

Choice of backfill material is usually based on availability and economics, though cohesionless soils tamped into place in 6 in (152 mm) lifts are preferable. When silts or clays are present in backfill, permeability is variable and added pore pressure due to moisture infiltration may impose significant load onto the structure if not properly drained.

Retaining walls must have an adequate system of drainage in order to avoid a buildup of hydrostatic pressure within the earth. A free-draining, granular material is suited best for backfill within a zone that is at least 24 in (610 mm) in width, placed adjacent to the wall for its full height. This filtering layer placed between the wall and cut face of an excavation also helps to prevent development of swelling pressures due to capillary rise or infiltration of water into a clayey-type backfill material during seasonal changes. A perforated drain pipe should be provided at the base of the foundation, running at a slope which allows water to be transported by gravity, surrounded by at least 12 in (305 mm) of pea gravel. A general drainage trench diagram that has been suggested by the U.S. Army Corps of Engineers is shown in Fig. 9-12.

An active soil pressure condition may be assumed for design when the wall is allowed to translate at the top by a distance of 0.1% of wall height during backfilling. This condition is difficult to prevent, except by either a very stiff floor or roof diaphragm or wall tie-back anchors, therefore most design applications can likely assume an active state of soil pressure (Amrhein and Vergun 1994, p. 59). If the wall is restrained at the top in any fashion prior to backfilling, an at-rest soil pressure condition must be assumed. The bottom of a wall can be designed with fixity from 0 to 100%, which helps in the distribution of restraining forces necessary for equilibrium in an effort to prevent failure through sliding or excessive rotation about the base with a safety factor of 1.5 (minimum). Passive pressure develops on the other side of the footing as the structure is pushed against the earth. The effect of this is seen in a loading and resistance diagram for a sheet pile wall in

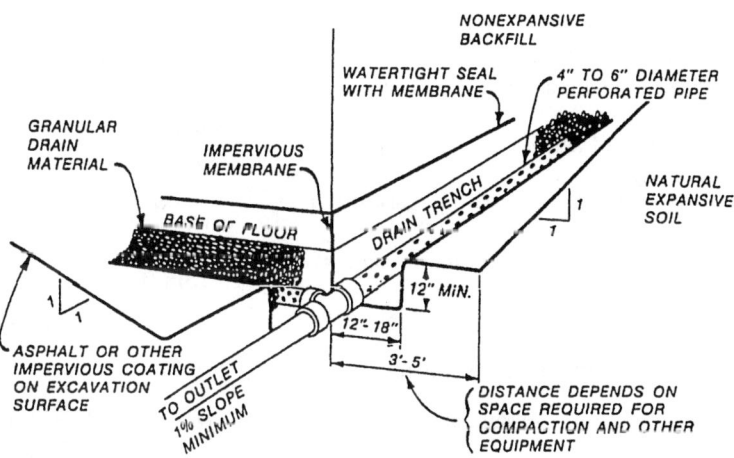

**Figure 9-12** Soil moisture drainage system behind a retaining wall. (*Courtesy U.S. Army Corps of Engineers*)

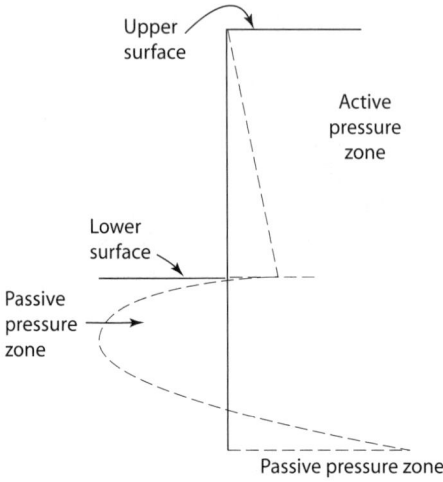

**Figure 9-13** Soil pressure terminology related to loading and support of sheet pile wall.

Fig. 9-13. Another strategy used successfully in retaining wall construction is the inclusion of buttress elements built into a wall to add strength. They are often detailed in such a way as to cause thinner wall segments to distribute loads as two-way plates.

For moist conditions of different nonexpansive backfill materials placed at optimum density, Table 3-1 of ASCE/SEI 7-05 lists active lateral soil loads that may be used for design purposes. Determination of active, at-rest, and passive soil pressures is relatively simple by calculation, however, once the internal angle of shear friction ($\phi$) and soil bulk density or unit weight ($\gamma$) are known. Resulting pressures for each case are distributed below the soil as a triangle which grows in magnitude linearly with increasing depth below the surface. For a cohesionless soil ($c = 0$), these relationships are as follows:

$$\text{Coefficient of active earth pressure} = K_a = \frac{1 - \sin\phi}{1 + \sin\phi} \quad (9\text{-}1)$$

$$\text{Coefficient of at-rest earth pressure} = K_0 = 1 - \sin\phi \quad (9\text{-}2)$$

$$\text{Coefficient of passive earth pressure} = K_p = \frac{1 - \sin\phi}{1 + \sin\phi} \quad (9\text{-}3)$$

$$\text{Active lateral pressure at a depth } h = P_a = K_a \gamma h \quad (9\text{-}4)$$

$$\text{At-rest lateral pressure at a depth } h = P_a = K_0 \gamma h \quad (9\text{-}5)$$

$$\text{Passive resisting pressure at a depth } h = P_p = K_p \gamma h \quad (9\text{-}6)$$

Due to the expense of excavation, extra material, and labor, basements are generally only constructed where the cost of land is excessive or a deep frost line exists.

Basements are also at a higher risk of flood damage than simple conventional foundation systems. A structural engineer must be aware of special loading in cases where the basement floor is located below the Base Flood Elevation as defined by FEMA's National Flood Insurance Program. The enclosed space will be subject to seepage and lateral hydrostatic and uplift pressures caused by high groundwater levels associated with regional flooding, so floors and walls should be designed for these effects. It is also strongly recommended that design proceed with the assumption that a subsurface drainage system is clogged and the full affect of hydrostatic pressure is realized in order to provide further assurance of protection against cracking.

## 9.5 FACTORS TO CONSIDER IN FOUNDATION DESIGN

Ali (1997) postulates that there are three types of foundation behavior possible when subjected to seismic loading on an elastic structure above: An elastic system, which promotes elastic behavior of the foundation; a ductile system, where energy received from the structure above will be carried inelastically; and a rocking manner, by which the structure and its foundation respond to load by rocking, assuming the overturning moment does not cause instability. The natural period of a structure is generally increased due to dynamic soil behavior when subjected to seismic loading. This interaction of structure and soil has been noted to reduce base shear design values and overturning moment, but increases the structure's lateral displacement and rotation.

### 9.5.1 Consequences of Poor Soils

An owner should be involved in the selection of a foundation system if there are several viable options available for sites with problem soils, such as expansive clays. Excessive foundation movement producing large differential settlement in buildings may be unacceptable (depending on the stiffness and sensitivity of the structure), and some systems have a greater potential for causing damage than others. The extent of movement is related to the amount of soil moisture change beneath the foundation system, which is often induced by seasonal changes in rainfall, excessive watering of landscaping, water pipe leakage, or extraction of water from the soil by large tree roots. Choosing from a variety of foundation options certainly does not establish any specific guarantee of structural performance, but the possibility of failure can be greatly reduced with a more robust foundation system that is well-suited to a particular site, and this sequence of thought may need to be presented to the owner for consideration.

Though soils located on a sloping hillside may not truly be of poor quality, they can be quite unstable and require maintenance to avoid future disaster. Measures that can be taken to avoid damage from falling rock include clearing of loose debris, addition of retaining structures, or altering the slope in some way. Surface failures occur during rainy seasons due to a reduction of shear strength from saturation, swelling, and seepage of water into fissures. It can be controlled to a certain extent by the presence of vegetation, geofabrics, lowering of the water table or content through drainage, or installation of stepped or tie-back retaining structures.

Measures to rehabilitate problem soils are used to improve bearing capacity, resistance to settlement, or protection against moisture intrusion. Different measures include compaction (discussed in previous sections) or chemical stabilization, where materials such as sodium silicate, lime, fly ash, or other is injected into a region of soil to improve mechanical properties. To accomplish this, pipes are driven to an appropriate depth and slowly removed once injection begins. Small quantities of cement can also be injected into a region of expansive soil to decrease capacity for volume change and to increase bearing strength.

### 9.5.2 Settlement

When load is applied to a soil mass, it is initially carried by pore water due to its incompressibility and is shifted to the soil structure when water drains from the pores. Total volume of the soil mass changes as a result, in equal measure to the volume of water drained, and is responsible for structural settlement. Total settlement is measured in three parts: Immediate, consolidation, and secondary compression. Immediate settlement occurs directly with the application of load and is exhibited primarily because of soil distortion. Consolidation and secondary compression settlement describe movement that occurs as a result of water being pressed out of voids in the soil mass and subsequent compression of soil particles together. The speed at which water escapes will obviously vary between different soil types and defines the time rate of foundation settlement. This occurs rather quickly in clean granular soils, therefore associated settlement is practically instantaneous upon application of load.

Most soil profiles are multilayered and nonhomogeneous, therefore total settlement cannot be determined or accounted for by review of a single variable. If the uppermost layer in a geotechnical profile is relatively thick compared to a foundation system's dimensions, it is practical to consider displacement as though the structure is uniformly supported atop a homogeneous layer of infinite depth and consideration of lower strata is not critical. If this upper region is thin by comparison, however, the effect of multiple layers and a nearby water table must be considered.

### 9.5.3 Risk

Geotechnical references include common warnings against lazy interpretation and use of engineering data. Behavior of soil under static and dynamic loading is far more difficult to predict than for well-established building materials like steel, concrete, masonry, or wood. Soil properties are naturally complex and depend not only on composition, but frequently on methods used in their determination. In fact, in-situ engineering properties of a soil type may differ greatly from that determined in the laboratory because of improper or inadequate sampling, different conditions of stress present than are accounted for during testing, and difficulty in evaluating the effects of disturbing geological layers.

One of the largest sources of construction-related claims for additional payment by a contractor is the encountering of unanticipated subsurface conditions, which occur due to a lack of definition and distinction between rock and other deposits that have great variance throughout the site. Excessive excavation needs, misrepresentation of the depth of soil profile types, unsuitable borrow fill material, misrepresentation of the groundwater level, undiscovered natural soil hazards (i.e., old fill deposits, large fissures, collapsible pockets), and unanticipated obstructions to pile driving or pier drilling have also been responsible for claims. Construction time delays and possible emergency foundation redesign or corrections add to frustration that can be experienced if site soil conditions are not properly evaluated or investigated.

Confusion of design methods sets up another area of unintentional risk. Consideration of foundations to support a structure which has been designed using strength-level forces becomes somewhat complicated, since soil resistance factors have not yet been fully studied nor even sufficiently understood. For allowable stress design, the building code, as well as most geotechnical reports, includes a safety factor in the range of 2 to 3.5 in the allowable bearing pressures noted. Different site conditions will dictate safety factor requirements and any uncertainties should be called to the attention of the geotechnical engineer, who will be able to provide guidance in applying appropriate statistical values to a strength design method. Additionally, prediction of ultimate consolidation settlement requires a basis in careful laboratory observation, as this condition corresponds to a complete loss of excess pore water pressure in a compressible medium, and the disturbance of in-situ samples must be considered in reported values.

Dynamic effects of earth movement can have a pronounced effect on buried structures. When codes or site conditions dictate, seismic pressure against an earth retaining structure is calculated based on the assumption that a wedge of soil bounded by the structure and the backfill shear failure plane moves as a rigid body under a percentage of peak ground acceleration. Type of soil, potential of seismic hazard, geologic formation, flexibility of the structure itself, and other factors are taken into account in determining the magnitude and character of dynamic soil movement.

## 9.6  CODES AND STANDARDS

The 2006 IBC allows soil to be classified according to ASTM D2487 in order to establish some expected engineering properties, such as allowable foundation pressure and lateral bearing capacities noted in Table 1804.2, in lieu of obtaining a complete geotechnical report for a site. Section 36.3.1.1 of NFPA 5000 allows this same practice with Tables 36.3.4(a) and (b). Properties for use in defining a soil's class can be determined by historical data or basic tests performed on samples, which save costs associated with a complete analysis. Though it is convenient to use allowable foundation pressure, lateral bearing, and lateral sliding resistance values listed in these codes, keep in mind that use of this table is also subject to approval by code enforcement officials within the region of interest.

Because underlying soil has a significant effect on seismic motion, site class based on an average of certain properties evident within the upper 100 ft (30.4 m) is to be determined according to 2006 IBC Table 1613.5.2. These properties include shear wave velocity within the stratum, standard penetration resistance, and the soil undrained shear strength, which can only be determined through proper geotechnical studies. If properties are not known in sufficient detail, the building official may allow assumption of Site Class D (stiff soil profile) for design purposes. Site Classes E and F represent special hazards as they may be vulnerable to failure under seismic loading in the form of liquefaction, collapse, or similar phenomena.

Other geotechnical considerations that are identified in model codes include excavations and general site preparation requirements, directions when encountering expansive soils or constructing within a flood-prone region, design and inspection of pier- or pile-type foundations, details for design and construction of foundation walls, and damp/waterproofing.

# 10 Understanding the Behavior of Concrete

Concrete is an engineering material that can be formed from a fluidic state to almost any solid shape or configuration that a person can dream of. Figure 10-1 shows the upper portion of a square reinforced column, waiting for tying of forms and further placement of a predefined mixture of cement, aggregate, water, and special admixtures. Reinforcing bars shown are lapped appropriately with those embedded in an earlier pour. This simple configuration represents a strong structural element for carrying load and further detailing adds to good ductile performance if used as part of a frame that resists seismic forces.

It is not the intention of this chapter to provide complete guidance on the design or specification of concrete mixtures, final elements and their type (plain, reinforced, prestressed), mode of formation (cast-in-place, precast), quality control measures (sampling, testing, observing), or means and methods of construction (tilt-up). Rather, background and general information is summarized from a variety of sources on the physical properties of elements found within a concrete mix, types of concrete elements used in construction, expected behavior of formed elements and assemblies, and usage of codes and standards. Because concrete is a complex building material, behavior patterns are also complex and sometimes difficult to grasp, however even rudimentary knowledge provides an engineer, whether new to the practice or a seasoned professional, with a solid basis for building effective problem solutions.

| **Figure 10-1** Concrete column reinforcing cage.

## 10.1 COMMON TERMS & DEFINITIONS

*Billet Steel*: Steel that is either reduced directly from ingots or continuously cast, made from heats of open-hearth, basic oxygen, or electric-furnace steel, or lots of acid Bessemer steel, with a specific chemical composition.

*Clinker*: Partially fused product from a kiln, consisting essentially of hydraulic calcium silicates burned to a temperature of 2600 to 3000°F (1420 to 1650°C), which is ground down to make cement.

*Darby*: A straight-edge measuring 3 to 8 ft (0.9 to 2.4 m) in length that is used during the early stages of concrete finishing to level the surface and supplement floating operations.

*Dobies*: Small concrete blocks in different sizes placed below reinforcing bars to maintain specified clearances or spacing. They are also manufactured with embedded wire ties in order to secure bars against movement during concrete placement.

*Finish*: Texture of a concrete surface after consolidation and finishing operations have been completed.

*Honeycomb/Rock-Pocket*: A porous, mortar-deficient portion of hardened concrete consisting primarily of coarse aggregate and voids, caused by improper consolidation, leakage of concrete from the forms, or separation during pour or difficult finishing operations.

*Hot Mud* (slang): Concrete that has been in a mixer for longer than an hour and not yet placed (mud is a term used for concrete in its fluid state).

*Kiln*: A furnace or oven for drying, charring, hardening, baking, calcining, sintering, or burning various materials.

*Rod Buster* (slang): Worker who sets, secures, and ties off steel reinforcing bars.

*Sintering*: A process by which particles are partially fused together, resulting in a porous mass, used in the development of lightweight aggregates.

*Slump Test*: Measure of the workability of a concrete mix. A 12 in (300 mm) high truncated metal cone is filled with fresh concrete, then it is lifted off and a measurement is taken of the distance that the top of the mass "slumps" down from its original position. The smaller the slump, the stiffer and less-workable the mix will be during placement. A slump of 3 to 5 in (76 to 127 mm) is common in building construction.

*Stirrup*: Reinforcement used to resist shear and diagonal tension stresses in a structural member, typically bent into a U-shape and placed perpendicular to longitudinal steel.

## 10.2 ELEMENTS OF CONCRETE

Plants that manufacture cement are typically located close to mines where raw materials are obtained in order to save transportation costs. Surface mining is the most common method of obtaining limestone, clay, silica, aluminates, and other material that will be fused together to make clinker, which is ground into cement powder. After the ore material has been excavated, it is crushed and screened into various particle sizes. It is then prepared and placed into a rotary kiln for heat processing, which produces the clinker.

### 10.2.1 Aggregate

Sand and aggregate are also mined, but may be dredged from the bottom of lakes, rivers, or streams. Aggregates affect the strength and durability of hardened concrete. They are graded as *fine* or *coarse*, which allows for a wide range of concrete mix designs depending on economic factors or desired physical properties. Recommended gradation limits are shown in the diagram of Fig. 10-2.

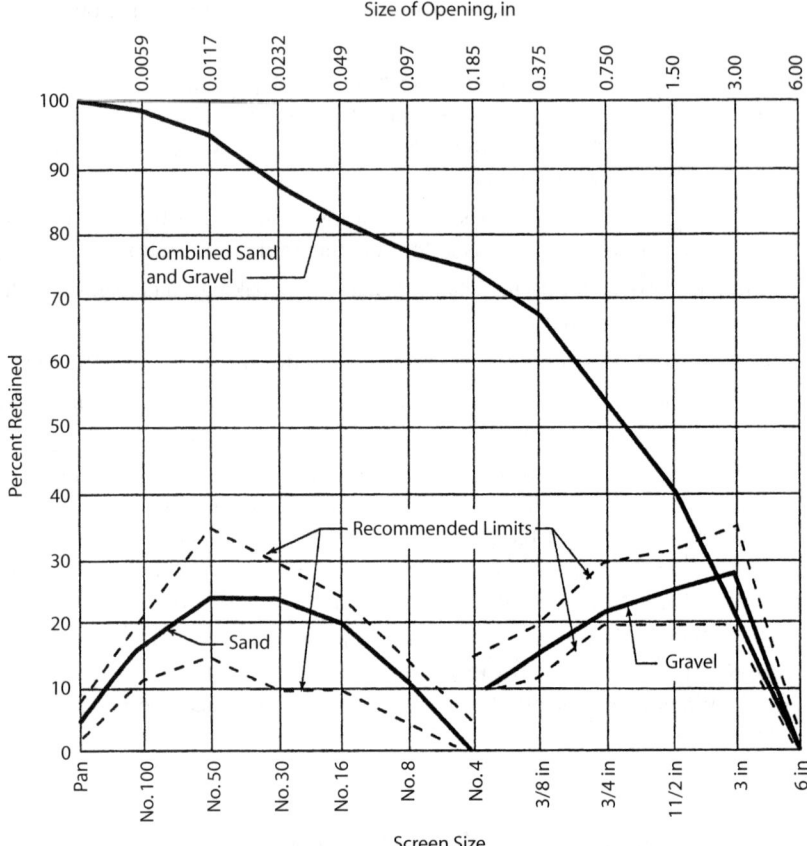

**Figure 10-2** Aggregate grading chart. (*Courtesy U.S. Army Corps of Engineers*)

The grading and quality of normal weight, natural stone aggregate is done according to ASTM Standard C33, resulting in a concrete unit weight of about 145 pcf (2320 kgcm), commonly taken to be 150 pcf (2400 kgcm) for design purposes when steel is incorporated within structural members. Fines are less than about 0.1875 in (5 mm) in diameter, consisting of natural and manufactured sand, whereas coarse aggregates are produced from gravel, stone, air-cooled blast furnace slag, or crushed hydraulic-cement concrete with a maximum size that is dictated by clearance requirements between forms and reinforcing bars. Quality control measures are incorporated into production to assist in preventing deleterious reactions in moderate to extreme weathering environments and include tests for measuring organic impurities, grading, soundness, lumps of clay, coal, bulk density, abrasion and reactivity, freezing and thawing, and chert (fine-grained siliceous rock) identification.

Lightweight materials conforming to ASTM C330 are prepared by expanding or sintering products such as blast-furnace slag, clay, diatomite, fly ash, shale, or slate. They may also be specified from materials in a natural state such as pumice, scoria, or tuff. Concrete unit weights of between 70 and 115 pcf (1120 and 1840 kgcm) may be expected. Samples are tested for conformance to the standard in regard to the inclusion of organic impurities, staining, grading, bulk density, and clay lumps.

Heavier-weight concretes are created for use in conditions where high density is required, such as for shielding against radiation, or where heavy ballast or counterweight is needed. Some of the higher-density materials (ores, minerals, or synthetics) have a tendency to powder or degrade during handling and batching, therefore production operations must be carefully monitored and documented. Unit weights between 200 and 350 pcf (3200 and 5600 kgcm) may be obtained for these purposes.

### 10.2.2 Hydraulic Cement

Cement is a chalky material that binds water, chemically combined in a process called hydration, with the aggregates. It consists primarily of silicates and aluminates of lime from limestone and clay or shale that is ground, blended, fused in a kiln, and crushed to a powder. Quality of cement produced in the United States is primarily governed by specifications from ASTM, AASHTO, and the American Concrete Institute (ACI).

Portland cement, so called because of its resemblance when hardened to Portland stone found near Dorset, England, is identified by conformance with ASTM C150 and is available in five different types. Type I is specified for general purposes, Type II possesses moderate resistance to sulfate attack, Type III is specified when high early strength is desired, Type IV produces a low heat of hydration which is critical during the construction of massive concrete structures, and Type V possesses high sulfate resistance. Compressive strength of concrete made with Portland cement is defined at 28 days for design purposes and typically requires 7 to 14 days to attain sufficient strength to allow removal of forms and imposition of some dead loading.

### 10.2.3 Water

Generally, water used in a concrete mix must be potable (drinkable) and ASTM D1129 helps to clarify some properties including hardness, acidity, and chemical content. The required amount of water increases as aggregate particles become more angular and rough in texture, but decreases as the maximum size of well-graded aggregate is increased or air entrainment is provided.

It has long been known that the lower the water-to-cement ratio (w/c), the higher the compressive strength of hardened concrete, though good concrete design also addresses durability, permeability, and wear resistance. Differences in measured strength for a given water-cement ratio certainly occur and are often attributed to changes in the maximum aggregate size; grading, texture, shape, strength, and stiffness of aggregate; differences in cement types; air content; and the use of chemical admixtures that affect the hydration process (ACI 211.1-91, Section 3.5).

## 10.2.4 Admixtures

An admixture is defined as any material other than water, aggregates, hydraulic cement, and fiber reinforcement used to enhance the properties of a concrete, and is added to the batch immediately before or during mixing. Some of the more common reasons for including admixtures are to increase workability without increasing water content, retard or accelerate set of the mix, reduce segregation, accelerate the rate of strength development, increase durability or wear resistance, increase bonding capabilities, decrease permeability, inhibit corrosion of embedded metals, and to provide color. Only a few of these will be discussed here.

*Cementitious materials*; including fly ash, blast furnace slag, silica fume, and natural pozzolans, are added to a mix in combination with normally specified cement for purposes of economy, reduction of the heat of hydration, improved workability, and durability. Pozzolans are siliceous and aluminous materials which chemically react with calcium hydroxide at ordinary temperatures to form compounds that possess fine cementitious properties. Fly ash is a residue left over from the combustion of ground or powdered coal with various properties. Blast furnace slag is a byproduct of pig iron production that possesses cementitious properties when it is rapidly quenched and ground. Silica fume is a by-product from the reduction of high-purity quartz with coal and wood chips in an electric arc furnace during the production of silicon metal or ferrosilicon alloys. It condenses from gases released from the furnace and may be used in production of low permeability, high-strength, or chemically resistant concrete.

*Air entraining chemicals* with qualities per ASTM C260 may be added to a concrete batch immediately before or during mixing to produce uniformly distributed air bubbles that will improve workability and resistance to freeze-thaw cycles. Water-soluble powdered agents are composed of a wide variety of salts or synthetic detergents. Solid particulates that have a large porosity, including a composition of hollow plastic spheres, crushed bricks, expanded clay or shale, may also be used for air entrainment. Important functions of air entrainment are derived from the total volume of bubbles produced, their size, and distribution throughout a mix.

*Water-reducing admixtures* are used to produce higher-strength concrete, obtain specified strength at a lower cement content, or increase slump of a mixture without increasing the water content. These products include a variety of salts that increase fluidity of the cementitious paste by reducing force between particles of cement. High-range water-reducing admixtures, also known as superplasticizers, are somewhat different from conventional products in that they do not affect the surface tension of water by a great degree and therefore may be used in greater proportions without excessive air entrainment.

## 10.3 CHARACTERISTICS OF A FINAL MIX

Ready-mixed concrete is prepared according to ASTM C94, which requires it to be batched by volume in a controlled manufacturing facility according to a specific recipe, mixed in a continuous mixer, and delivered to the purchaser in a condition that is ready to place. The purchaser may either assume responsibility for the proportioning of the concrete mixture or may delegate this task to the producer, giving consideration to desired compressive strength, workability, placeability, durability, surface texture, and density. Concrete batches are produced by stationary or truck mixers. Delivery tickets are provided with each batch of concrete that list the ready-mix company name, serial number, delivery date, truck number, purchaser, job information, concrete designation for use on the job, amount of concrete and the time it was loaded, water added by receiver, reading of revolution counter at first adding of water, cement and cementitious material information, type and quantity of admixtures or fibrous reinforcement, aggregate size and mass provided, and official signatures. These tickets become a critical asset to any quality control plan in creating a history trail.

The paste created of cement and water is the portion of a mix that shrinks as the concrete hardens and dries out. When tighter controls are desired on the amount of shrinkage that will occur, it is necessary to reduce the volume of paste created in a mixture—randomly reducing the water-to-cement ratio is not the answer. To accomplish this, the size and quantity of voids between aggregates, and thus the total surface area that will be covered by paste, should be reduced by using greater quantities of coarsely graded aggregate. *Shrinkage-compensating cement* can also be used instead of ordinary Portland cement to help control drying shrinkage, though it is not guaranteed to prevent all cracking. This special cement works by expanding slightly during early stages after placement of the mix to add a compressive force throughout, partially counteracting internal tensile stress of the concrete and resulting in fewer drying shrinkage cracks. Some restraint is necessary to develop these forces and can be provided with a minimal amount of steel reinforcement or subgrade friction for slabs-on-grade.

*High-performance concrete* (HPC) has simply been defined as a special mixture where constituents are carefully controlled in order to produce a special combination of performance and uniformity that cannot always be achieved by the use of conventional materials and methods. Some of the critical characteristics sought in HPC mixtures include ease of placement and consolidation, early attainment of strength, reduced permeability, greater density, controlled heat of hydration, toughness, or added durability in severe environments. Codes and standards of different countries recognize the use of higher strengths through HPC formulas and research continues to quantify knowledge about behavior, limitations, and benefits. Though the cost of materials, mixing, placing, and curing of HPC may be significant compared to that of normal concreting operations, financial benefits will likely be realized in terms of maintenance and long-term performance, which is especially critical for longer-span bridges or those where access is limited. Careful proportioning of constituents and attention to their respective characteristics are obviously of critical importance.

If conditions of exposure are such that air voids within hardened concrete become blocked, durability against freezing damage is lessened. Rock used to make coarse aggregate also contains a certain volume of pores that absorb water and saturation can lead to damage in wet, freezing environments. Testing and observation of actual service conditions have shown that entrained air within a concrete mixture renders the cement paste essentially immune to damage from freeze-thaw activity of pore water. Accumulation of water atop concrete structures in service where freezing temperatures are experienced should also be controlled by good design practice, such as providing positive drainage and preventing the formation of baths or dams. Air-entrainment also has the benefit of reducing segregation and surface bleeding, which controls formation of water pockets beneath aggregates and embedded steel.

## 10.4 BEHAVIOR OF CONCRETE ELEMENTS

Strength and serviceability are key criteria reviewed in the process of designing any concrete member under all possible loading scenarios. The term *strength* may be defined as the point of imminent failure under a condition of overload, whereas *serviceability* describes acceptable performance under ordinary loading conditions. A proper review of concrete member serviceability involves control of unsightly cracking, which may allow the intrusion of water and lead to corrosion of reinforcing steel or deterioration of surrounding concrete, limit of deflection so that adjacent or supported nonstructural elements may be protected from damage, and a check on vibration effects.

### 10.4.1 Plain Concrete

In reviewing the behavior of concrete as a structural element, it is helpful to first consider mechanisms that concrete itself sets up within a structure before launching into the contribution of reinforcing steel. Concrete does exhibit a certain amount of resistance to tensile stress, but it is only recognized for limited applications. The primary restrictions to using plain concrete as a structural element are where ductility is a necessary factor in design or where cracking will affect structural integrity. A basic stress-strain curve for plain concrete is shown in Fig. 10-3.

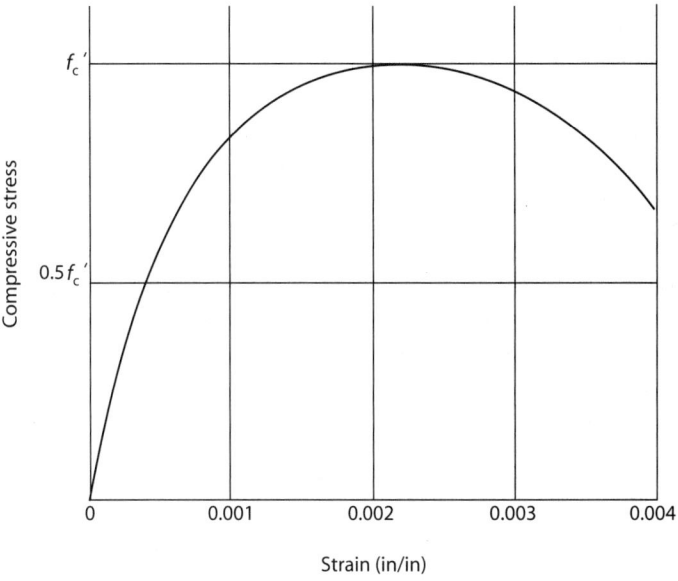

**Figure 10-3**  General stress-strain curve for plain concrete.

Structural plain concrete, whether cast-in-place or precast, is recognized by ACI 318-05 only for members that are continuously supported by soil or some other means (footings, slabs-on-grade), members by which arch action provides continuous compression for all load combinations, or for walls with continuous vertical support and pedestals with a ratio of unsupported height to average least lateral dimension less than or equal to three. Plain concrete columns, defined as members that primarily support longitudinal compression and are taller than pedestals, are restricted against structural uses.

Plain concrete resists shear from transverse loading by means of aggregate interlock occurring tangentially along a formed crack. Since cracking occurs when the tensile strength of concrete is exceeded, Chapter 11 of ACI 318-05 recognizes the fact that a member's shear strength increases when axial compression is present, as the opening of shear cracks are delayed, but decreases in the presence of significant axial tension.

Construction and contraction (control) joints are of particular importance for plain concrete to relieve tensile stress from creep or temperature and shrinkage movement. A construction joint is created by two successive concrete pours and sometimes includes a metal (or other) keyway or crossing reinforcement. A control joint, on the other hand, is formed, sawed, or tooled into concrete to regulate the location and amount of cracking that occurs during hydration or that due to movement from changes in ambient temperature.

## 10.4.2 Reinforced Concrete

Reinforced concrete is a perfect example of how strength characteristics of two different building materials can work together in creating a structure: Plain concrete provides resistance to high compressive forces and steel reinforcing provides the needed strength in tension, adds compressive strength to axially loaded members, aids in reducing creep and shrinkage deformation when placed in the compression region, and provides needed ductility for good performance when subjected to lateral forces. These two materials work well compositely due to development of a strong bond and resistance to slippage provided by deformations that are rolled onto the bar surface. Plain steel bars, which do not have surface deformations, may only be used in assemblies as spirals, confining reinforcement at longitudinal bar splices, or prestressing steel (ACI 318-05, Section 3.5.1) and have also been used as dowels at isolation or contraction joints. Additionally, concrete and steel have very similar rates of thermal expansion, so changes in temperature will introduce negligible force differentials between materials.

Reinforcing steel supplements the strength of concrete in shear through dowel action (a force transverse to the longitudinal bars), with direct shear steel in the form of stirrups, and through arch action whereby a rib of concrete arches over the neutral axis and is tied together by the longitudinal bars. Stirrups may be placed either perpendicular to or inclined 45° or more with longitudinal steel.

Reinforcing bars are identified in terms of eighths of an inch when using inch-pound terminology (see Fig. 10-4). For example, a bar that measures 5/8 in (nominal diameter) is called a #5 bar. Using metric terminology, this same bar measures 16 millimeters (nominal diameter) and is therefore called an M #16 bar. Since 1997, reinforcing bar producers have been issuing soft-metric products, where the actual nominal dimensions of bars are produced per inch-pound specifications, but identified in terms of equivalent metric units. Produced sizes include #3 (M #10), #4 (M #13), #5 (M #16), #6 (M #19), #7 (M #22), #8 (M #25), #9 (M #29), #10 (M #32), #11 (M #36), #14 (M #43), and #18 (M #57). The Concrete Reinforcing Steel Institute (CRSI) issues many helpful publications to assist engineers and contractors in the proper design and installation of bars, including information on bar characteristics, bend and splice details, placing, support, and inspection. Steel reinforcement is commonly specified per the following designations, depending on the application and environment:

*ASTM A185 (plain steel welded wire reinforcement).* This standard covers an assembly of cold-drawn wire placed in longitudinal and transverse directions, resulting in an essentially square or rectangular mesh, welded at their intersection points by electrical resistance welding (fusion combined with pressure). The fabrication is also known as welded wire fabric or mesh and is produced in sheets or rolls. The purchaser must specify wire size and spacing, length and width of sheets or rolls, yield strength requirements for the wires as noted in ASTM A82, and possible inspection needs. Mechanical property tests of the finished assembly include

Understanding the Behavior of Concrete | 255

**Figure 10-4** Reinforcing steel with identification stamp (16-mm bar shown).

tensile either across or between welds (not less than 50% may be taken across welds), reduction of area, bending performed on a specimen taken from between the welds, and weld shear strength.

*ASTM A497 (deformed steel welded wire).* Sheets or rolls are fabricated similar to those for plain wire, yet are primarily of deformed material with strength properties per ASTM A496.

*ASTM A615 (deformed and plain carbon-steel, billet).* Available bar sizes range between 0.375 and 2.257 in diameter (9.5 and 57.3 mm) and can be ordered in three yield strength levels of 40, 60, and 75 ksi (280, 420, and 520 MPa), identified as *grades*. Grade 40 (280) is only available in #3 through #6 (M #10 through M #19) and Grade 75 (520) may only be ordered for #6 through #18 (M #19 through M #57). No specific provisions are included to enhance the *weldability* (see discussion for ASTM A706 bars below) of this type of bar, therefore a special welding procedure would be necessary to indicate proper methods and materials to complete the work (ACI 318-05, Section 3.5.2). The engineer must specify bar size, grade, and whether deformed or plain without assuming the contractor is aware of code limitations regarding the use of these types. Bars are rolled from mold or strand cast steel using the electric furnace, basic oxygen, or open hearth process and the average distance between deformations cannot exceed seven-tenths of the bar's nominal diameter. Testing is performed on a sample from every heat produced to check properties of ductility and chemical composition.

*ASTM A706 (low-alloy steel, deformed and plain).* These bars are intended for applications requiring tight controls on tensile properties or improved weldability and are only available as Grade 60 (420). The weldability of reinforcing steel is based on its carbon equivalent (CE), calculated by Equation 10-1 using the chemical composition shown in a mill test report, and may not exceed 0.55% (AWS D1.4-05, Section 1.3.4.2).

$$CE = \%Carbon + \%Manganese/6 + \%Copper/40 + \%Nickel/20 \\ + \%Chromium/10 - \%Molebdenum/50 - \%Vanadium/10 \quad (10\text{-}1)$$

Preapproved welding processes include shielded metal arc (SMAW), gas metal arc (GMAW), and flux cored arc (FCAW).

*ASTM A767 (zinc-coated steel for corrosive environments).* Steel bars that have been produced according to ASTM A615, A706, or A996 are prepared and dipped into a molten bath of zinc according to this standard. Bars are shaped either before or after galvanizing, but any damage to the coating is repaired prior to delivery. The coating's quality is checked on a sample from every lot produced, where a *lot* is defined as all bars of one size furnished to the same steel bar specification that have been galvanized during a single production shift.

*ASTM A775 (epoxy-coated steel for corrosive environments).* Steel bars that have been produced according to ASTM A615, A706, or A996 receive a protective epoxy coating by the electrostatic spray method according to this standard. Steel bars are prepared to receive epoxy material by sand-blasting and application of a coating that promotes adhesion of finishing material, reduces metal-coating reactions, improves corrosion resistance, and increases resistance to blistering. The finished coating is fusion-bonded to steel and contains pigments, thermosetting epoxy resins, cross-linking agents in the resin to improve strength, and other additives. It is applied as a powder and difficulties occur where sharp edges are present, as well as at deformations and bar marks.

*ASTM A955 (stainless-steel for corrosive environments).* This reinforcement is specified for applications where special corrosion protection or controlled magnetic permeability is necessary. Standard bars are plain or deformed and available in the same sizes and grades as A615, manufactured from stainless steel alloys noted in A276 (identified by chemical composition) that are to be selected for appropriate suitability in the constructed environment. All bars are furnished with heat treatment either by annealing, hot rolling, or strain hardening.

*ASTM A996 (rail-steel and axle-steel)*. Available bar sizes range between 0.375 and 1 in diameter (9.5 and 25.4 mm) and are designated with a "rail symbol," Type R (produced from rail steel), or Type A (produced from axle steel). Type "rail symbol" and Type R members can be ordered in two yield strength levels of 50 and 60 ksi (350 and 420 MPa), whereas Type A bars are available as Grade 40 (280) and Grade 60 (420). All bars are rolled from standard tee-rail sections or carbon steel axles taken from retired railway cars and locomotives. The manufacturer is to determine carbon content of axle steel and stock the material for rolling in separate lots according to carbon range.

**Flexural Effects** Flexural effects in reinforced concrete elements are evident through resistance to bending, shear, and also for consideration of deflection. It is to be remembered from design courses that the distribution of compressive stress in concrete on one side of the neutral axis forms a sort of parabolic shape, though other geometries like Whitney's rectangular stress block have been assumed for simplification. Steel may also be placed within the compression region, though this practice is not commonly done to augment strength but rather in an effort to reduce creep and shrinkage deflection.

Nominal flexural strength in design is based on a maximum compressive stress when concrete crushing strain is reached at a value of 0.003 (ACI 318-05, Section 10.3.3). A member's behavior under this condition depends on the amount of reinforcing present. After a properly reinforced concrete beam cracks, it will continue to deform in a linear fashion until the steel yields at a moment equal to $M_y$, identified in the idealized moment-curvature diagram of Fig. 10-5. Beyond this

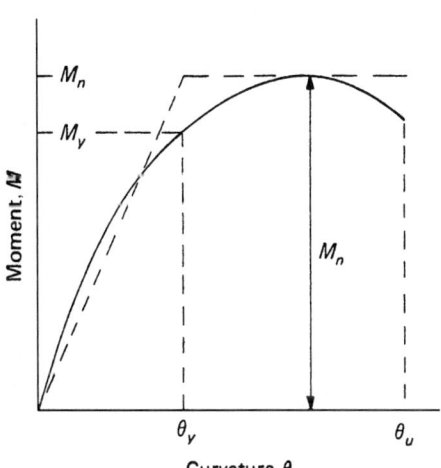

**Figure 10-5** General reinforced concrete moment-curvature diagram.

point, inelastic deformation occurs until the nominal bending strength, $M_n$, is reached. When a large amount of tension steel is present, it will remain in an elastic range of stress and the structural member will fail in a brittle fashion, whereby concrete crushes prior to yielding of the steel. This is undesirable and design standards are written to control against this phenomenon by specifying a percentage of steel low enough to assure a ductile mode of failure.

Openings present special challenges in terms of understanding reinforced concrete behavior, partially due to observations that resistance of members with small openings exhibit characteristics of a *beam*, thus adhering to the same familiar flexural design procedures as a solid member, whereas those which contain large openings resist load much like a *frame* or Vierendeel truss. A simple beam-type of analysis for such a member will still be valid, as long as the opening through the member's web does not reduce the area of concrete necessary to develop the flexural compressive stress block (Mansur 1999). Solid sections above and below a large penetration may not have the strength to resist imposed forces, leading to a frame-type failure. Action carried by these portions can be determined by distributing vertical shear force at the centerline of the large opening in proportion to relative flexural stiffness of the solid regions through maintenance of deformation compatibility. The upper segment will essentially play the part of a top chord member, resisting compression due to global bending moment (when loading is applied in a downward direction), and the lower portion will likewise resist tension forces. Stirrups and diagonal reinforcement assist in crack control and transfer of shear forces between the solid regions.

Flexural behavior of a reinforced concrete column is assessed from a moment-curvature analysis of the member's cross section. Under lateral loading, behavior is defined in terms of a linear elastic response to the point of initial reinforcement yield, then entering into nonlinear behavior until the ultimate moment is reached, identified as a reducing slope of the moment-curvature line. Once the column cracks, the concrete shear resisting mechanism is provided mainly by interlock of aggregate. After this occurs, transverse reinforcement activates to carry the tensile portion of an established truss mechanism (with concrete playing the role of compression "web elements"), contributing to the overall shear capacity of the column.

Shear and torsional forces produce diagonal cracking in flexural members as indicated in Fig. 10-6, causing not only tension in the stirrups, but in the longitudinal steel as well. Stirrups can be visualized as similar in function to tension web elements of a truss as they transfer shear forces across inclined cracks. The optimal amount of transverse reinforcement is determined through code limitations setup such that both the stirrups and the compression zone of concrete continue to carry shear after formation of inclined cracks, until the point where this reinforcement yields, assuring ductile behavior.

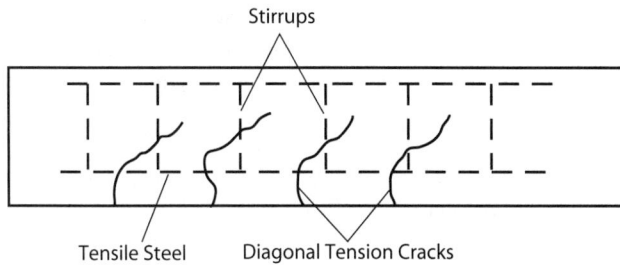

**Figure 10-6** Pattern of shear stress cracking in reinforced concrete bending member.

An important shear transfer mechanism described in Section 11.7.4 of ACI 318-05 is shear-friction, where concrete bond to perpendicular or inclined bars on each side of a crack sets up a compressive force that is not only necessary to produce friction between surfaces, but also prevents or delays the opening of that crack. A coefficient of friction is used which corresponds either to concrete placed monolithically, along the surface of previously hardened concrete that has either been left untouched or intentionally roughened, or along the plane between newly placed concrete and structural steel joined by headed studs or reinforcing bars.

When stirrups are not present in larger beams, shear strength is dependent on the width, spacing, and roughness of cracks in conjunction with straining of longitudinal steel (Collins and Kuchma 1999). The relationship between shear and bending moment along the length of a concrete beam indicates that internal shear is resisted by a combination of *beam action*, based on the internal moment arm length, and *arch action*, identified when bond between longitudinal steel and concrete is lost. Experiments have shown that beam action tends to govern flexural behavior when a member's span-to-depth ratio exceeds 2.5, whereas arch action is significant when this ratio is less than 2.5 (Kim, Kim, and White 1999).

Deflection that a flexural member is allowed to experience is to be calculated based on service-load conditions and it may be governed by performance of the framing system or by tolerances relative to supported or interacting components of the structure. Limitations on deflection due to live loads as stipulated in Table 9.5(b) of ACI 318-05 are historical and somewhat arbitrary, commonly limited to L/360 for floors and plastered ceilings. Service load moment usually exceeds that which causes concrete to crack on the tension face, therefore elastic properties of the cracked member are used to compute instantaneous deflection, including the effective moment of inertia which is based on the gross section as well as the transformed cracked section (both tension and compression steel is transformed into equivalent concrete).

Total long-term deflection of a reinforced concrete beam includes an instantaneous elastic deformation coupled with creep and shrinkage effects. *Creep* is defined as inelastic deformation with the passage of time and is affected by properties of the constituent elements (cement, grading and mineral content of aggregates, admixtures), the water/cement ratio, quality of curing operations, member volume and geometry, as well as magnitude and duration of applied loading. Wang and Salmon (2002, p. 566) indicate that after about 1 year of service, 90% of shrinkage will have been realized, whereas 5 years are necessary for 90% of creep deformation to occur, otherwise it is difficult to tell the two phenomena apart. *Shrinkage* causes warping of a reinforced member as bond between steel and concrete restrains movement along that face. Both creep and shrinkage can be controlled to a certain extent by good curing, assurance of high strength at the time of initial loading, inclusion of reinforcing steel within the compression region, and appropriate measures for shoring, but there will always be some level of uncertainty.

**Behavior Under Axial Loading** Multiple combinations of nominal axial load strength ($P_n$) and nominal moment strength ($M_n$) for a reinforced concrete column are defined along the curve of a strength interaction diagram, such as that shown in Fig. 10-7. Radial lines that project from the origin of the axes represent different values of axial load eccentricity, with $e = 0$ along the vertical axis and $e = \infty$ along the horizontal axis. A balanced strain condition occurs at $e = e_b$ along the interaction curve, which divides the *compression controls* region (concrete compressive strain reaches 0.003 prior to yielding of the steel) from the *tension*

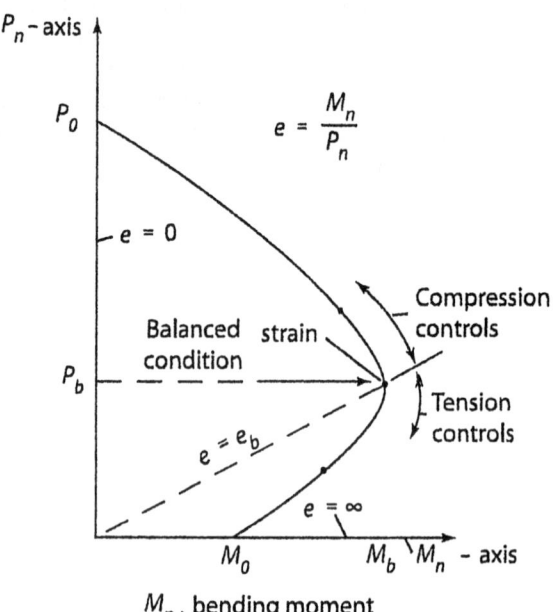

**Figure 10-7** General concrete column strength interaction diagram.

*controls* area of the diagram (strain in tension steel is greater than yield at the time concrete compressive strain reaches 0.003). When $P_n$ exceeds $P_b$, the member behaves more like a column than a beam (compression controls) and conversely, when $P_n$ is less than $P_b$, beam behavior may be expected as crushing of the core transforms to a more ductile response. As a logical consequence, the strength reduction factor $\varphi$ gradually increases from that for a column to that of a beam.

For design purposes, the axial load strength of compression members is limited to 80% (spirally tied elements) or 85% (common lateral ties) of concentric nominal strength, stipulated by ACI 318-05 (Section 10.3.6) in recognition of the fact that it is extremely rare to find a column that only carries concentric axial forces. Initial imperfection and support conditions are partially responsible, but long-term effects of carrying a compressive force also play a role in the uncertainty of exact load alignment. For example, when concrete and longitudinal reinforcement act together in resisting compression, the proportion of load carried by each material changes over time as creep and shrinkage causes steel to gradually absorb more load than the initial elastic portion.

### 10.4.3 Precast and Prestressed Concrete

Ordinary or prestressed concrete that is precast in a special facility or directly on site prior to placement has inherent benefits of tighter quality control measures, making the selection an attractive one to consider. The design process can at first be unfamiliar, however, as it not only involves loading that is experienced during the life of the structure, but an engineer must also consider the manufacturing process, shipping restrictions, handling, and erection operations. As a rule-of-thumb, the average depth of a prestressed beam will be about 75% of the depth of a standard reinforced concrete beam under the same loading conditions and that depth will be about 0.6 in (15 mm) for each foot of span.

Cost savings are generally realized in the process of repeatability and standardization, including use of the following:

1. Modular dimensions for the overall plan and individual members
2. Commonly available shapes and sizes without too many different products on one project
3. Predictable reinforcing patterns and prestressing strand placement
4. Preferred connection types using common grades of steel
5. Element weights that can reasonably be produced, shipped, stored, and handled
6. Design and fabrication measures that allow for common skill levels of workmanship, including allowance of tolerances

Connections between precast concrete elements are often made by welding plates or bars to embedded steel elements. Thermal expansion and steel distortion from welding operations can cause failure of bonding, induce cracking, or precipitate spalling of the surrounding concrete. These effects can be reduced by the use of appropriate welding procedures, low-heat welding rods, staggered weld patterns instead of continuous runs, or smaller weld sizes. Other connection types include simple bearing pads that distribute member reactions over a defined area and allow limited horizontal and rotational movement for relieving stress (see Fig. 10-8). Bearing material in common use include neoprene pads that are reinforced with randomly oriented fibers, chloroprene pads reinforced with alternating layers of bonded steel or fiberglass, multimonomer plastic bearing strips, tempered hardboard strips, and different proprietary products.

*Prestressing* is an operation where a concentric or eccentric force is applied in the longitudinal direction of a bending member in such a way that tensile stress at midspan (for simply supported elements) is either eliminated or greatly reduced, thus allowing full use of the compressive strength of concrete across the entire depth of an element. A prestressed structural system can be made either flexible or rigid without influencing strength through proper control of applied longitudinal force, developed by high strength steel in order to maintain the necessary prestressing force after normal losses occur, which can be in the range of 35 to 60 ksi (241 to 414 MPa). *Pretensioning* is the most common method for applying this axial force, where steel is tensioned prior to placement of concrete within forms and then cut once concrete has encased the tendons and hardened sufficiently to take up the compressive force transferred from tension in the steel.

**Figure 10-8** Bearing conditions of precast concrete sections in a parking garage.

*Posttensioning*, on the other hand, is done by threading tendons through a duct or conduit and applying stress after concrete hardens. After all of the force is imparted to the tendons from jacking or other means, ducts are typically grouted to provide permanent protection of steel and to develop bond with surrounding concrete in cored ducts. There are some cases where grouting of tendons is unreasonable or difficult, such as in areas of high steel congestion or in sections where concrete thickness is somewhat small, and other means of tendon protection may need to be considered. Whether sheathed or cored, it is critical that ducts remain clear of deleterious material and be properly supported or formed without constriction.

In most situations, concrete remains in the elastic, uncracked range of behavior and stress analysis is thereby simplified. Different geometries have been adopted for use according to need and historical performance in service, as well as economy, and include tee-shapes, double-tees, I-sections, standard rectangular forms, hollow-core slabs (see Fig. 10-9), and hollow box girders commonly used in large span bridges.

Special steel tendons (bars, strands) are necessary for prestressing because of the need to carry high, continuously applied tension. Figure 10-10 illustrates the difference in behavior of prestressing strands and mild steel bars. Reinforcement used in common prestressing applications may be constructed of wires, strands, or bars in accordance with the following standards:

*ASTM A416* (*uncoated seven-wire steel strand*). Low-relaxation strands are considered to be standard, but stress-relieved (normal-relaxation) material may be specially ordered. Stress-relieving is achieved by heating steel to about 1202°F (650°C) and maintained at that level for a period of time, after which the steel is slowly cooled. Either type of strand is available with a minimum ultimate strength of 250 or 270 ksi (1725 or 1860 MPa), where a *strand* is defined to be a group of wires having a center wire enclosed tightly by six helically placed outer wires. Finished strand sizes are available in diameters of 0.25 to 0.60 in (6.4 to 15.2 mm) and are designated by the nearest whole number in millimeters (i.e., a strand measuring 15.2 mm in diameter may be identified as #15).

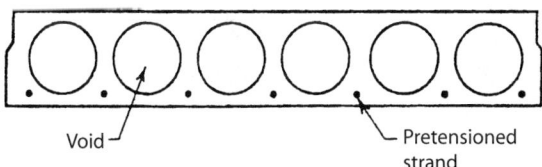

**Figure 10-9** General hollow-core prestressed concrete slab section. (*From Gaylord, Structural Engineering Handbook, McGraw-Hill, 1990*)

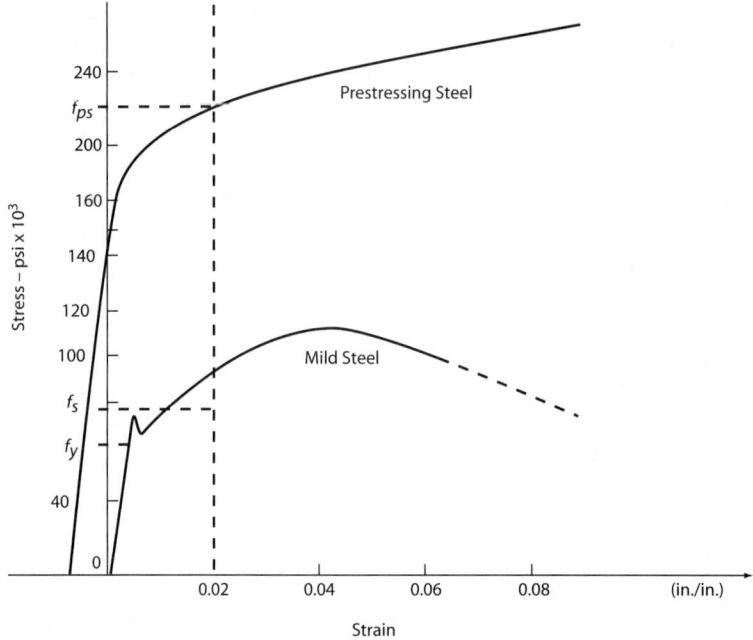

**Figure 10-10** Stress-strain comparison between mild steel and prestressing steel.

*ASTM A421 (stress-relieved, cold-drawn wires).* These are high-carbon, uncoated wires produced as either Type BA (cold-end deformation, or button, anchorage) or Type WA (wedge-end anchorage). They are available in nominal diameters of 0.192, 0.196, 0.250, and 0.276 in (4.88, 4.98, 6.35, and 7.01 mm) and tensile strengths between 235 and 250 ksi (1620 and 1725 MPa).

*ASTM A722 (high-strength, stress-relieved alloy steel bars).* Bars are available as plain or deformed with a minimum tensile strength of 150 ksi (1035 MPa). Plain bars are available in nominal diameters of 0.75 to 1.375 in (19 to 35 mm) and deformed bars are available at 0.625 to 2.5 in (15 to 65 mm). Steel is cold-drawn to raise the yield strength and stress-relieved to improve ductility.

Transfer of force in pretensioning occurs gradually along the length of a strand, whereas this load is delivered to a small area at member ends when posttensioning is employed (see Fig. 10-11). Adequate conventional reinforcement, therefore, is critical within tendon anchorage zones to prevent longitudinal cracking or even bursting of concrete below a tendon anchorage plate. Conventional reinforcement also plays an important role in the design of partially prestressed members. Tests performed on partially prestressed hollow beams subjected to bending and torsion,

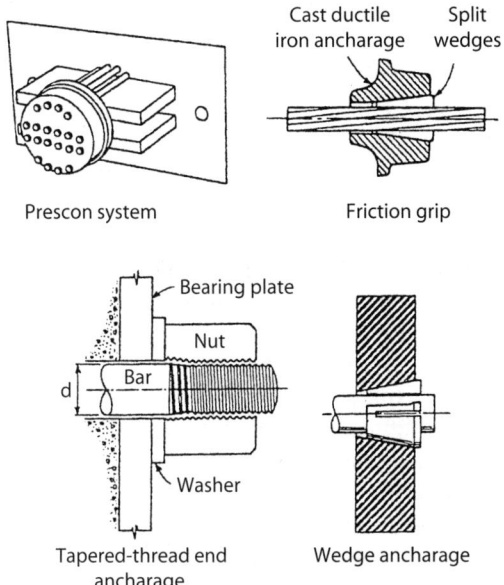

**Figure 10-11** Sample prestressing steel anchorage connections. (*From Gaylord, Structural Engineering Handbook, McGraw-Hill, 1990*)

in an effort which proved the conservatism of the general method defined in the AASHTO LRFD specification, showed evidence that longitudinal steel contributes significantly to the ultimate strength for large bending moments, whereas transverse steel dictated strength at intermediate and low moments (Rahal 2005).

Both methods of prestressing require consideration of prestress force loss by elastic shortening of concrete, relaxation of tendons, creep, and shrinkage, but posttensioning also requires consideration of force loss through slippage at anchorage locations and conduit friction losses. If jacking is done at one end of a member, the maximum frictional stress loss occurs at the opposite end and varies linearly along the span; it is a function of tendon or form alignment and accidental local deviations in alignment. Time-dependent losses, such as creep and shrinkage, continue over a period of about 5 years (Nawy 2003, p. 73). Concrete shrinkage, which is approximately 80% complete within the first year of service, is affected by proportions of materials in the mix, type and size of aggregate, type of cement, curing time, elapsed time between final curing and imposition of prestress force, member size, and environmental conditions. Prestress force can also be lost from corrosion of the strands, both by intrusion of environmental elements and formation of microscopic stress cracks that can lead to brittleness and failure.

## 10.5 BEHAVIOR OF CONCRETE SYSTEMS

Extreme loading events provide engineers with useful data concerning differences between predicted and actual in-service building performance. An earthquake of magnitude 7.2 on the Richter scale rumbled through Kobe, Japan on January 17, 1995 with both lateral and vertical ground motions. Scientists noted that the event forced structures to behave in much the same way that would be expected in a laboratory, except on a much larger scale, thereby providing valuable information by reconnaissance (Sarkisian 1996). In reinforced concrete multistory buildings with an age of 20 or 30 years, instability of first story columns accounted for observed collapses. Many of these columns contained smooth reinforcing bars and minimal ties at a larger spacing than would be required by modern codes for confinement during cyclic loading. Collapses were also attributed to stiffness variations between heavy concrete partition walls (nonstructural) and reinforced concrete moment frames, in some cases causing a torsional response not adequately detailed for transfer of load. Buildings with a sudden geometrical change in plan or elevation behaved poorly, as well as those structures which exhibited "weak column-strong beam" action.

Collapse of a nine-story concrete building from an earthquake in Vina del Mar, Chile, on March 3, 1985 was primarily attributed to the fracture of flexural reinforcement at the base of the walls in a brittle fashion because of a low steel ratio (Bonelli, Tobar, and Leiva 1999). Tests conducted on a scale model of this building indicated that a larger reinforcement ratio would have developed a greater number of thinner cracks along the story height and reduced yielding at the base of the wall, thus avoiding the concentration of nonlinear deformation at a single section. This translates into a greater ability to absorb and dissipate energy.

A general description of the construction, mechanics, and behavior of structural systems, including frames and diaphragms, has been presented in Chapter 8 and the reader may need to review that information before proceeding.

### 10.5.1 Rigid Frames or Cantilevered Columns

Concrete frames are assumed to behave within an elastic range prior to cracking and the usual methods of indeterminate structural analysis apply to force distribution. In regions of moderate to high seismic activity, frames must be able to sustain drifts far greater than elastic limits, with beams providing the bulk of this postyield deformability by rotation in the plastic hinge region. The temptation to design beams that are stronger than necessary must be resisted, as ductility will be negatively affected by a haphazard increase of reinforcing steel. Englekirk (2003, p. 124) lists the factors that are to be considered in conceptual sizing of reinforced concrete frame beams as beam-column joint capacity, reinforcing details, beam shear demand, column shear strength, and available ductility. Frame systems can also be created using precast elements, provided the connections are capable of sustaining the postyield deformation discussed here.

Testing of 2-story and 3-story scale models by El-Attar, White, and Gergely (1997) revealed surprising lateral load resistance in concrete buildings that have been detailed per ACI 318-89 to primarily resist gravity loads, being characterized by discontinuous positive moment reinforcement in beams at column locations, a low reinforcement ratio in columns, little confinement reinforcing in joints, and lap splices positioned immediately above the floor level. Neither of the models collapsed even under the highest level of dynamic loading (0.75 g), though a significant reduction in stiffness was also recorded. As a result, P-$\Delta$ effects were greatly pronounced. Most damage, deformation, and energy dissipation occurred within the first story columns with the formation of plastic hinges.

Tests performed by Cheok and Stone (1990) on scale models of circular bridge columns defined a pattern of failure when subjected to cyclic loading that has been noted by other experimenters. Flexural cracks initiate within the plastic region near the column base, where they gradually extend around its body. Damage continues with spalling of surface concrete away from the reinforced core, followed by yielding and subsequent fracture of spiral reinforcing, leading to eventual buckling and fracture of the longitudinal steel. Cyclic energy-absorption capacity decreases significantly when confining reinforcement fractures, leading a designer to the conclusion that ties or spirals are critical to assuring good flexural behavior during a seismic event. Whether by column, beam, connection, or joint, adequate absorption of cyclic energy is critical to structural success and failure can be catastrophic, as shown in Figs. 10-12 and 10-13.

The design of bridge piers or building columns includes consideration of certain characteristics based on load history such as stiffness, ductility, strength, and energy

**Figure 10-12** Concrete bridge damage from the Loma Prieta earthquake, California 1989.

**Figure 10-13** Concrete bridge failure during the Northridge earthquake, California 1994. (*Courtesy UCLA Department of Earth and Space Sciences*)

absorption (Esmaeily and Xiao 2005). Seismic detailing of columns demands a ductile response to these forces, achieved by providing adequate lateral reinforcement to resist shear forces, to confine the core, and to inhibit premature buckling of the longitudinal steel within the plastic hinge regions. Higher axial forces will thus require greater lateral reinforcement, which is recognized in the building and bridge codes. Confining reinforcement affects a column's ductility, whereas most of the energy absorbed during a seismic event is dissipated by flexural response, not through shear deformation (Pilakoutas and Elnashai 1995).

## 10.5.2 Shear Walls

Cracking is a necessary phenomenon in concrete shear walls for resisting high cyclic forces, which accounts for behavior within the inelastic range of response. Imposed vertical load along the center of a shear wall causes an increase in stiffness as the opening of cracks is delayed, though the effect is difficult to quantify. For purposes of system and structure analysis, this initial stiffness can be approximated by assuming the effective moment of inertia of the shear wall is 50% of the gross moment of inertia (Englekirk 2003, p. 374). Characteristics of concrete shear wall deformation are shown in Fig. 10-14.

Enlarged and heavily reinforced boundary elements, which are often required in regions of high seismic activity, are expensive to build and create architectural challenges. Many reinforced concrete shear wall buildings constructed with uniformly thick panels that are tilted up into position do not contain obtrusive chord elements and have performed satisfactorily during large earthquakes, though they can still contain a large quantity of reinforcement as shown in Fig. 10-15. Boundary zone geometry and reinforcing requirements are determined in Section 21.7.6 of ACI 318-05 in response to compression related to tributary gravity loading plus the added resultant associated with overturning. Since this loading condition is repeatedly applied, confinement of longitudinal bars is critical to good performance when calculated compressive stress exceeds a critical index value.

The continuity of a multistory, unbonded posttensioned precast concrete wall assembly is handled by posttensioning strands between individual panels across the construction joints at floor levels, which also happens to define the system's behavior

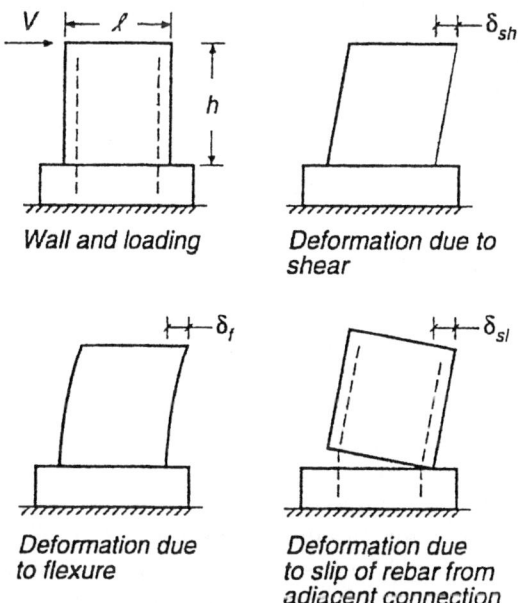

**Figure 10-14** Mechanics of concrete shear wall deformation. (*Courtesy Federal Emergency Management Agency*)

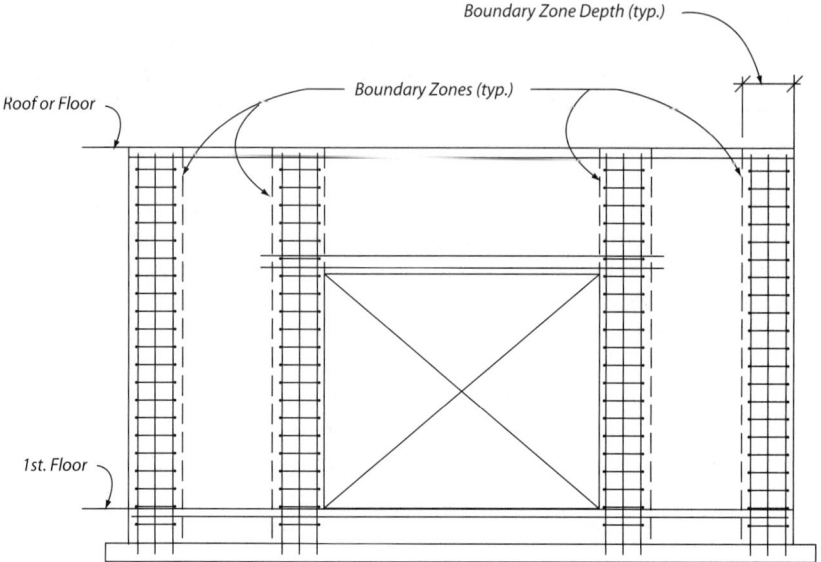

| **Figure 10-15**  Boundary zone reinforcement in concrete shear wall.

(Kurama et al. 1999). In the case of overturning, the force in the strands work together with applied axial load to maintain a panel's stability and close the construction joint gap which occurs from this action. However, there is not a mechanism present to reverse the effect of panels slipping along this joint and special consideration is necessary to control the magnitude, including such remedies as lengthening the wall, increasing the area and initial stress of the posttensioning steel, and laying out the framing system in such a way as to increase gravity loads applied to the wall.

### 10.5.3  Horizontal Diaphragms

Concrete horizontal diaphragms, whether plain or reinforced, carry both in-plane and out-of-plane loads as a plate element, subject to the load combinations of ASCE/SEI 7-05. Square horizontally placed plates with simply supported edges subjected to out-of-plane loading uniformly distributed over the top surface have been shown to exhibit cracking to both faces. Cracks first occur on the bottom, being subjected to flexural tension, then diagonally from each corner toward the middle of the plate. The concrete surface above these cracks can be expected to fail by crushing due to an opposing compressive component. When a restraining effect is applied to resist uplift or rotation at the corners (a semirigid boundary condition), the upper surface has been shown in tests to crack near the corners in a direction perpendicular to the plate's diagonal axis (Aghayere and MacGregor 1990). Figure 10-16 shows a series of concrete floor slabs that must have rigid boundary conditions, with support being traced completely to the foundation, in order to safely support code-required loads.

**Figure 10-16** Concrete cantilevered slab elements may be designed as special plates using charts or simple computer analysis methods.

A typical moment-interaction diagram for a reinforced concrete column indicates that a moderate amount of axial force increases the moment capacity, but testing has shown that this relationship also holds true for flat plates, where inplane loading leads to increased capacity and flexural rigidity, though it depends greatly on shape and slenderness (Ghoneim and MacGregor 1994). Plates with large out-of-plane deflections tend to resist loading by creating a tension field near the middle, held stable by a compressive field around the perimeter, resulting in increased capacity due to combined bending and tensile membrane actions.

Design and detailing of concrete diaphragms in resisting in-plane lateral forces in regions of moderate to high seismic activity is described and regulated by Chapter 21 of ACI 318-05. In regions of low seismic activity, or where wind forces govern element design, the provisions of Chapters 1 through 18 and 22 are intended to provide sufficient strength and ductility without relying on the detailed terms of Chapter 21.

### 10.5.4 Shell-Type Structures

Many different geometrical structures can be created using a thin concrete slab, barrel-type shells or arches, catenary-type membranes, folded plates, and other arrangements that provide three-dimensional support of load. Elastic behavior is commonly assumed to determine stress, displacement, and for analysis of stability, and any method of analysis used must satisfy compatibility of strains in the resulting shape (ACI 334.1R-92, Section 2.2). Milo Ketchum, the structural engineer who served as editor of *Structural Engineering Practice* magazine from 1982 to 1984 and wrote *Handbook of Standard Structural Details for Buildings* (Prentice-Hall, 1956), was arguably one of the greatest leaders in the design of shell structures as he continually presented seminars across the nation on their design and construction (Mr. Ketchum passed away in 1999). Shell structures truly provide an opportunity for an engineer to think outside the box and exercise the ability to visualize load paths and structural relationships, as is evidenced by advice offered by Mr. Ketchum in his writings and lectures:

1. Shell structures can usually be understood as a set of beams, arches, and catenaries and can be analyzed as such.
2. For any shell structure, there will be a simple method of analysis that can be used to check a more precise analysis.
3. Stiffest path concepts are useful in understanding shell structures.
4. Do not throw away all intuition when designing shell structures.
5. In case of doubt, reinforce. Shell structures are very complex and carry forces by many paths.
6. Shell structures get their strength by shape and not by high strength of materials.

Shells represent a true harmony of disciplines, as a project architect must understand the potential for great designs, yet also realize the practical limitations of the resulting structural systems. An engineer, in turn, must realize the important contribution a contractor brings to the phase of erection, dictation of construction methods, and insight of economics. Problems that may arise during design, bidding, or construction phases will best be solved by a team approach, since everyone has a stake in the final product.

Membrane forces in a shell structure develop in response to deformation under load and those which are tensile act to return the shell back to its original shape, thus enabling the system to carry greater load than anticipated by simple buckling theory. If both principle membrane forces are compressive in nature, however, they increase deflection and negatively influence stability to a certain degree. In this case of dual compression of principle membrane forces, there is poor correlation between theory and experimental results (ACI 334.1R-92, Commentary, Chapter 5).

## 10.6 CONSTRUCTION

When a concrete mix is placed, it must be properly consolidated to assure uniform distribution, removal of entrapped air, and adequate coverage of steel reinforcement. It is done by vibration for most work, but can be suitably accomplished for some simple work through the external tamping of forms, and is especially

**Figure 10-17** Close placement of concrete shear wall jamb reinforcement calls for careful consolidation measures.

important under congested reinforcing conditions (see Fig. 10-17). Slabs are commonly vibrated by surface equipment applying static pressure or centrifugation, whereas concrete shapes and elements are worked through by vibrating inserts with a range of power capabilities depending on the work involved. Vibration by inserting wands or other equipment obviously becomes a difficult chore if reinforcement and other inserts or embedments are heavily congested, which call for an easily flowing or self-consolidating concrete. A self-consolidating concrete mix is possible through the use of increased fine content, admixtures to increase viscosity, and a superplasticiser.

Finish work involves the smoothing, compacting, or treating the surface of fresh concrete to produce a desired appearance by use of a trowel, float, or other equipment such as is shown in Fig. 10-18. *Screeding* is the act of striking off concrete lying above the desired plane and can be accomplished by use of a wood or metal straight-edge or by mechanical means, usually performed immediately after placement of

I **Figure 10-18** Concrete finishing equipment.

fresh concrete. *Floating* is the act of consolidating and compacting an unfinished concrete surface in preparation for final finishing operations and is done in two stages: Initial floating (bull floating) is done right after screeding, prior to the appearance of excess moisture or bleed water on the surface. After waiting for the concrete to stiffen a bit, or dewatering has been employed, secondary floating is then performed after most of the bleed water has evaporated, using power trowels with special shoes attached. Concrete edges are then straightened or formed, and the operation is followed by troweling in such a way that a relatively hardened surface results. If noted on the project drawings, contraction joints are then cut into flat surfaces and coatings are applied to complete the finishing process.

After finishing operations are completed, hardened concrete must be properly cured, which is a process intended to slow the loss of moisture so that the process of cement hydration may be as complete as possible prior to the imposition of load or final surfacing materials. *Curing* is done either by continual application of moisture by sprinkling or ponding, by wetted coverings such as burlap, by moisture-retaining coverings, or by the use of special chemical compounds. The duration of curing operations will vary according to method used, ambient temperature and humidity, type of cement, and whether admixtures have been used.

On-site precast two-way floor slab systems may be installed by use of the *lift slab* method, which can be hazardous to workers and the structure itself if not properly done by an experienced crew. A ground level slab is placed first and also serves as the casting bed for successive floors, separated by a membrane or an appropriate debonding agent. All other slabs are cast with steel collars providing clearance around columns to allow lifting by jacks that have been secured to the tops of these columns, connected to collars with threaded rods or other means. Jacks must be triggered in unison to avoid imbalance of the heavy plates until they are secured at target floor elevations.

Adequate performance of a concrete slab-on-grade depends heavily on how the subgrade is prepared. The subgrade must be uniformly compacted to proper specifications to remove incident soft spots. It is then overlain by a granular base to better control the slab's thickness and retain initial moisture to aid in curing. If a vapor barrier or retarder is provided, which serves an important function where slabs are to be surfaced with moisture-sensitive products, an additional granular subbase may also be placed. Steel reinforcement (properly set on chairs or dobies) or proprietary fibers (added to the concrete mix) are often placed within nonstructural slabs to control the width of shrinkage cracks.

### 10.6.1  Risks in Design and During Service

Simplifying assumptions are made in the design of concrete members, as exact distribution of flexural and shear stresses within a member cannot be known for certain. These assumptions are determined through a multitude of tests, sound principles of engineering mechanics, and some statistical analyses in order to justify a level of design peace of mind.

The actual compressive strength a concrete mix will produce is not exactly predictable and usually shows considerable variability between specimens. In order to guard against this being a serious factor in a constructed facility, a concrete mix is to be designed with an average compressive strength that exceeds requirements. Adequate control of strength can only be achieved by sampling and comparing values with a standard deviation, calculated by first computing the average strength of a specified number of compression test results, taking the absolute value of the difference between each record and the calculated average, then obtaining the square root of the average of the squares of the deviations. Small standard deviation indicates better strength consistency in a mix.

An *alkali-silica reaction* occurs between amorphous silica or cryptocrystalline quartz present in sand used in a concrete mix and in the alkali component of the cement, by which a gel-type material forms that absorbs water, causing excessive expansion within the concrete and resulting in cracks. This chemical reaction has been shown to affect the tensile strength of concrete, leading to a significantly reduced bearing capacity when load is applied eccentrically (Ahmed, Burley, and Rigden 1999).

Steel corrosion is an electrochemical process that requires an anode (point of oxidation, where ferrous ions are produced) and a cathode (point of reduction, where hydroxyl ions are freed), by which rust is produced when ferrous ions combine with oxygen or the hydroxyl ions. Corrosion can result if the concentration of oxygen, water, or chloride differs at various locations along a steel bar and sometimes by the coupling of different metals. Cracks that have occurred as a result of internal steel corrosion will be directed along the length of reinforcing bars, allowing for further moisture and oxygen infiltration and thus greater damage. The load carrying capacity of flexural members will be reduced in proportion to the reduction of steel cross-sectional area due to corrosion (Maaddawy, Soudki, and Topper 2005). Under ordinary conditions, the alkaline environment of surrounding Portland cement paste provides adequate protection of embedded steel, though quality depends on the amount of cover, construction detailing, degree of exposure to chlorides from concrete-making components, and the environment.

## 10.7 QUALITY CONTROL

Familiarity with the responsibilities of all trades involved in meeting the concrete and assorted accessory stipulations of the contract is the first step toward establishing any form of quality control system. Notes and details on a set of construction documents are to be created in conformance with current industry standards, using familiar and consistent terminology in order to avoid confusion. Concrete work for any project typically involves a primary subcontractor, a steel reinforcing fabricator and detailer, a concrete supplier who may also serve as the batcher and mix designer, prestressed element fabricators and their own support staff, those who place and secure formwork, those who place and secure reinforcing bars, and those who place and finish the concrete itself.

Chapter 5 of ACI 318-05 defines measures to assure a level of quality in construction and materials, but model building and bridge codes impose further regulation depending on type of construction and importance of the structure. ASCE/SEI 7-05 contains quality assurance provisions in Appendix 11A, which describe the generation of a quality assurance plan, special inspection and testing of materials and installations, and observations for more detailed work. Chapter 17 of the 2006 IBC contains detailed testing and inspection requirements, paralleled to a degree by Chapter 40 of NFPA 5000. Confirmation of material and placement quality includes slump testing, sampling and breaking of concrete cores or cylinders, and possibly a chemical analysis of aggregate or other constituent.

Both design drawings and a subcontractor's placing drawings should clearly identify the condition and placement of reinforcing bars, including size and location, anchorage length, location and length of lap splices, and type and location of welded splices or mechanical connections. A fabricator of reinforcing bars is responsible for supplying all materials shown within the project documents, including the proper quantity of the correct sizes that are already cut and bent accordingly, unless field bending is allowed by the design professional and so stipulated within the contract.

### 10.7.1 Crack Control

Concrete cracks when internal tensile stress exceeds the modulus of rupture through movement by drying shrinkage, plastic shrinkage, temperature changes, settlement, chemical reactions, frost action, or by the imposition of load. *Drying shrinkage* occurs as the cement paste loses moisture and potential change in volume is somewhat reduced by the aggregate. Cracks form if something restrains the section from movement. In massive concrete sections, tensile stress develops due to differential shrinkage between the surface and interior. *Plastic shrinkage* cracks occur when temperature, humidity, or wind movement cause moisture to evaporate from the surface of freshly placed concrete too quickly and restraint that is provided by concrete below the surface layer causes tensile stresses to develop.

Cracks can either indicate structural problems or merely affect appearance, but they also provide an opening for moisture and oxygen intrusion that can corrode reinforcing steel. Flexural cracks are mainly vertical (perpendicular to lines of principal stress) in the middle third of a simply supported beam, or over interior supports for continuous beams, and begin developing at about 50% of the failure load (Nawy 2003, p. 229). As external loading increases, cracking extends toward and beyond the neutral axis with pronounced member deflection. Depending on a member's span-to-depth ratio, flexural cracks can turn into diagonal tension cracks as the effect of shear begins to dominate behavior, followed by a loss of bond between reinforcing steel and the surrounding concrete near the support, resulting in sudden failure when cracks join the zone of crushed concrete in the top compression fibers.

## 10.8 CODES AND STANDARDS

Basic requirements for construction, design, and quality of materials for concrete construction are found in Chapter 19 of the 2006 IBC or Chapter 41 of NFPA 5000 with different levels of description. Both model codes, however, make reference to *Building Code Requirements for Structural Concrete (ACI 318-05)* for detailed design, construction, and quality assurance guidelines. It is formally adopted as a standard of ACI and written in such a way that it may be given the same legal status as a model building code when referenced therein. Section 1908 of the 2006 IBC contains modifications to ACI 318-05, many of which are also listed in Chapter 14.2 of ASCE/SEI 7-05 related to seismic design and detailing, that govern over the original text. In terms of hierarchy, the model codes take precedence over applicable sections of ASCE/SEI 7-05, which takes a position in front of ACI 318-05, followed by other standards or design manuals.

Provisions for design and detailing found within ACI 318-05 are intended to improve the redundancy and ductility of structural systems in an effort to resist collapse or failure and to cause reasonably expected behavior patterns when subjected to load. Design conditions are also to include effects of prestress forces, crane loads, vibration and impact, shrinkage and temperature changes, creep, expansion of shrinkage-compensating concrete, and unequal settlement of supports (ACI 318-05, Section 8.2.4) This is certainly not an exhaustive list, but serves as an important reminder that a structure may be subjected to a wide variety of loading conditions and not just the familiar dead, live, wind, and seismic effects.

# 11 Understanding the Behavior of Masonry Construction

Advancements in the manufacture of materials and development of construction techniques have greatly increased the scope of uses for masonry. A typical single-wythe concrete masonry wall is shown in Fig. 11-1, having been laid up around steel reinforcing to a prescribed height and awaits either further placement of block or grout to be pumped into the cavity. Design of a completed assemblage of masonry, cementitious mortar and grout, and reinforcing steel proceeds under very similar methods that are used for reinforced concrete, but equations and assumptions are expanded to include the appropriate behavior characteristics considering how each portion functions together as a whole.

It is not the intention of this chapter to provide complete guidance on the design of masonry, as there are already many fine references available for that purpose. Rather, background and general information is summarized from a variety of sources on the origin of materials used for constructed masonry systems (mortar, grout, masonry units, reinforcing steel), as well as expected behavior of elements and assemblies. Because masonry construction can be a complicated engineering endeavor, knowledge of how elements function together is a big step toward discovering safe problem solutions for engineers of all experience levels.

## 11.1 COMMON TERMS AND DEFINITIONS

*AAC Masonry*: A masonry assembly made of autoclaved aerated concrete (AAC) units, manufactured without internal reinforcement, and bonded together with thin or thick-bed mortar. The product is based on calcium silicate hydrates,

**I Figure 11-1** Concrete masonry unit wall construction.

attaining a low density through the use of an added agent that creates macroscopic voids according to ASTM C1386. This material was developed and patented by a Swedish architect in 1924 who was searching for a material with properties similar to wood, but without inherent problems of combustibility, decay, and boring insect damage. These porous blocks provide excellent fire resistance, thermal and sound insulation, and weigh about 50% less than a standard concrete masonry unit.

*Bed Joint*: Horizontal layer of mortar in masonry construction.

*Bond Beam*: A continuous row of grouted, reinforced masonry units. A bond beam that spans across an opening is called a *lintel*.

*Bonder/Header*: A unit of masonry that ties two or more wythes of a wall assembly together by overlapping.

*Cleanout*: Opening at the bottom of a masonry cavity into which grout will be poured, created by the removal of a single brick or face shell of a hollow unit, in order to allow for the removal of debris which fell to the bottom of the assembly from mortaring units into place and setting reinforcing steel, typically occurring at the same spacing as vertical reinforcement.

*Collar Joint*: Space between wythes of masonry, or between a veneer and backup construction, that may be filled with mortar or grout.

*Grout Lift*: Height of grout placed in a single continuous operation prior to consolidation.

*Grout Pour*: Total height of masonry to be grouted prior to erection of additional masonry.

*Head Joint*: Vertical layer of mortar placed when units are stacked together.

*Pilaster*: Column built integrally with a wall and may project beyond the wall's thickness.

*Shell*: Outer portion of a hollow masonry unit.

*Web*: Inner solid portion of a hollow masonry unit.

*Wythe*: Continuous vertical section of a wall, measuring one masonry unit in width.

## 11.2   ELEMENTS OF MASONRY ASSEMBLIES

A masonry assembly is truly a curious thing, in that part of a completed building, such as a wall or beam, is itself composed of individual pieces: Masonry units, grout, mortar and steel reinforcement. The behavior of the whole must be understood in light of individual component properties. Common terminology for a masonry wall is illustrated in Fig. 11-2.

### 11.2.1   Masonry Units

Individual units come in the form of bricks or blocks and may be solid or hollow, stacked in different patterns depending on code requirements and intended aesthetics of the structure. Intuition, judgment, and experience directed early builders to rely on mass and special arrangements for stability, such as cellular-type construction, by which masonry walls held each other in place for most loading conditions. Today, masonry units are formed of clay or shale, concrete of different weights, calcium silicate, stone, and glass.

**Clay or Shale**   Raw material for making clay units is mined directly from different layers of earth, of both old and more recent geological formations. Lumps and stones are crushed within a mixture of clays and the final material is ground up prior to storage. Over the years, technology has made it possible to monitor variability in these materials and has strengthened the ability to combine elements from different sites for additional quality control.

In the manufacturing process of better clays, water is added to clay or shale materials in order to improve plasticity, typically to a moisture content of 12 to 15% of the beginning dry weight. Small sections are extruded from a shredder plate as the plastic material is pushed through a machine, after which a vacuum chamber removes the entrapped air. The final journey of the undefined mass occurs through a cylinder that reduces in size, adding pressure as the clay exits

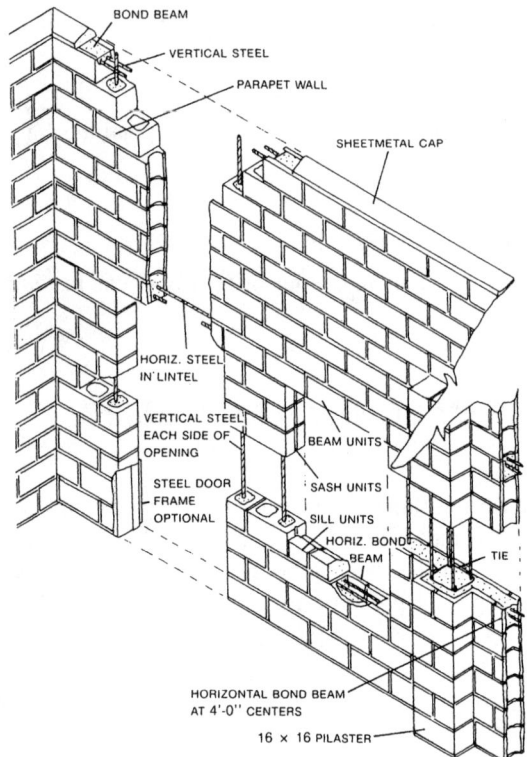

**Figure 11-2** Terminology of masonry wall construction elements. (*Courtesy Masonry Institute of America*)

a rectangular aperture or die. Holes are formed and colors or other finishing is incorporated into the rectangular mass before it enters a wire cutter which slices the extrusion into individual bricks. Modern kilns are fired continuously at 194 to 212°F (90 to 100°C) to completely dry out the new specimens, which will expand slightly upon exiting due to absorption of moisture from the air. Clay will shrink during both firing and drying periods and it is next to impossible to predict the exact finished dimensions, though variation is not large enough to be of structural concern. Cooling of the units must be carefully controlled in order to avoid cracks and checks and is usually maintained for 48 hours in tunnel kilns or 48 to 72 hours in periodic kilns.

ASTM C62 (solid) and C652 (hollow) regulate the manufacture of structural and nonstructural clay or shale bricks based on desired compressive strength, absorption characteristics, and resistance to freeze-thaw activity. If surface appearance is a design issue for solid units, the purchaser is to consult ASTM C216 for facing brick. Grades used in construction include SW, which offers a high resistance to freeze-thaw damage and are used in regions where units may be frozen when saturated with water; MW, used in areas where only moderate weathering protection is necessary and when units are not expected to be saturated when freezing occurs; and NW (for solid brick), solely for interior uses as protection against environmental damage is not provided. Grade SW brick is recommended for any region where they will be placed in contact with the ground, in horizontal surfaces, or in any position where they would become permeated with water.

Durability of clay units is directly linked to compressive strength as defined in manufacturing standards. Solid and hollow units of Grade SW are required to have a minimum individual compressive strength of 2500 psi (17.2 MPa), translating to a maximum water absorption rate of 20% after a 5-hour boiling test. For comparison, individual MW bricks have a minimum compressive strength of 2200 psi (15.2 MPa) and a maximum water absorption rate of 25%, revealing less durable characteristics than the aforementioned grade of unit. In order to lend validity to property tests, five specimens are used to obtain an average and provide information on the extent of variability. If a purchaser does not specify the grade of brick desired, the manufacturer is to provide the durability requirements for Grade SW.

Measured compressive strength can be as high as 35,000 psi (244 MPa) depending on the clay itself, firing conditions, coring pattern, and unit geometrical properties (Drysdale, Hamid, and Baker 1999, p. 125). Because actual environmental conditions will extend beyond the limitations of standard laboratory tests, it is important to consider historical behavior of clay units in situ when specifying for purposes of durability. Though relatively stable, clay brick can expand or contract up to 22% more in the vertical direction due to thermal effects.

Clay brick is porous and will absorb moisture from mortar and grout, which can cause problems if the rate of penetration is great enough to create a dry layer of mortar on the brick surface or lower than what is required to draw the mortar into a sufficient bond. Units that do not bond properly with mortar will not achieve desired flexural strength and water tightness will be questionable. Clay brick that exhibits such behavior requires special care, such as the use of compatible, adjusted mortars or prior wetting.

**Concrete** A type of hydraulic cement, aggregate, and the process of hydration with added water define the strength of concrete brick and block. Most contractors associate clay products with the term *brick*, whereas the term *block* is reserved for larger sized concrete elements. These units can be subjected to the same property enhancing admixtures as a standard concrete mix, so great variety is possible. Cementitious material may be Portland cement, blended or ordinary hydraulic cement, hydrated lime, ground granulated blast furnace slag, or a type of pozzolan (i.e., fly ash).

The concrete mixture used to create units has nearly zero slump and is vibrated and pressed into steel molds of the desired shape. Once the concrete has hardened enough to remove molds, shapes are immediately transported to an area for curing which is accomplished by accelerated hydration or steam methods at either atmospheric or higher pressure (autoclaving), each requiring less than a day's time. Special AAC may also be used to form units according to ASTM C1386, for which special design and construction specifications are provided in ACI 530-05.

Concrete masonry unit quality is controlled through the requirements of ASTM C55 (structural or facing brick), C90 (solid or hollow structural block units), and C129 (solid or hollow nonloadbearing block units suitable for exterior applications). The greatest amount of resistance to moisture penetration and severe freeze-thaw activity is offered as Grade N brick, which also maintains a higher level of strength than Grade S brick. Depending on the aggregate used in a mix, concrete block is classified as either lightweight with a density of less than 105 pcf (1680 kgcm), medium weight between 105 and 125 pcf (1680 and 2000 kgcm), or normal weight of greater than 125 pcf (2000 kgcm). The height and length of commonly used concrete blocks in construction are manufactured in nominal increments of 8 in (203.2 mm), whereas the actual dimensions are 3/8 in (9.5 mm) less to account for a standard mortar joint on both sides and the top of each course. Therefore, an 8 × 8 × 16 nominal precision unit actually measures 7 5/8 in × 7 5/8 in × 15 5/8 in (193.7 mm × 193.7 mm × 396.9 mm). Common block terms are indicated in Fig. 11-3.

Like clay brick, the durability of concrete units is intimately related to compressive strength, and behavior of individual shapes will be similar to poured concrete. Shrinkage of units must be controlled, both in the short term during the drying process and for extended periods of use, in order to limit cracking and restrict moisture penetration. Higher compressive strength and lower absorption rates assure a more durable concrete masonry unit. The more dense the aggregate, the less the final product will shrink, although the amount is highly variable.

A typical concrete masonry element stress-strain curve is shown in Fig. 11-4. It is evident that nonlinearity under applied load will be displayed at 35 to 50% of unit strength, partially related to creep which continues in proportion to applied stress. After about one year of constant stress, most of the creep will have occurred.

**Calcium Silicate (sand lime)**  Autoclaving is also used to produce calcium silicate masonry units, which consist of sand and hydrated lime pressed tightly together. Lime and silica react with water during this process, creating calcium silicate hydrate, which is the same strength-giving product as in hydrated Portland cement, producing a more fine-grained product than standard concrete brick

Understanding the Behavior of Masonry Construction | 285

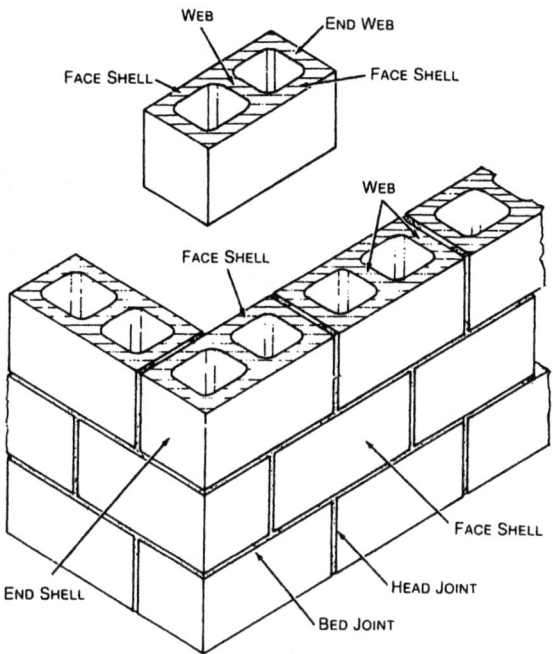

**Figure 11-3** Concrete masonry unit terminology. (*Courtesy Masonry Institute of America*)

(Hendry 1991, p. 28). Limestone is a sedimentary rock consisting mostly of calcium carbonate ($CaCO_3$), typically existing in the form of calcite or dolomite (a carbonate of calcium and magnesium) with a granular texture. These units have primarily been used in European markets, though they can be seen worldwide.

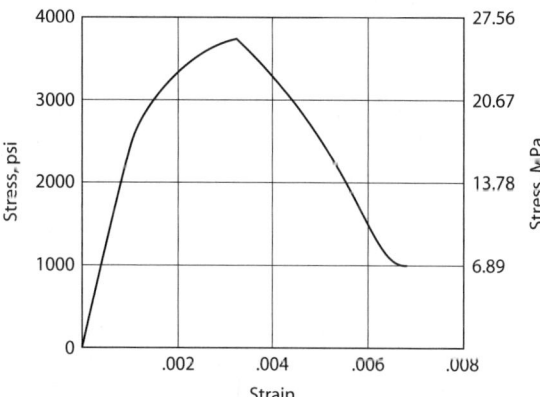

**Figure 11-4** General stress-strain curve for concrete masonry units.

Manufacture is regulated by ASTM C73, which defines two grades of unit. SW bricks are intended for use in environments where freeze-thaw occurs with a high concentration of moisture. Grade MW units may also be subject to freezing temperatures, but must be protected from becoming saturated with water. Measured compressive and tensile strength depend on the amount of binder present, which is the lime, as well as pressure exerted to the mixture and conditions used in autoclaving.

**Glass** Architectural expression by the use of glass block cannot be understated, as the public can enjoy the supernatural activity of looking through a solid wall, though the image on the other side is typically distorted and unclear in order to maintain a certain level of thermal and ultraviolet resistance, as well as privacy (see Fig. 11-5). The main raw materials used to create these special units are silica sand, limestone, and soda ash, melted together and pressed into blocks that most commonly measure 8 in (203.2 mm) square. Units may be completely solid or hollow, the latter being formed of two halves fused together in such a way as to produce a partial vacuum within the void. Underwriters Laboratory is responsible for inspecting and monitoring quality control operations in production of units.

Manufacturer's literature often includes restrictions on use, handling, and placement of these units. Additionally, the International Building Code (IBC) disallows its use in construction of fire or party walls, fire barriers or partitions, or in load-bearing applications (2006 IBC, Section 2110.1.1). Other code provisions related to glass block units and constructed panels are empirical, based on historical performance, older code requirements, and manufacturer's recommendations. In terms of normal forces, applied wind or seismic loads must be transferred to the main structure by use of embedded panel anchors, spaced at a maximum of 16 in (406 mm) on center per Section 7.3.3.1 of ACI 530-05, or confining channels and angles. Lateral supports are required to transfer a minimum service-level lateral force of 200 pounds per lineal foot (2919 N/m) to the building.

**Stone** Building stone is typically selected by an architect according to color or texture, intended to be somewhat variable throughout a single structure to accentuate charm. It is also used in a structural capacity, though practical applications are somewhat limited due to the fact that the assembly must remain unreinforced. A strong bonding of the joints, therefore, becomes a crucial element of a completed member's behavior.

Five main types of stone are regulated by ASTM in terms of absorption, density, measured compressive strength, modulus of rupture, and abrasion resistance. *Marble* for general building and structural purposes is covered by ASTM C503 and includes stone that is cut, sawed, split, or otherwise finished into blocks, slabs, or tiles. Marble is defined as a crystalline rock composed predominately of

**Figure 11-5** Glass masonry walls provide beauty with structural strength, provided they are properly anchored to supporting elements.

calcite and dolomite and further categorized by soundness depending on repairs that may be necessary prior to or during installation.

*Limestone* and *sandstone* are sedimentary by nature, having been formed by mineral deposits from water. *Granite* is known as igneous rock produced by solidification of volcanic material, whereas marble and slate are classified as metamorphic, being produced from igneous or sedimentary rock by pressure, heat, or chemical reaction. ASTM C568 addresses material characteristics, physical requirements, and sampling appropriate to limestone dimension stone, classified into three categories based on density. Granite, the most dense of quarried stones, is covered by ASTM C615 for use as exterior and interior cladding, curbstone or paving, structural components, retaining walls, or monuments. Determination of physical characteristics are widely based on historical performance in terms of strength, abrasion resistance, durability, permanent volume change, thermal expansion, and modulus of elasticity. Quartz-based material, also called sandstone, is defined according to free silica content per ASTM C616, whereas *slate* (ASTM C629) is simply classified according to either exterior or interior use.

Blocks of stone are cut at special quarries directly from a natural rock face and are delivered for final shaping by saws and grinders. The preparation and finishing of stone pieces suitable for construction is very labor intensive and commonly more expensive than common brick-type veneers, but the final product has a somewhat human aspect to it, inviting people to enjoy the building it embraces and sharing a spiritual element that cannot be reproduced with other materials.

Design, selection, and installation of anchors used to secure architectural stone veneers to a solid support system are to be done according to ASTM C1242. Smooth dowels, wire hairpins, or threaded rods are used to anchor cladding panels to precast concrete structural wall elements with the use of adhesives at a minimum depth of two-thirds of the stone's thickness. Support may also be provided by face anchors, blind anchors, or liners (pieces of stone mechanically connected to the back of the veneer stone that transfer loads to anchors).

## 11.2.2 Mortar

Mortar is a cementitious paste that bonds masonry units together and provides a uniform bearing surface, which is important in compensating for manufacturing tolerances of block or brick. The most convenient form is prepackaged in distinct proportions of lime (improves workability, increases water retention) and Portland cement (adds strength), easily mixed with sand and water on the jobsite. Sand used for creating mortar is graded more finely than that used for making concrete. The mason will mix components together, preferably by mechanical means for a period of between 3 and 5 minutes, though this may vary depending on equipment efficiency, place the completed paste onto a board, and spread a $3/8$ in (9.5 mm) thick layer between each successive masonry unit on both the horizontal and vertical faces. The very first layer of mortar upon which the entire wall will be constructed is allowed a range between $1/4$ to $3/4$ in (6.4 to 19.1 mm) in thickness (2006 IBC, Section 2104.1.2.1).

A product called quicklime, $CaO$, is created when limestone is heated in a kiln to drive off water of crystallization and carbon dioxide. Hydrated lime is produced by placing the quicklime into water, which can then be dried and ground into a white powder for use in mortar mixes. ASTM C207 defines four types of hydrated lime: Types S and N contain no air entraining admixtures, but Types SA and NA will typically produce more entrained air in the final mortar mix than is allowed by code. Type N does not control unhydrated oxides, leaving Type S as the only hydrated lime suitable for masonry mortar (Amrhein and Vergun 1994, p. 17).

ASTM C270 describes three different types of mortar—cement-lime, mortar cement, and masonry cement—each of which may be specified in terms of proportion or property. When defined according to property, the standard includes requirements for average 28-day compressive strength, water retention, air content percentages, and aggregate ratio. For all families of mortar, Type M requires a compressive strength of 2500 psi (17.2 MPa), 1800 psi (12.4 MPa) for Type S, 750 psi (5.2 MPa) for Type N, and 350 psi (2.4 MPa) for Type O. It is commonly stated that only the weakest mortar that will perform adequately should be specified for any project, not by virtue of being the strongest (ACI 530.1-05, p. SC-10, Section 2.1).

Workability and rate of mortar hardening play important roles in economic construction of a masonry assembly and the importance of a mason's experience is immediately apparent. Bed joints must be firm enough to support increasing wall weight as units are quickly stacked, without shifting, yet must also be soft enough to allow enough material to squeeze out for a clean, weatherproof finish and assure proper bonding. Slump is expected to be in the range of 5 to 8 in (127 to 203.2 mm). Water may be added at the site to improve workability, but must be carefully controlled. If mortar hardens too quickly, it must be discarded and some delay in the project's completion will be experienced. Unused mortar for typical masonry construction must be discarded within $2^1/_2$ hours after initial mixing, except that which is used for glass block must be used within $1^1/_2$ hours.

Bond of mortar not only depends on its own proportions, but is also influenced by quality of the masonry units in terms of water absorption and surface texture. The mechanics and scientific behavior of mortar bond with masonry units are difficult to describe in absolute terms but in general, mortar with a higher percentage of Portland cement will exhibit stronger bonding tendencies whereas that with a higher percentage of retained air will be weaker. Initial setting of the units in paste is also delicate as primitive early stages of bonding cannot be disrupted by a mason's excessive movement.

When mortar is first placed, the masonry unit below absorbs some water from the mix and it is important that sufficient moisture be retained in order to create a solid bond through the same contact between units that share this joint. Bond can also be disrupted by layers of dust, loose dirt, mud, or other deleterious material that stands on the surface of units to be joined together. Variation in bond can occur as a result of raked back bed or empty head joints.

### 11.2.3 Grout

After masonry units have been mortared together to a specified height, highly flowable grout is placed in the cavity between wythes or within hollow block according to structural design or purposes of durability. Grout is mixed according to defined proportions of cementitious materials, aggregate, and water that result in a high slump paste of 8 to 11 in (203.2 to 279.4 mm). This consistency is important since the space into which it must flow is usually quite narrow and water is quickly absorbed into the units during placement. As is the case with mortar, this absorption is essential for creating a sound bond between materials. Grout is not only necessary for binding reinforcing steel and masonry together into a system, but it also adds sound transmission, fire resistance, improved energy storage, and better resistance to overturning.

Before grout can be placed, the masonry cavity must be cleared of debris and solid mortar that has squeezed out into the space between wythes or inside hollow block. Excess material is knocked to the bottom of the constructed assembly, often by use of a reinforcing bar from above, and removed from clean outs (see Fig. 11-6). After they have been inspected, holes are covered by mortaring brick or a face shell back into place. Depending on design requirements, grout either completely fills the void or is restricted to particular cells in a vertical or horizontal pattern by installing dams and screens during erection. Since it is a form of concrete placed around embedded items, grout must be properly consolidated within its tight space by the use of paddles for small lifts or thin vibrators for larger lifts.

Aggregate used in a mixture are classified as fine or coarse, prescribed in ASTM C476, and typically depends on the space available for placement. Fine grout is

**Figure 11-6** Clean out in concrete masonry wall construction for inspection and removal of mortar and other debris.

used in very small spaces and will typically contain mortar sand, but no pea gravel. Coarse grout is added to spaces of at least $1^{1}/_{2}$ in (38.1 mm) in horizontal width, or where minimum block cell dimensions measure $1^{1}/_{2}$ in $\times$ 3 in (38.1 mm $\times$ 76.2 mm). Approved aggregates are limited to a maximum size of $^{3}/_{8}$ in (9.5 mm), but a coarse grout mixture may use gravel up to $^{3}/_{4}$ in (19.1 mm) in size with the appropriate space. One of the advantages to using larger aggregate is a reduction of Portland cement for equivalent strength, leading to less shrinkage and a more manageable slump.

Grout is delivered to larger projects from a concrete batch plant that has prepared the mix based on a design meeting tabulated proportions or special combinations specified by the project engineer, therefore size and condition of aggregate is controlled far ahead of construction. The four most common admixtures for grout include a shrinkage compensator, used to counteract the loss of water and shrinkage of cement by creating expansive gases; plasticizers, used to obtain a high slump without adding extra water; cement replacements, such as fly ash; and accelerators, used primarily in cold climates to reduce the amount of time needed to protect a wall against freezing by decreasing set time and speeding up strength gain.

### 11.2.4 Reinforcement

A masonry assembly may be strengthened by the addition of reinforcement in the form of smooth steel wire, commonly #9 in size [0.144 in (3.66 mm) diameter], embedded in mortar along bed joints or by metal ties that hold multiple wythes together (see Fig. 11-7). This form of joint reinforcement is effective in otherwise unreinforced masonry construction for controlling cracks and increasing flexural strength in the horizontal direction, though with unstable hysteretic behavior due to a lack of ductility at higher lateral loads (Drysdale, Hamid, and Baker 1999, p. 323). Wire reinforcement must be stainless steel, hot-dipped galvanized, or epoxy coated when used in masonry exposed to earth or weather and be protected by at least $^{5}/_{8}$ in (15.9 mm) of mortar cover. All other wire still requires mill or hot-dipped galvanizing or stainless composition (ACI 530-05, Section 1.13.4).

Wall ties are used to connect two or more wythes of brick together or to attach masonry veneer to a backing wall. Due to environmental conditions, wythes located on a building's exterior experience movement at a greater rate than inner layers, which requires ties to be sufficiently flexible to accommodate this differential movement. Ties are therefore made into sections having a low flexural stiffness, yet are strong enough to transfer lateral forces between wall layers. During construction of a double-wythe wall, ties must resist the tendency of wythes to separate if the cavity is to be filled with grout. Ties are placed within mortar joints and are commonly produced in the form of corrugated strips for anchoring veneer to wood backing, as Z or rectangular wires to anchor veneer to a structural masonry wall, or

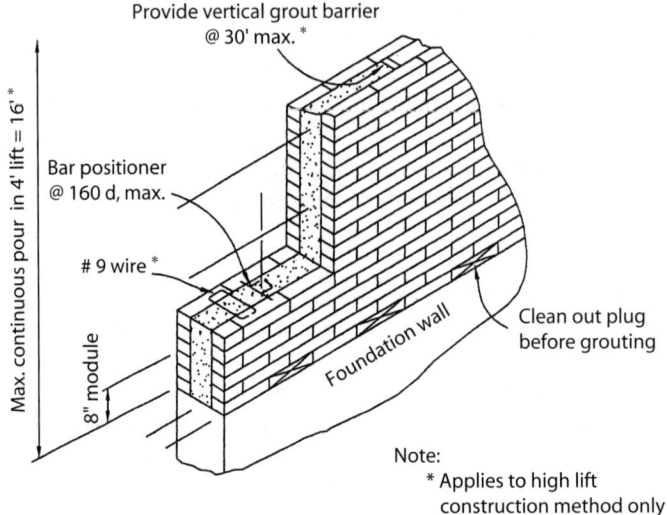

**Figure 11-7** Metal ties for brick masonry construction. (*Courtesy U.S. Army Corps of Engineers*)

as ladder or truss joint reinforcement that is commonly used to tie separate wythes together.

Steel reinforcing bars, which have already been discussed to a certain degree in Chapter 10, are included in grout spaces for the purpose of resisting tensile and shear forces, increasing axial load carrying capacity, reducing the occurrence and magnitude of thermal cracks, and providing ductility when subjected to seismic forces. These are the same bars used in concrete construction, but ACI 530-05 restricts available sizes to #11 (M #36) and smaller or one-half the least clear dimension of a masonry unit's cell or an assembly's collar joint. Corrosion of bars is prevented by good grout consolidation and minimum coverage of 2 in (50.8 mm) for bars larger than #5 (M #16) or $1^1/_2$ in (38.1 mm) for #5 (M #16) and smaller when the completed assembly will be exposed to earth or weather.

Masonry also lends itself to prestressing using the same steel and methods as in concrete construction. Precompression delays crack initiation, thereby enhancing flexural capacity as well as serviceability and water-tightness. Bars or strands in wall construction are placed within open cells or cavities, anchored to the foundation, and stressed through a properly sized bearing plate and mechanical connector on top. Once tensioning has been completed, steel may then be surrounded by grout or left uncovered. As with concrete, design of the entire assembly needs to account for losses in prestressing force.

## 11.3 BEHAVIOR OF MASONRY ASSEMBLIES

A designer is permitted to use Allowable Stress Design (ASD) per ACI 530-05, Chapter 2, the Strength Design (SD) method of ACI 530-05, Chapter 3, or an empirical method described in ACI 530-05, Chapter 5 [not allowed in Seismic Design Categories (SDC) D, E, or F] for all elements present in a masonry structure. Different block sizes may be placed together to create specified dimensions as shown in Fig. 11-8.

For design purposes, compressive strength of the entire masonry assembly, $f'_m$, is determined either by tabulated values or through testing of properly constructed prisms (2006 IBC, Section 2105.3; NFPA 5000, Section 40.3.9). Tables 2105.2.2.1.1 and 2105.2.2.1.2 of the 2006 IBC specify net area compressive strengths assumed for different types of masonry and mortar in both clay and concrete. For example, when clay units having a compressive strength of 3350 psi (23.1 MPa) are used with Type M or S mortar, the completed assembly may be assumed to attain a minimum net area compressive strength of 1500 psi (10.3 MPa). When concrete masonry units having a compressive strength of 1900 psi (13.1 MPa) are mortared together using Type M or S, design of the final element may proceed using a net area compressive strength of 1500 psi (10.3 MPa). This method is quick and simple, whereas the taking and testing of prisms present expensive difficulties and it is recommended that they

**Figure 11-8** Masonry walls may be constructed using blocks of differing widths for structural or architectural purposes.

only be required when absolutely necessary, such as for justification of questionable construction or under special cases of design.

As with concrete, linear elastic behavior may be assumed for a completed masonry system before cracking occurs. Steel reinforcing therefore does not influence shear strength of the members, whereas once cracking has occurred and stresses are redistributed, steel plays a significant part in the overall element shear stiffness.

## 11.3.1 Beams and Columns

Masonry lintels are not required to carry the total load distributed above an opening, as arching action tends to direct loading that occurs above a triangle's apex (see Fig. 11-9) away from the opening. Any load, including the weight of the masonry itself, that is located above the triangular area shown does not need to be included in the design equations. One cautious observation, however, is the fact that there must be sufficient masonry along each side of the opening to counteract the arch's thrust, as well as above the apex to support the arch's compressive force. Concentrated forces are assumed to be distributed along a wall length of the bearing area plus $4t$, where $t =$ wall thickness (ACI 530-05, Section 2.1.9.1), and the portion along this distribution which falls within the triangular area above the lintel is to be added to the loading pattern. This is also demonstrated in Fig. 11-9. Standard lintels are limited in length by threat of cracking

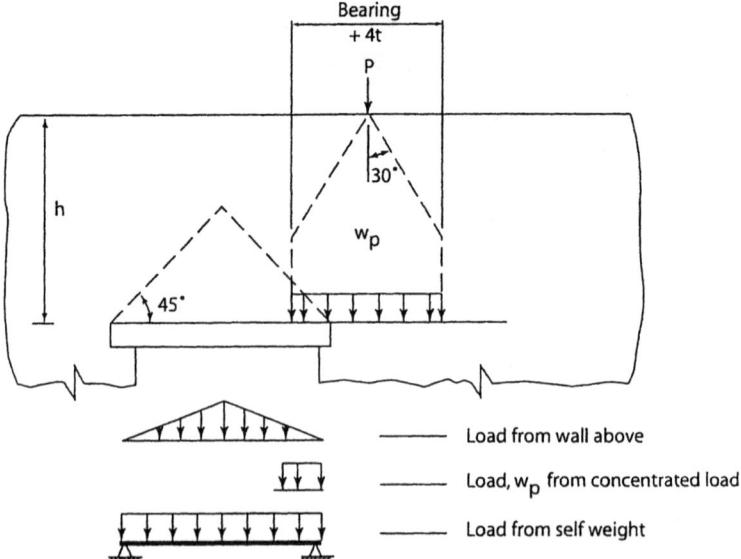

**Figure 11-9** Effect of load distribution and combination over masonry lintel.

and by practical constraints of placing reinforcement within a confined unit space, but the maximum span can be increased by prestressing or by incorporating a steel beam within the wall space to carry imposed loads.

Because the modulus of rupture for masonry is small compared to measured compressive stress, flexural tensile forces are only nominally resisted by masonry components (unit, mortar and grout) and steel must be introduced into the system to increase flexural strength. Elastic behavior will continue until masonry compression strain exceeds 0.0025 (concrete units) or 0.0035 (clay units) in over-reinforced members or tension steel yields in underreinforced assemblies. Yielding of tensile steel is the preferred failure mode since it produces large deformation and opening of cracks across the member instead of the brittle, sudden failure of a compression-controlled section. For both ASD and SD, the maximum percentage of longitudinal reinforcement in a member is therefore limited in order to assure the masonry will not crush before the steel yields for the purpose of maintaining a ductile response to loading (ACI 530-05, Sections 2.3.2.1 and 3.3.3.5).

The main function of beams is to distribute forces from the roof and floors across an opening and to tie the lateral force resisting elements together. Depending on anchorage and construction, they may also resist pressure applied normal to the face. One of the most basic assumptions about the behavior of reinforced masonry is that there is strain continuity between reinforcement, grout, and masonry such that all applied loads are resisted by composite action (ACI 530-05, Section 3.3.2). Additional assumptions used to simplify the design of masonry beams include:

1. Internal stress at any location along the length is in equilibrium with the effects of external load.
2. Strain in masonry due to flexural compression and that developed in reinforcing steel by flexural tension is directly proportional to computed distance from the neutral axis.
3. After cracking, flexural tension will be entirely resisted by reinforcement. There is still some contribution by masonry to full tensile strength, but it is ignored at this level of stress.
4. Linear elastic behavior may be assumed with the use of ASD methods.

Deep beams, those with a span to depth ratio less than about 2.5 for continuous beams and 1.5 for simply supported members, display a somewhat different behavior than the assumptions listed above. Plane sections do not remain plane, resulting in a nonuniform distribution of internal stress about the neutral axis and anchorage of the longitudinal reinforcement plays an important role in flexural strength. Additional reinforcing should be provided in both directions to assist in crack control and shear capacity.

Shear stress occurs in a beam along the neutral axis and is displayed in a diagonal pattern, either toothed along mortar joints or directly through weaker masonry units, being evident near the supports where shear stress is highest (see Fig. 11-10). If not properly reinforced by a sufficient amount of stirrups, shear tension cracks can propagate up into the compression region of a beam and cause sudden failure before the flexural capacity is realized.

As one would expect, the presence of confining reinforcement greatly improves the load carrying capability of masonry columns. Test results have shown that embedment of ties within the grout space caused a column's failure mode to change from vertical splitting of the entire section to spalling of face shells. Additional experiments demonstrate elastic behavior of reinforced block masonry columns under concentric axial loading up to about 75% of ultimate with similar results for eccentrically loaded brick columns (Drysdale, Hamid, and Baker 1999, p. 406). The axial load capacity of masonry columns is determined in a manner similar to that of reinforced concrete, using relationships of equilibrium and strain compatibility, regardless of whether loads are applied along the member's neutral axis or in an eccentric fashion. Sections 3.2.3 and 3.3.4.1.1 of ACI 530-05 set maximum nominal axial capacity of masonry columns (with or without compression reinforcement) at 80% of the theoretical value under concentric loading in order to account for section variability and inevitable eccentricity of load path.

Masonry columns are regulated differently depending on whether they are fully or partially embedded into a wall (pilaster), they have a width-to-thickness ratio (b/t) less then or equal to 3 (column), or their height is equal to or less than 5t (pier). Required prescriptions for columns do not necessarily apply to piers unless specifically noted in the reviewed code section. If specific provisions are met, pilasters may be considered to include part of the wall on each side as flanges for design purposes (ACI 530-05, Section 2.1.7.1).

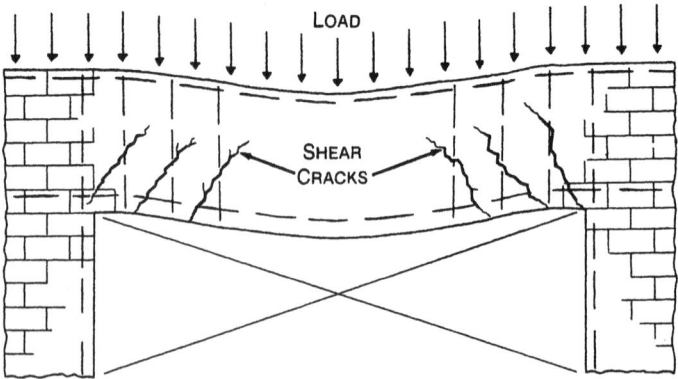

**Figure 11-10** Shear cracking pattern in masonry lintel from in-plane forces. (*Courtesy Masonry Institute of America*)

## 11.3.2 Walls

Out-of-plane loading applied to reinforced masonry walls is resisted by both vertical and horizontal action, as evidenced by cracking patterns shown in Fig. 11-11, especially if horizontal span between cross walls or load resisting frames is not more than 3 times the wall's height. The orthotropic stiffness and strength characteristics of masonry impose limitations on simplified methods of elastic plate analysis and it would be more precise to model openings, geometric relationships, and boundary conditions with finite elements to determine the reaction pattern from applied loads. However, a simple and fairly common, albeit conservative, method for designing reinforced masonry walls with openings is to divide the panel into strips that run both vertically and horizontally to transfer loads around the opening, tracing loads and reactions across each strip, and placing reinforcement according to need.

When load is applied normal to a wall's surface, the masonry unit itself is subjected to compression and grout may not actually be stressed at all from flexural. It is important to remember that the weight of a wall and carried vertical loads provide great assistance in maintaining stability when subjected to lateral forces, especially in the case of unreinforced masonry assemblies, in addition to delaying the onset of masonry cracks due to a precompressive effect. Amrhein and Vergun (1994, p. 65) comments that the inherent flexibility of masonry walls perpendicular to their plane allows them to tolerate a significant amount of out-of-plane bending

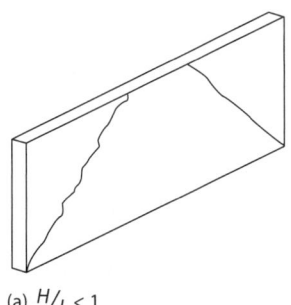

**Figure 11-11** Cracking patterns that may be seen in masonry walls subjected to out-of-plane pressure.

(a) $H/L < 1$

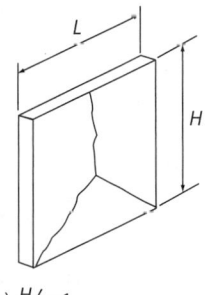

(b) $H/L \approx 1$

and translation due to movement of a horizontal diaphragm without negatively affecting their shear resisting capacity in the parallel-to-wall direction.

Because of the need for structural elements to perform in a ductile fashion under seismic loading, it is important to identify brittle behavior patterns and detail the construction in a way that will effectively produce the desired result. As an example of appropriate detailing, a higher capacity for energy absorption is available by staggering vertical steel about the wall's centerline.

Masonry shear walls resist in-plane lateral forces through modes of flexure as well as shear. Failure in shear generally dominates capacity and is evidenced by diagonal cracking, which is brittle by nature. A ductile, flexural-type of failure, on the other hand, occurs as a result of yielding the reinforcing steel and is the preferred response under strong seismic loading. As a result, researchers have suggested that an ultimate strength design approach be used for the design of masonry shear walls in order to detail the assembly such that a flexural mode of failure will govern.

Shear walls are designed as cantilevered elements fixed at the base when connected floors or other horizontal diaphragms are not stiff enough to create a coupling effect. Methods of design consistent with an assumption of linear elastic behavior provide acceptable results, whereby the total lateral force is distributed to individual shear walls according to their combined flexural and shear rigidities. These equations were introduced in Chapter 8. Openings modify shear wall stiffness in relation to their placement and size. In general, when the combined area of penetrations is less than about 10% of total wall area, and they are relatively small and distributed in a somewhat regular fashion, rigidity of the entire wall may be calculated as though the openings were not present.

Masonry walls are not homogenous, therefore different patterns of behavior occur in perpendicular directions. Compressive strength is commonly associated as being normal to bed joints, but a system that spans horizontally to resist lateral pressure must rely on strength characteristics parallel to these same joints. It is interesting to note, however, that the anisotropic nature of masonry construction also reveals a higher tensile strength parallel to bed joints of ungrouted assemblies than in a normal direction, almost by a factor of 2. When grout is added to a wall system, strength properties become more uniformly distributed because load paths affected in the weaker directions, normal and diagonal to bed joints, now must pass through the column of grout. Regardless of the presence of grout, masonry walls have exhibited greater horizontal flexural strength than vertical which assures plate-type action for most walls with a reasonable aspect ratio.

### 11.3.3 Frames

Masonry shear walls that contain large central openings will behave as a type of rigid frame, where the critical region for design and attention to detailing is at the beam-column (pier) joint. Rigidity of this joint is what distinguishes this configuration from a simpler pair of coupled shear piers. Detailing of beam reinforcement is intended to cause plastic hinge formation away from the pier joint in order to develop a ductile failure mechanism as opposed to reinforcement bond or masonry diagonal shear failure, both of which exhibit brittle characteristics.

### 11.3.4 Prestressed Assemblies

Pretensioned wall reinforcement, referred to as tendons, is easily placed within a cell of hollow units or in the space between a multiwythe configuration and is stressed after the masonry assembly has gained sufficient strength. Bars, strands, or wires are anchored into the foundation and threaded through a bearing plate at the top which allows stressing operations to take place.

One of the ideas behind prestressed masonry is that a wall will remain uncracked when subjected to service loads, assuring ductile behavior. Even after brittle cracks have formed from a condition of overload, prestressing steel provides the wall with a ductile tensile component, and after load is removed the cracks close up. When cracking does occur, it usually takes the form of spalling due to flexural compression, tensile cracking due to flexure alone, and that due to the combined effects of bending and shear.

As in prestressed concrete, reduction in the initial tension applied to the system is expressed in design as *prestress loss*. It occurs as a result of elastic deformation, slip of tendons at anchorage during jacking, friction, moisture expansion and shrinkage loss, relaxation of steel, creep of masonry, and possible thermal effects. Elastic deformation occurs as an element compresses and shortens along its length when prestress force is released. Strain in the masonry will be compatible with strain in the reinforcing steel. Loss of force due to locking off tendons after the jacks are released is commonly determined using anchorage manufacturer's data since the magnitude is difficult to determine otherwise. Some forms of applying force will not result in loss of force, such as a threaded rod-type system. Concrete masonry will creep to a greater extent than will fired clay units and is time-dependent with constant stress, therefore further prestress losses will be experienced from this characteristic.

A prestressed wall behaves differently at ultimate flexure than a standard reinforced masonry wall because the tendons apply an additional tensile component to the internal resisting moment. Testing has also shown that wall behavior will be affected by the level of prestressing and whether tendons are guided against lateral translation, which can occur when a wall deflects under lateral load. Hendry (1991, p. 130) asserts that the effectiveness of prestressing masonry walls can be assessed by a simple comparison of the ratio of wall section modulus to cross-sectional area: As this ratio increases, the benefits obtained through prestressing (as opposed to using standard bar reinforcement) also increase.

## 11.4 CONSTRUCTION

A mason has several options for completing a vertical assembly, including low-lift and high-lift grouting. Masonry units are set to a height based on the anticipated completion of a day's work or $1^1/_2$ in (38.1 mm) to half a block height below a construction joint that will occur in the grout. For grout pours less than 5 ft (1.5 m) in height, clean outs may not be necessary since inspection can be easily done from above with a flashlight and mortar droppings will not greatly impact structural capacity. Table 1.16.1 of ACI 530-05 indicates allowable grout placement heights based on coarseness of the mixture and size of the masonry cavity, primarily based on experience and some test observations. After placement, grout requires consolidation in order to eliminate voids and to assure proper flow around reinforcement. It may be done by the use of puddle sticks for shorter lifts [those less than 12 in (304.8 mm)] or mechanical vibrators. Minimal effort is necessary since there is generally a small volume of grout to be consolidated within a cell. Over-vibration can cause blow-out of face shells or movement of elements beyond their necessary positions.

Tolerances during construction are important, yet difficult to maintain. Section 3.3 G of ACI 530.1-05 lists allowable deviations in joint thickness, cavity/grout space width, placement of reinforcement, alignment of horizontal and vertical surfaces, and position of assembly as noted on the plan or elevation. Individual ASTM standards regulate manufacture of components, so their physical properties are not typically in dispute as long as they were created as anticipated with proper measures of inspection and certification. As with other materials, construction that requires specific architectural appearance must be properly identified on the plans and the designer must include a schedule of tighter tolerance requirements if different from standard practice.

Masonry walls that are built under normal environmental conditions do not require special steps for curing, but even a calm rainstorm can create havoc with material bonding properties as cementitious components are washed out of the mortar and the absorption rate of units is reduced. Strong winds can cause premature drying of the mortar, reducing bond strength, and present a danger to stability of the assembly.

Under cold conditions, typically defined as below 40°F (4°C), a higher percentage of water is retained within the mortar and bond will thus be weakened. This condition would be worse in the presence of frozen masonry units with blocked pores. Low temperature construction can be done effectively if careful attention is given to the process of hydration of mortar cement and appropriate water contents are balanced. Hot weather construction is loosely defined as extremes of 85°F (30°C) and negative effects during these periods are experienced by not only temperature but also humidity, the drying effects of warm wind and direct sunshine. Walls should be constructed with the protection of shades or screens and carefully cured through maintained dampness.

A masonry beam, pilaster, or wall that is cantilevered or spans in a horizontal or vertical direction is physically connected at the ends to the rest of the building structure that will provide support. These reaction forces or moments are carried to the soil through a well-defined load path which includes horizontal and vertical diaphragms, frames, drag struts, and ties that must be properly noted and detailed on the project drawings.

Placement of conduit, pipes, and sleeves must be carefully coordinated and given consideration in design prior to assembly of construction documents. An engineer is required by ACI 530-05 to discount voids created by all embedments in determining an assembly's capacity. Unfortunately, the needs of relevant trades may not be readily available until after structural design is almost complete, so the engineer will typically proceed with certain assumptions: Large concentrations of piping are typically found in restroom areas and conduit or special sleeves often occur near clusters of mechanical or electrical equipment. Restrictions on placement of embedded items should be described in the project drawings, though it is not possible to anticipate every condition. ACI 530-05 requires a minimum of three pipe or sleeve diameters on center when placed in groups and displacement of column cross-sectional area is restricted to 2%.

Openings through masonry walls must also be planned in advance, such as might be required for ductwork or larger fire suppression piping, in order to account for the lost material during design. Mechanical plans often identify ductwork in schematic form and are not terribly specific in regards to the effect these pathways have on the structure, so questions must be asked in the early stages of design. Typical details on a set of structural drawings that show added reinforcing at the head, sill, and jambs may suffice for most work, as shown in Fig. 11-12, but openings will affect a completed wall's stiffness and seismic load resistance.

## 11.4.1 Constructability

Masonry construction begins and ends with units that come in predefined dimensions, commonly in increments of 8 in (203.2 mm), which includes a $^3/_8$ in (9.5 mm) mortar joint. Although they can be easily cut with modern tools, the best approach is to define the plan, elevation, and openings of a building in common increments to avoid additional labor expense. A structural engineer should make this suggestion

| OPENING WIDTH | "A" BARS | "B" BARS | "C" BARS |
|---|---|---|---|
| 2'-0" OR SMALLER | 1 - #5 | 1 - #5 | 1 - #5 |
| 2'-1" TO 4'-0" | 2 - #5 | 1 - #5 | 1 - #5 |
| OVER 4'-0" | 2 - #5 | 2 - #5 | 1 - #5 |

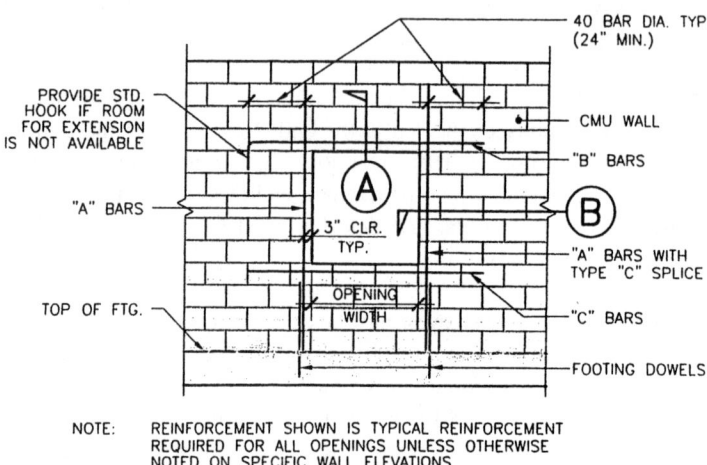

**Figure 11-12** Sample distribution of reinforcing steel around opening in a masonry wall.

early in the design process so the architect can plan accordingly and make adjustments.

Rebar congestion can also cause a mason headaches and take great pains to install properly. It is important to remember that these are not concrete elements: Reinforced masonry units have face shells and webs that often prevent ideal placement of bars and if an engineer does not account for them in his design or notes, he will deal with a field fix or have a finished assembly that will not perform as intended. If the grout cannot be properly vibrated around all bars in a group, inadequate bond will be the consequence, resulting in a lower member capacity. In an effort to help limit congestion potential, the maximum bar size should be limited to either the wall thickness or wall thickness minus 1. For example, an 8 in (203.2 mm) nominal concrete masonry wall should preferably not have bars greater in size than #8 (M #25).

Placement and spacing of confinement reinforcing for columns is critical in regions of high seismic activity. Section 2.1.6.5 of ACI 530-05 allows placement of ties at 6 in (152 mm) clear from a laterally supported bar within a mortar joint under general loading cases, but ties should be placed in contact with longitudinal bars within the grouted space in SDC D, E, and F.

## 11.4.2 Risk in Design and During Service

Major disasters tend to reveal problems that engineers once believed were taken care of through design methods and construction prescriptions. In areas of high seismic activity, for instance, unreinforced masonry construction has proven to be greatly limited in its ability to absorb energy. Grout and mortar with strength levels appropriate to the type of block or brick used, as well as reinforcement that is properly placed within masonry units or between wythes, all play an important role in the adequacy of the finished product. This interdependence can lead to uncertainty and a few surprises, but careful attention during the design process and monitoring of material and construction quality can help remove some of the mystery.

It is easy for a structural engineer to find himself happily crunching numbers and sketching details, then upon presenting the results of his analysis discover that the architect needed a special type of system to resist thermal or acoustical effects of the surrounding environment. A standard reinforced masonry wall might be suitable, but specific requirements must be confirmed. Thermal resistance can be increased by reducing the density of the materials, such as using lightweight aggregates or aerated concrete, but these considerations will impact design. Better to ask questions ahead of time than to design something twice. Find out what plans the architect has for the masonry, including thermal or acoustic resistance, texture (such as split face block), color (some types of admixtures can have a slightly negative effect on bonding), additional insulation, level of fire rating required by the governing code, and finishing (such as adding a layer of plaster over the surface).

Flexural tension failure normal to a bed joint begins with debonding of mortar from the masonry unit and continues with further cracking in regions where concentration of grout is lowest. When stresses are directed parallel to bed joints, cracks normally begin at adjacent head joints and may pass completely through alternate courses, leaving intervening units intact except in a stack bond configuration where vertical cracks may propagate further. Relatively strong units may force cracking to alternate between head and bed joints in a toothing pattern, whereas weaker units lack the strength to displace stress and crack themselves in a diagonal pattern.

The most common cause of crack appearance in masonry related to structural movement has been attributed to inadequate site drainage (Cooper 2003). Poor drainage causes a weakness in the soil support capacity related to infusion of moisture resulting in uneven sinking or lifting of the foundation system, leading to distress in masonry elements. Imperfections in masonry units themselves can also result in cracking as applied compressive forces attempt to transfer around voids, creating regions of tension above and below which cause new cracks to form.

## 11.5 QUALITY CONTROL

Assurance of quality construction begins with a well-defined set of construction documents that not only comply with applicable codes, but are easy to follow and understand, including identification of embedded items as shown in Fig. 11-13. Quality of many masonry components can be ascertained by proportion, testing, or other means specific to a production facility. Even after a masonry assembly is completely erected, prisms can be sawn or core samples drilled from the wall and tested in a laboratory to confirm or refute suspicions. The resulting cavity is grouted back solid in order to prevent the intrusion of moisture. Inspectors try to take cores at inconspicuous locations, but when appearance is an important issue, the affected face shells are sawn out and replaced by new ones that are mortared back into the wall.

Mortar that is proportioned according to guidelines set forth in ASTM C270 is deemed to meet the physical requirements of strength and durability intended for various types (M, S, N, and O), due in part to the historical performance of these mix designs, and does not require further testing. In fact, ASTM C1586 clarifies that strength results obtained in a laboratory per ASTM C270 will typically exceed results of field tests due in part to a higher amount of water present in actual mortar used on a construction project, therefore field tested specimens cannot be expected to attain strength levels tabulated for mixes in ASTM C270.

Compressive strength of a masonry assembly, defined as the maximum compressive force resisted per unit of net cross-sectional area of masonry, may be determined by taking and testing a prism according to ASTM C1314, but construction and transportation of specimens need to be given careful attention. A prism specimen for testing purposes is made up of masonry units, mortar, and sometimes grout. It is expensive and can be difficult to control properly. The unit strength method of ACI 530.1, Section 1.4B.2 explains that an assembly's design 28-day compressive strength, $f'_m$, may be assumed as a tabulated value based on masonry unit strength and mortar type, using Table 1 for clay masonry unit assemblies [4000 psi (27.58 MPa) maximum], Table 2 for concrete masonry [3000 psi (20.69 MPa) maximum], or knowledge of unit strength for AAC masonry assemblies. For example, when units meeting the requirements of ASTM C90 are provided and have been sampled and tested by the manufacturing facility per ASTM C140 to meet a compressive strength of 1900 psi, an assembly design strength of 1500 psi may be assumed when these units are laid up with Type S mortar. Prism or core testing is not required for assemblies of this proportional type, unless specifically required by the local enforcement agency.

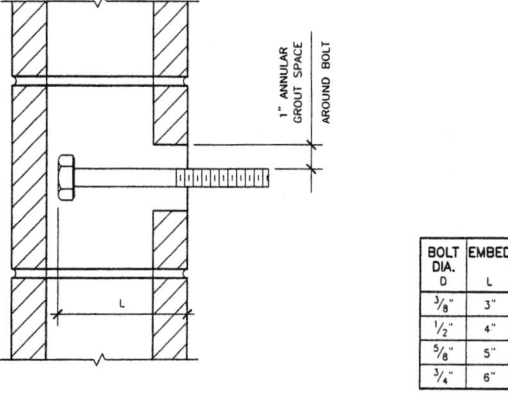

**Figure 11-13** Typical anchor bolt placement in the face of a concrete masonry unit wall.

## 11.6 CODES AND STANDARDS

Basic requirements for construction, design, and quality of materials for masonry construction are found in Chapter 21 of the 2006 IBC or Chapter 43 of NFPA 5000 with different levels of description. Both model codes, however, make reference to *Building Code Requirements for Masonry Structures* (ACI 530-05/ASCE 5-05/TMS 402-05) for detailed design, construction, and quality assurance guidelines. It is formally adopted as a standard of the publishing organizations and written in such a way that it may be given the same legal status as a model building code when referenced therein. The 2006 IBC contains modifications to the standard in different areas, mainly regarding ASD in Section 2107. Chapter 14.4 of ASCE/SEI 7-05 contains additional modifications for seismic design and detailing, which would govern over the original text. In terms of hierarchy, the model codes take precedence over applicable sections of ASCE/SEI 7-05, which takes a position in front of ACI 530-05, followed by other standards or design manuals that may be available for different aspects of construction.

When masonry elements are combined with other building materials, or are used in a special form of construction, other portions of the model codes become applicable to the complete design picture and it is easy to miss them. Section 49.2.5.3.3.5 of NFPA 5000, for example, requires special sealing measures by grouting or provision of a concrete beam within an otherwise hollow block foundation wall to control the passage of radon from the soil. Section 1604.8.2 of the 2006 IBC requires a minimum strength-level anchorage force for out-of-plane seismic loads of 280 plf (4.10 kN/m) along the length of the diaphragm providing lateral support, though this requirement is repeated in Section 12.11.2 of ASCE/SEI 7-05 with additional minimum force level checks. The provisions of a duly adopted model code are intended to govern the choice of similarly worded passages found in a referenced standard or code (2006 IBC, Section 102.4; NFPA 5000, Section 1.3.2).

# 12 Understanding the Behavior of Structural Steel

As a construction material, structural steel is versatile and reasonably predictable, as it has historically been formed into many different shapes and tested under most conceivable conditions, while maintaining a certain simplicity in the design effort. One such example is pictured in Fig. 12-1, where a closed rectangular beam is connected to the corner of what is to function as a compression ring for a skylight. The connection is made by use of simple, normal-strength bolts and component portions have been reviewed for local buckling, tensile, and shear forces.

It is not the intention of this chapter to provide complete guidance on design of structural steel elements or systems, nor to cover the vast amount of great resources available on the subject. Rather, background and general information is summarized from a variety of sources on the origin and manufacture of steel as a construction material, expected behavior of elements and assemblies, and the effect of loads on connections. Knowledge of these things adds to a good steel-problem solving base for engineers on all levels of experience.

## 12.1 COMMON TERMS AND DEFINITIONS

*Ingot*: A solid produced by the casting of molten metal into a mold.

*Killed Steel*: Steel that has been deoxidized either by the addition of an agent or by a vacuum treatment that reduces the oxygen content to such a level that no reaction occurs between carbon and oxygen during solidification of the molten product.

| **Figure 12-1** Simple steel beam connection.

*Quenching*: Rapid cooling of steel from high temperatures of about 1650°F (900°C) to about 400°F (200°C) by immersion into water, brine, oils, polymer solution, molten salts, or forced gas in order to increase hardness and strength.

*Residual Stress*: Stress that exists in a member after it has been formed into a finished product, caused by cold bending, finishing, straightening, flame cambering, oxygen cutting, welding, cooling after rolling, or quenching during heat treatment.

*Strain Hardening*: A phenomenon that describes the occurrence of strength gain as a metal deforms plastically.

*Tempering*: Quenched steel is reheated to a suitable temperature below some critical level to improve toughness and ductility, though there will also be some reduction in strength and hardness. It is done to a temperature of around 1100°F (600°C).

## 12.2 WHERE DOES STEEL COME FROM?

The best place to begin our exploration of structural steel is at the atomic level, where it can be seen that chemical properties (interaction between elements, atomic particles, and the environment) define other material characteristics that engineers use to design structures, such as physical properties (those related to the physics of a metal, such as density, thermal and electrical conductivity, and the like) and mechanical properties (those which reveal a material's elastic and inelastic behavior under load, such as modulus of elasticity, tensile strength, hardness, and the like.)

In basic terms of chemistry, steel is a ferrous material of iron combined with low or medium amounts of carbon. Other elements are added to the molten mixture of iron and carbon to create steel alloys exhibiting a variety of different properties including surface hardness (due to added aluminum), corrosion resistance (due to added chromium, molybdenum, silicon, or nickel), and wear or abrasion resistance (due to added manganese, tungsten, vanadium, or cobalt). As molten metal cools and solidifies, many different crystals forms that include atoms from constituent elements arranged in a regular, repetitive pattern. As these crystals move and join together in the formation of a solid, they create many larger entities, called *grains* (see Fig. 12-2), whose respective planes are not typically in alignment. Formation of these grains always involves some complication, as small irregularities develop that have an interesting effect on mechanical properties.

One such imperfection is known as a *dislocation*, which ordinarily moves easily through a crystal when external load is applied to a steel system. However, when an obstacle occurs to disrupt their movement—such as the abnormal boundary that is formed when different grains come together—these dislocations build up and cause *strain-hardening* of the metal, which was defined earlier. See Fig. 12-3 as an illustration of this phenomenon. Plastic (or inelastic) deformation of the material corresponds to the movement of a large number of these dislocations. *Ductility* is perhaps one of the most attractive mechanical properties of structural steel and is defined by this ability to deform plastically before failure. Measurements of ductility obtained

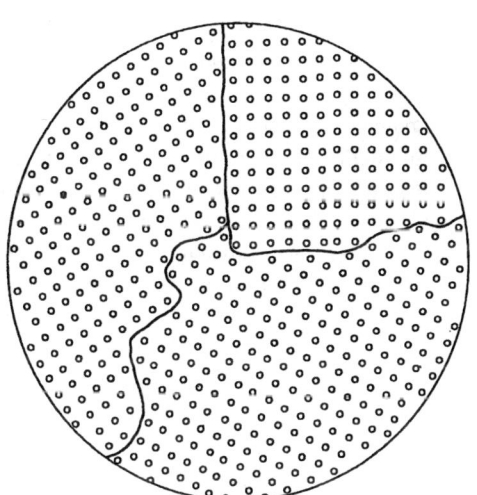

**Figure 12-2** Defined grain boundaries in the microstructure of steel. [*From American Welding Society (AWS), Linnert, G.A. 1994, Welding Metallurgy-Carbon and Alloy Steels, Volume 1, 4th Edition, Miami: AWS*]

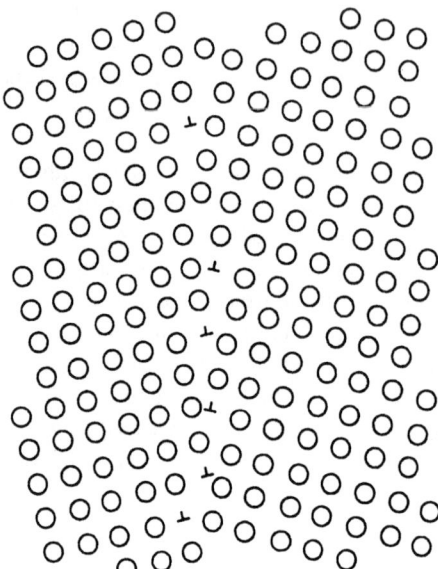

**Figure 12-3** Impeding of dislocation movement by steel grain boundary (Dislocations are shown as "⊥") [*From American Welding Society (AWS), Linnert, G.A. 1994, Welding Metallurgy-Carbon and Alloy Steels, Volume 1, 4th Edition, Miami: AWS*]

by testing are not used directly for design purposes, but the results are usually studied to compare the reaction of different metals to identical test conditions. In cold-forming operations, the measure of ductility can be useful in determining how a metal will behave during brake bending, deep drawing, and straightening. Ductility of a tension specimen is indicated by elongation, measured both prior to and during necking to the breaking point, and by the reduction of cross-sectional area at the fracture location within the necked region.

Other important mechanical properties of metals defined by chemical structure include the elastic moduli (bulk modulus, shear modulus, and Young's modulus), determined primarily through interatomic binding forces. They are highly structure insensitive, which means their values remain somewhat constant from one metal sample to another of the same kind and are unaffected by grain size or the addition of alloying compounds. These moduli, however, appear to be affected to a certain degree by temperature changes. Young's Modulus (elastic modulus), $E$, for example, is a reasonably constant $29 \times 10^6$ psi (200 GPa) at room temperature but will decrease by about 10% for each 400°F (200°C) increase in temperature (Linnert 1994, p. 140).

### 12.2.1 Mining and Refining

Although ore can be excavated from the earth's crust in the form of oxides, sulfides, and carbonates of iron, most of the iron and steel manufactured is reduced from iron oxide-type ore. Selection of suitable ore is important, as iron content may be questionable (a minimum of 50% iron is commonly required for commercial production), and control of impurities, mostly silicon oxide, aluminum oxide, and phosphorous pentoxide, is necessary to achieve predictable behavior of the final product. To extract iron from the ore, carbon obtained from coal, coke, oil, natural gas, or other material is introduced as a reducing agent, which combines with the iron oxide to form carbon monoxide and carbon dioxide gases. This process introduces a large amount of carbon into the molten iron, close to 5%, which must be reduced to create a product for most structural uses. Hydrogen has also been used as a reducing agent, thus avoiding the problem of high carbon content in the iron product, but operational costs are typically higher.

Most of the metallic iron is extracted from ore through blast furnaces, a process that has been in use worldwide for more than 500 years. Raw materials are introduced through the top of the furnace and pressurized air that has been heated to about 1200°F (650°C) enters the bottom. The air burns a portion of the carbon-containing element (coke, and the like), causing the necessary chemical reactions for reduction of iron. Limestone is added with the raw materials to assist in the removal of impurities from the ore. The product is a highly carbonous liquid iron, which also contains about 1.0% manganese, 0.5% phosphorous, 0.05% sulfur, and 1.0% silicon, which is then poured into molds and given the term *pig iron*.

Pig iron from the blast furnace is then transferred into a more sophisticated type of furnace to create liquid steel, at which time scrap steel and cold pig iron can also be mixed in depending on demand and availability. More than 60% of steel used throughout the world is created through basic oxygen furnaces which have a refractory lining not ordinarily equipped with a source of heat for melting raw materials. Rather, hot metal from the blast furnace is poured in and further heated by a high-pressure stream of purified oxygen that reacts chemically with silicon, manganese, and carbon to reduce impurities, bring temperatures well above iron's melting point, and reduce carbon content to an acceptable level. Steel is drawn from the furnace when it is at the right temperature and composition while remaining slag (waste product) is floated off the molten metal and stored for use in other construction industries (cementitious material for concrete mixes or aggregate in road building). An electric-arc furnace is also used for creating steel from scrap and iron and offers better control over temperature and chemistry of the molten product than the basic oxygen furnace.

A ladle refining, or secondary refining, station is sometimes introduced into the process for making final adjustments to molten metal, though not always necessary for metal produced from an electric-arc furnace. After being refined, molten steel is then either cast directly into a mold of desired shape or poured into a basic form called an ingot that can then be hot or cold worked to produce the desired shape. The former is typically done in a foundry, whereas the latter is more common in steel plants because of the large initial capital necessary for rolling, extruding, and drawing equipment. In addition to casting individual ingots, molten steel may also be placed into the top of a special machine that pushes out a continuous bar or slab that can be cut into desired lengths.

## 12.2.2  Mills and Suppliers

Before steel is shaped into its final form, it is tested in a laboratory to determine the quality of a particular heat of product. Different properties are determined to qualify usage under both ordinary and special conditions of service, and though properties of interest will be common in all nations, adopted testing methods for measurement may differ. Scientists must be very familiar with the chemical and physical regulations that are intended to be met for any basic or alloyed steel produced (these will be discussed later).

Roughing mills produce a semifinished product from ingots without a close watch on tolerances because they are intended for further working to obtain a final shape. Hot rolled products of this type are called blooms, which are square or rectangular sections with rounded corners of at least 36 in$^2$ (225 cm$^2$) in cross-sectional area; billets, which are square sections with rounded corners of at most 36 in$^2$ (225 cm$^2$); and slabs, which are rectangular sections at least $1^1/_2$ in (38.1 mm) in thickness and a width of at least twice the produced thickness. These shapes are further worked in a finishing mill to meet specified tolerances and produce a smooth surface that is reasonably free of defects. The final element takes the form of a bar, plate, structural shape (I-beam, channel, angle, and the like), rod, or strip.

Hot or cold rolling of an ingot involves the passage of metal between rolls with a large amount of pressure. Hot rolling of a material typically joins deep-seated blowholes, reduce grain size of the steel, improve toughness and ductility, create better material homogeneity, and produce a type of directionality in the final shape. Cold-rolling or drawing has the effect of creating better uniformity in size and shape throughout a cross-section, as well as producing a smoother surface finish and increasing strength (though losing some ductility). Extruded shapes are created by pressing a metal mass through dies which contain openings of miscellaneous configurations, whereas the process of drawing involves a pulling of metal through the die.

Once these shapes are produced according to demand, they are shipped from mills to distributors and fabricators. Larger distributors must find a workable balance between purchasing, processing, and distributing materials in order to maintain a competitive edge in the steel market, which can be somewhat volatile depending on the health of many different industrial nations. Many smaller distributors remain competitive by specializing in a particular type of steel or processing operations they offer. Because a distributor's health depends on their position within the market, owners want to have a reputation for quality and careful attention to customer needs, which begins with knowledgeable purchase of products that conform to recognized standards and regulations for the type of steel or alloy sold.

### 12.2.3 Regulations

The American Society for Testing and Materials International (ASTM) regulates the production of different types of steel and sorts them into categories based on physical and chemical properties with stress-strain behavior as shown in Fig. 12-4. ASTM A6 covers most requirements that are common within specific product standards such as identification, tolerances or permitted variations, standard shape profiles, terminology, testing, and general information on treatments. For design purposes, the yield stress of steel will be used the most and it may either be defined as the *yield point*, which is a well-defined instance of strength that occurs once increasing stress moves the material from a perfectly elastic condition into a slightly plastic range, or *yield strength*, which is most often the stress defined at 0.2% offset strain for steels that do not have a well-defined yield point.

Though all types of steel contain some carbon, the most basic steel is classified generically as *carbon steel* and includes, among others, ASTM A36 (shapes, plates, and bars), A53 (pipe, available in two different grades of strength), A500 (cold-formed round, square, rectangular, or special shape tubing available in four different grades), and A501 (hot-formed round, square, rectangular, or special shape

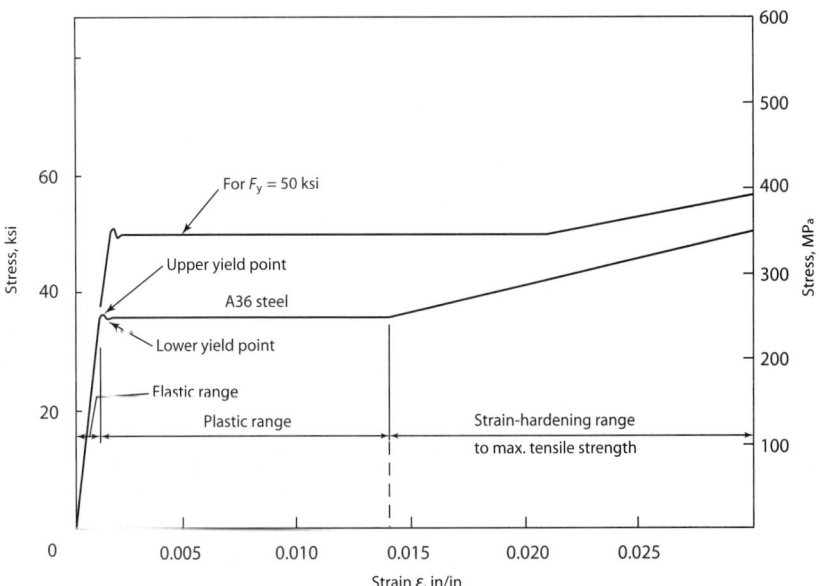

| **Figure 12-4**  General stress-strain diagram for different steels.

tubing). These are further classified according to carbon content: Low (< 0.15%), mild (0.15 to 0.29%), medium (0.30 to 0.59%) and high (0.60 to 1.70%). In addition to iron and carbon, these steels contain a small percentage of manganese, copper, phosphorous, sulfur, nickel, chromium, molybdenum, vanadium, and silicon.

High-strength low-alloy steels, including ASTM A572 (shapes, plates, sheet piling, and bars) and A992 (rolled shapes), possess increased strength due to the presence of one or more alloying agents, including chromium, nickel, vanadium, or greater amounts of silicon and copper, normally limited in solution to 5% or less of the total. They are also more corrosion resistant than carbon steels and may be further enhanced by adding more copper to the mixture of elements. A572 members are available in five different grades and four types of alloy combinations for different purposes, therefore it is important for the purchaser to determine which type is economically produced in the locality of a construction project.

High-strength low-alloy steel that is produced according to ASTM A242, A588, or A709 is recommended where additional durability is necessary, depending on its local availability, and has a resistance to atmospheric corrosion better than most other steels. It would thus serve a unique purpose for structures where a painting and maintenance schedule for corrosion resistance may be difficult to maintain, such as for transmission towers and bridges.

Quenched and tempered alloy steels, including ASTM A514 (plates) and A913 (shapes), round out this family with the highest strength of all, contain alloying elements in greater quantities than lower classifications. A514-type plates are strengthened by heating to a temperature of not less than 1650°F (900°C), quenching in water or oil, and tempering at not less than 1150°F (620°C). Shapes created according to A913 are quenched and self-tempered at a minimum temperature of 1100°F (600°C). These steels do not exhibit a well-defined yield point and have reduced ductility. Their use may be governed by serviceability limit states including deflection and vibration, at which point the very high strength characteristics may not help the final member selection much.

High-alloyed steels are those which contain at least 10% of chromium, manganese, or nickel and sometimes a large amount of carbon, therefore welding operations must be done according to special procedures. These steels commonly include stainless, austenitic manganese, or heat-resisting. ASTM A276 regulates the manufacture of stainless steel bars and shapes for common structural use, not including free-machining types for high-temperature service. Bars are furnished in a condition that is either annealed, low or high temperature hardened and tempered, or strain-hardened with light or severe cold working. A multitude of available types can be ordered depending on quality of corrosion resistance needed and availability. Chromium in the mixture creates an invisible surface film that resists oxidation, thus making the material resistant to corrosion.

The American Society of Mechanical Engineers (ASME) has also published many different design and product manufacturing standards for steel that is to be used for high-temperature or high-pressure piping and other services sensitive to mechanical systems. The society was founded in 1880 and has produced codes and standards for boilers and pressure vessels, cranes and hoists, HVAC equipment, industrial gas, welding and brazing, and others. Material standards for specialized uses have also been produced by the American Petroleum Institute (API) and the Society of Automotive Engineers (SAE).

## 12.3 BEHAVIOR AND CHARACTERISTICS OF STEEL SHAPES

For purposes of design, steel shapes are chosen not only for grade of strength, but also for economy and benefit of the geometrical cross section in resisting imposed forces. For example, flexural elements are typically chosen based on bending stresses or allowable deflection. Shear is a concern that is often only a concern with short, heavily loaded spans. They are also chosen to be symmetrical about the plane of loading, which is a step toward reducing torsion. Tension members are carefully chosen in relation to their method of connection in order to minimize shear lag (length of connection and location of the shear plane relative to member cross-section) and other unwanted, unsymmetrical effects. Compression members are usually chosen with due regard to individual slenderness, based on effective length and radius of gyration, to account for buckling effects under load. Behavior of structural members such as beams and columns will be discussed further.

### 12.3.1 Hot-Rolled Shapes

Steel billets or ingots are rolled into different shapes, primarily wide-flanges, channels, and angles, through a hot process. Tolerances in resulting camber, sweep, thickness, weight, length, straightness, and others are closely monitored in order to meet the general requirements of ASTM A6. Heating the steel mass first softens it for work through a rolling machine, where stresses above yield are imposed in order to create the desired shapes. Residual stress that is always present in metal shapes after hot-rolling is distributed according to member cross-section, rolling temperature, conditions under which cooling takes place, procedures used for straightening, and properties of the metal itself.

Common structural shapes with a wide range of uses include wide-flanges ("I-beams"), channels, angles, and sections that are built up of a combination of these basic shapes. M-shapes are also wide flange elements that may have a sloped inside flange face, similar to the sloping surface provided with S-shapes (particularly suited for crane runway beams) that is at about 2 in 12. HP-shapes, yet another type of wide-flange are known as bearing piles and have flanges of the same thickness as their webs, making them particularly suitable for driving (thus the name they are known by). Wide-flange shapes are particularly suited for resisting loads in flexure with high strength and good lateral stiffness for their weight.

American standard channels, C-types, also have sloping inner flange surfaces to match that of S-shapes. Miscellaneous channels (MC-type) are also fabricated to a different set of dimensions, usually with wider flanges than American standard channels. Channel shapes have good flexural strength, but are relatively weak in lateral stiffness, therefore buckling is of primary concern and may dictate a need for added bracing elements. Structural tees are not specially rolled, but rather are cut from existing rolled wide-flange shapes and take the same initial letter in its title. For example, a "WT" is cut from a W-section, an "ST" is cut from an S-section, and so forth. Tees and angle shapes are usually reserved for lighter loads in flexure, but may be quite capable of carrying high tensile or compressive loads depending on stiffening requirements or connection layout.

### 12.3.2 Plate Girders

Plate girders are simply built-up beams, where their shape can be defined according to stress patterns and applied load and, since the bulk of the section is placed where needed, these built-up cross sections, such as is shown in Fig. 12-5, can be quite economical in carrying large forces. Plate girders can become very deep, resulting in slender webs that sometimes require stiffener plates spaced at varying increments along the length in order to provide postbuckling shear strength through tension field action. This can be visualized as a form of truss action, where flanges serve as top and bottom chords, the web resists diagonal tension forces, and transverse stiffener plates act as compression web elements.

When subjected primarily to bending stress, plate girder beams will fail by lateral-torsional buckling, local buckling of the compression flange, or yielding of one

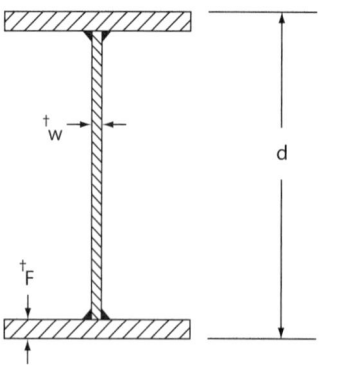

**Figure 12-5** General plate I-girder section.

or both of the flanges. Though web buckling does not usually lead to complete loss of capacity, it redistributes internal stress as the web becomes less effective in resisting load. Longitudinal web stiffeners also increase bending strength by controlling lateral deflection of the web, though the increase may not prove to be significant with thin webs. Critical stress for shear buckling, however, can be greatly increased by the installation of a longitudinal stiffener at mid-depth of the web. In general, unstiffened webs are more economical for depths less than 50 in (1270 mm) and it may be more economical to simply design for a thicker web.

Plate girder geometries and individual parts should be chosen such that special fabricating requirements can be avoided, which may not necessarily result in a member of the lowest weight, but reduced time and effort during fabrication can save a lot of the final cost. Member depth is commonly set at 1/25th of the span as a minimum and web stiffeners are only provided where needed to prevent buckling in regions of high shear stresses. Flange plates are ordered as bars rolled to the proper width and thickness and may be flame cut with a horizontal curve when specified for such use. Very thin and deep webs require special protection during fabrication to prevent premature buckling and to assure proper alignment with flanges for welding.

### 12.3.3 Tubular and Pipe Sections

Hollow Structural Sections (HSS) are available in round, square, or rectangular shapes. It is important to remember that the actual design thickness is only 0.93 times the listed nominal thickness of a section, due to tolerances listed in ASTM A500, A501, A618, and A847. Finished shapes may be produced by electric resistance welding (A500 only covers this type), which is a process by which a flat steel strip is cold-formed about its longitudinal axis by moving the strip through a progressive set of rolls (round pipe sections), or progressively shaped by driven dies which produce corners through subsequent forming stations (square or rectangular shapes), then forged together by welding rolls to create a continuous longitudinal weld without filler metal. Shapes may also be formed by a hot-rolled process, defined by ASTM A501 and A618, which are more commonly available in round sections in the United States (Stine 2005).

HSS columns tend to be sensitive to horizontally applied loads directed perpendicular to the flat side of a wall, as opposed to along the axis of the wall (Shneur 2006) and consideration of wall stiffening or a redirection of stress away from thin regions may be necessary. Additionally, ASTM A500 explains that cold-formed tubing products per that standard may not be suitable as dynamically loaded, welded structural elements where low-temperature notch toughness is required. On a more positive note, however, closed HSS sections are far more efficient in resisting torsional loading than shapes with an open cross-section, such as a channel or wide-flange. One of the reasons for this is that normal and shear warping stresses are of insignificant value in closed sections, so the total torsional moment is assumed resisted by pure torsional shear stresses.

Pipe sizes that are manufactured according to ASTM A53 are ordered in terms of nominal outside diameter denoted as *nominal pipe size*, or *diameter nominal* in metric units, and are formed by furnace butt-welding (Type F), electric-resistance welding (Type E), or in a seamless fashion (Type S) by piercing through a prepared billet. They come in standard, extra-strong, and double-extra-strong gauges, each being more expensive than the thinner brother. In terms of resistance to imposed concentrated forces and torsional action, they react similar to rectangular or square HSS sections. ASTM A500 regulates the manufacture of ordinary cold-formed round HSS sections, A501 covers hot formed sections, and both A618 and A847 regulate high-strength, low-alloy sections for different uses.

### 12.3.4 Composite Members

The mechanics of composite elements have been briefly discussed in Chapter 10. When steel is forced to deform in unison with another material, predominantly concrete, special design considerations are necessary. Connections between materials transmit shear flow in order to cause adequate sharing of load and thereby produce economical assemblies. A fully composite section is designed to completely share the total shear flow forces, whereas a partially composite section will only develop a percentage. A brief discussion of composite sections has already taken place in Chapter 5 (as related to bridge construction), where the neutral axis in either case shifts upwards as the slab becomes part of the compression flange and the bottom flange of the steel beam becomes more effective in resisting tension.

Steel columns can also be economized in some cases by the addition of concrete to enhance mechanical behavior. In order for a column to qualify as *encased composite*, the steel core must be at least 1% of the total composite area and concrete encasement must contain longitudinal reinforcing bars and ties or spirals (ANSI/AISC 360-05, p. 79). A *filled composite* column must have a steel HSS outer shell that comprises at least 1% of the total composite cross-section with a limit placed on width-to-thickness (rectangular) or diameter-to-thickness (round) ratios. In terms of flexural stiffness, measured EI for a concrete-filled steel tube will be influenced far more by the wall of the steel section than by the concrete, whereas the effective stiffness of a concrete-encased steel member will be influenced more by the concrete itself.

## 12.4 BEHAVIOR AND CHARACTERISTICS OF STEEL CONNECTIONS

The type of joint that best suits a particular situation is dependent on the size and shape of members to be connected, type and direction of loading, and relative cost of available fastening means. Whether bolts or welds, the effect of an applied force is to be determined based on its magnitude, position, and direction with respect to center of action of a fastener group. The center of action may be defined by geometrical arrangement or by distribution of individual rigidities (weld lines).

## 12.4.1 Bolts

For structural applications, the most common carbon steel bolt is created in accordance with ASTM A307 (carbon steel bolts and studs), although threaded bars created from rolling, cutting, or grinding A36 steel are also in common use. Both types are referred to as machine, unfinished, or rough bolts. Carbon and alloy steel nuts are to be manufactured according to ASTM A563 and unhardened, plain, flat steel washers for general use are produced based on ASTM F844. Strength limit states for bolted connections include shear or tension failure of the bolt and shear, tension, or bearing failure of the connecting parts.

There are three general methods for cutting holes in steel. Straight drilling is the most expensive, though it is common for joining thicker pieces, by which holes are made to a diameter of 1/32 in (0.8 mm) larger than the bolt diameter. The most common and least expensive method involves punching a standard hole that is 1/16 in (1.6 mm) larger than the bolt diameter. A third method involves punching a hole that is 3/16 in (4.8 mm) in diameter smaller than the bolt, then reaming to a finished size.

Simple bolted beam end connections are often assumed in design to resist shear reactions only (a "simply supported" reaction). Cyclic testing of these nonrigid connections proves, however, that there is some moment capacity available, even though it is not typically considered in the design of structures (Liu 2000). The shear tab connecting the beam to a column dictates the joint's behavior, which is first marked by bolt slippage and yielding of the plate, followed by deformation of the bolt holes and out-of-plane warpage of the shear tab and beam web. Girders tested included W18 × 35 and W24 × 55 sections which developed an average of 15 and 20% of each beam's plastic moment capacity, respectively. The existence of a concrete floor slab atop the beam added to the capacity of simple beam connections on the order of 30 to 45% of plastic moment capacity, though crushing of concrete also resulted in loss of some composite action and therefore of the carrying capacity of those members.

Actual distribution of force within a connection involving multiple fasteners is complex, but it has long been common practice to simply assume that equal sized bolts share an equal percentage of applied force when they are arranged symmetric to the member's neutral axis. This greatly simplifies design and adds incentive for creating symmetrical, well-defined joints. It is therefore assumed that frictional resistance to joint slip, plate deformation, and tensile stress concentration at holes may be neglected, shearing stress is uniform over the bolt cross section, bearing stress is uniform over a surface of contact, and fastener bending may be ignored.

## 12.4.2 Welds

A welded connection is one in which two pieces of metal are heated to a plastic state and fused together in a process that may or may not include the application of pressure or the use of filler metal (also called weld metal). Basic joint types are shown in Fig. 12-6 and include the butt, by which two plates or members of similar thickness are carefully aligned and may be beveled to receive weld metal; lap, which is the most common and easily suited for shop or field welding and a variety of member combinations; tee, used most often for creating built-up shapes such as I-beams or columns, plate girders, stiffeners, brackets, or other pieces aligned at right angles; corner, used to form box-type shapes with high torsional resistance; and edge joints, which are not typically used for structural purposes, but rather for holding pieces together to maintain proper alignment for completed fastening. An electrode is used in an electrical circuit that terminates at the arc, slag, or base metal, and may be filler metal (metal arc) or graphite (carbon arc) in the form of a wire or rod.

The interface between base metal and weld metal contains discrete zones with different microstructures produced by heating and cooling cycles. These are shown in Fig. 12-7. The weld (mixed) zone is a mixed production of base and weld metals. The adjacent unmixed zone consists of a boundary layer of melted base metal that did not get incorporated with the weld zone and is produced by a flow of excess heat from that requiring melting as the source travels along beyond the previous region. A partially melted zone follows the unmixed zone, but is not always present during fusion depending on the chemical composition of the base metal. The heat-affected zone (HAZ) is a much broader area, sometimes thought to include the partially melted zone, defined as the portion of base metal where mechanical properties or microstructure have been changed in some way by the heat of welding. The unaffected base metal is sometimes labeled as a final zone in the creation of a weld.

In the construction of buildings and bridges, electrical energy in the form of an arc is the most common source for creating welding fusion heat. The arc is a large current discharge measured in terms of amperes between an electrode and the base metal and is conducted through an ionized gaseous column. Special arc

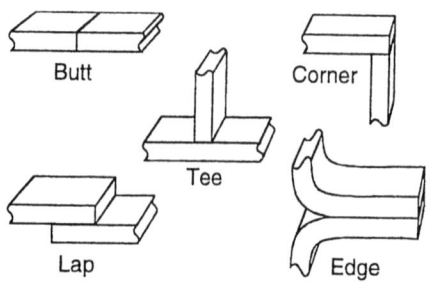

**Figure 12-6** Welded joint types (*FEMA*).

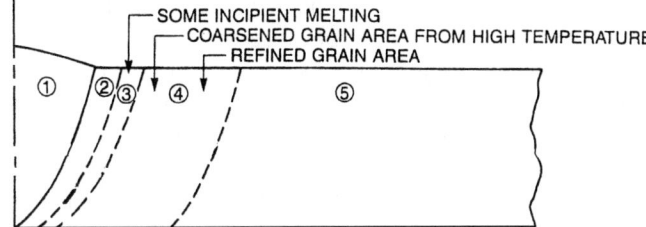

① - WELD ZONE (MIXED ZONE)
② - UNMIXED ZONE (SUPERHEAT MELT-BACK)
③ - PARTIALLY-MELTED ZONE
④ - HEAT-AFFECTED ZONE (OF BASE METAL)
⑤ - UNAFFECTED BASE METAL

**Figure 12-7** Distinction of weld zones. [*From American Welding Society (AWS), Linnert, G.A. 1994, Welding Metallurgy-Carbon and Alloy Steels, Volume 1, 4th Edition, Miami: AWS*]

welding machines supply the electric current and operate at relatively low voltage, commonly stepping the primary line voltage from generating stations down to between 15 and 35 V. Voltage is essentially a measure of the pressure pushing electrons through a circuit, which regulates the length of the arc and thereby the amount of protective shielding required. Different *arc welding* processes involve consumption of the electrode in a specific manner, and material must be selected for economy, availability, and application to the work. The most common forms of arc welding include shielded metal arc welding (SMAW), submerged arc welding (SAW), gas metal arc welding (GMAW), and flux cored arc welding (FCAW).

Another family of welding processes is designated as *resistance-type*, which uses a current channeled by an electrode through solid work pieces that are in contact with each other to generate heat, along with the application of pressure to complete the joint. Methods of this type include resistance spot welding (RSW), resistance seam welding (RSEW), and electroslag welding (ESW). Other methods of generating heat and completing joints have been studied and employed in the form of electron beam welding, laser beam welding, oxyfuel welding, and solid-state processes.

Heat input is proportional to welding amperage multiplied by arc voltage, divided by speed of equipment travel. More heat added to the system than is necessary will slightly affect yield and tensile strength of the weld metal and will generally lower notch toughness, so caution is important for work requiring larger welds and thicker materials. Preheat and interpass temperatures are intended to control the formation of cracks, primarily in the base metal. If the base received little or no preheat, rapid cooling of the work may lead to a reduction in notch toughness and cracking. If equipment travel speed is extremely slow, the weld puddle may actually roll on ahead of the arc, resulting in reduced penetration of the work.

Four basic types of welds are groove, fillet, slot, and plug, each requiring a different level of precision regarding joint preparation and execution. The most economical type of weld is the fillet, primarily due to the fact that it is particularly suited for many different types of joining, fit-up of connected pieces does not require an excessive degree of accuracy, and edges of base metal seldom require special preparation. Fillet welds may or may not be designed to transfer the full strength of joined pieces. Experiments have shown that as the angle of loading applied to a weld increases from 0° (along the weld's longitudinal axis) to 90°, the strength of a fillet weld increases but overall connection ductility decreases (Deng, Grondin, and Driver 2006). The predominant use of slot and plug-type welding is to transfer shear between lapped parts when the connection size does not allow for long enough fillets, but they have also been used to prevent buckling between overlapping plates that transfer structural load by fillet welding along the edges.

Groove welds, on the other hand, are applied to structural elements whose edges have been prepared in some fashion to receive weld metal. Different forms that are used in construction are shown in Fig. 12-8.

Welding that completely penetrates the joint is intended to transfer the full strength of connected parts, whereas welding that only partially penetrates the base metal may be designed to achieve a range of capacities. Complete joint penetration groove welds (CJP) are more expensive than partial joint penetration (PJP) or fillet welds due primarily to the labor involved in preparing the metal and depositing weld. Fillet welding should be specified wherever possible and it may even be advantageous to reconfigure a connection so it allows for them.

When subjected to load, welds, and joined pieces must deform together and the sharing of stress to accomplish this is somewhat complex. The presence of residual stresses due to weld cooling, poor welding procedures, and high joint restraint further complicate this phenomenon. Despite these complications, an engineer should remember that the weld metal itself has a higher strength than the materials joined. This is due to the fact that the core wire used in the electrode is a premium grade of steel held to tight specifications, and the shielding of molten metal and certain ingredients in the electrode coating provide a uniformity of crystal structure and physical properties (Blodgett 1966, p. 1.1–3).

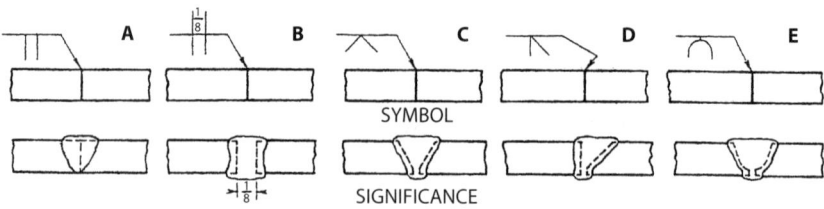

**Figure 12-8** Groove welds that are in common use.

**Welding Procedure Specifications (WPS)** To make a proper weld, specific procedures need to be followed in accordance with an approved manual or standard. The American Welding Society (AWS) was founded in 1919 and is the largest and best-known organization that produces standards, codes, and guidelines for procedures as well as welded construction and materials. The more common publications include the structural welding codes for steel (D1.1), sheet steel (D1.3), reinforcing steel (D1.4), and bridges (D1.5); recommended welding practices for steel hulls (D3.5), underwater operations (D3.6), and railroad cars (D15.1); and material specifications for carbon steel electrodes used in SMAW (A5.1), SAW (A5.17), GMAW (A5.18), FCAW (A5.20), and others.

All welded joints, whether prequalified or specially engineered, are to be created in strict conformance to a detailed procedure. Welding Procedure Specifications (WPS) are a mystery to most beginning engineers, even for many professionals, which may be due to a tendency to avoid designing any type of connection that cannot be done using a prequalified weld with existing, prequalified procedures. Prequalified welds are those which have been shown to be adequate for a particular situation due to their historical track record, but do not require special testing to confirm the adequacy of their WPSs as long as they meet all of the applicable requirements of Section 3 in the structural welding code for steel (AWS D1.1-06). A WPS contains special directions to the welder in addition to specification of an electrode, description of the process, electrical characteristics, base metal requirements, minimum preheat and interpass temperatures, type of shielding gas and rate of flow, operating current, arc voltage, travel speed, position of welding, heat treatment needs after completion of the weld, and design details of the intended joint (Miller 1999).

A prequalified WPS includes the following conditions: The welding process itself must be SMAW, SAW, GMAW, or FCAW, the base/filler metal combination must be found in Table 3.1 of AWS D1.1-06, preheat and interpass temperatures are used as prescribed in AWS D1.1-06, Table 3.2, and specific positions for different types of welds are necessary. A prequalified weld itself may, in fact, require special procedures if some of these conditions are not met, such as when introducing unlisted steel or a different welding position.

Generally, a special WPS requires testing when project specifications ask for it or when any of the conditions listed in a prequalified WPS are deviated from. Fabricators are required to complete their own WPSs for special work and qualify them by independent testing, but it is the contractor's responsibility to verify the suitability of a particular WPS and confirm working conditions. This is where communication between parties is essential to prevent the generation and testing of WPSs that cannot be used on a site in the first place, such as for a weld that can only be placed in an overhead position due to physical constraints, but has been qualified in a different orientation.

Labor is easily the greatest cost associated with a welded connection, often reaching 75% of the total, with energy costs, filler metal and shielding materials making up the remainder. The most cost-effective weld will be the one which requires the least amount of time to prepare and produce.

### 12.4.3 High Strength Bolted Connections

Two of the more familiar high strength bolts for structural purposes include ASTM A325 (quenched and tempered, with or without alloying elements) and A490 (quenched and tempered, alloy steel) and are most commonly specified at 3/4 or 7/8 in (M20 or M22) for buildings and 7/8 or 1 in (M22 or M24) for bridges. The principle of using high-strength bolts is to provide sufficient tightening to impose a prescribed clamping force on the joint, though some connections are more dependent on reduced slippage and are therefore designed with certain restrictions in mind.

Three methods of pretensioning high-strength bolts have been used successfully over the years including calibrated wrench, turn-of-the-nut, and miscellaneous direct tension indicators. In the calibrated wrench method, torque or power wrenches are adjusted such that they cease operation at a specified torque corresponding to a target tension value, typically 70% of the minimum tensile strength. Tension actually achieved by this method may be at variance from what was originally intended, sometimes as high as 30% (Salmon and Johnson 1996, p. 112), so accurate calibration with due consideration to environmental conditions is essential. The turn-of-the-nut method was developed in the 1950s as a simple procedure that does not require special equipment. Necessary pretensioning force is achieved by specified nut rotation beyond an initially snug-tight condition ("a few impacts of an impact wrench or the full effort of a man using an ordinary spud wrench"). Compressible-washer-type direct tension indicators are manufactured according to ASTM F959.

O'Leary and Zoruba (2006) define a slip-critical joint as one in which service loads are transferred through a joint without slip where all faying surfaces have a Class A or better mean slip coefficient (0.33). Fabrication and preparation of these special connections can cost three times as much as a joint that is only snug-tightened, so they should only be specified on construction drawings where absolutely required. Resistance of bolts to slip is a serviceability limit state rather than strength and requires investigation according to service-level loading. Section 4.3 of the *Specification for Structural Joints Using ASTM A325 or A490 Bolts* (RCSC 2004) states that slip-critical joints for shear or combined shear-tensile forces are only required where fatigue loading with reversal of direction is applied, oversized or slotted holes are used (except where loading is applied between 80° and 100° to the direction of the long dimension of a slotted hole), and those in which slip at faying surfaces would be detrimental to the structure's performance.

## 12.5 BEHAVIOR OF STEEL-FRAMED SYSTEMS

Steel structures are generally divided into three categories: Those that are framed with elements that act as tension members, beams, columns, or those under combined loading directions; shell-type structures that resist loads by a combination of elements dominated by axial forces; and suspension-type structures that are dominated by axial tension as a supporting mechanism. One of the great benefits of steel as a construction material is this inherent versatility, high strength, and relatively light weight.

### 12.5.1 Stability of Beams

Lateral-torsional buckling of a beam is said to have occurred when deflection changes from being predominantly in the plane of applied load to a combination with lateral deflection and twisting (see Fig. 12-9), as the capacity for carrying load drops off because of significant lateral movement and yielding. Because of this complex action, bracing will only be effective if it restrains both lateral movement and twisting of the cross section. Lateral-torsional buckling strength is dependent on distance between braces and their stiffness, type and placement of transverse load, restraint of member at ends and interior supports, cross sectional geometry (some sections are more stable than others), inclusion of stiffening elements that restrain warping, material properties, residual stress, applied compressive loading,

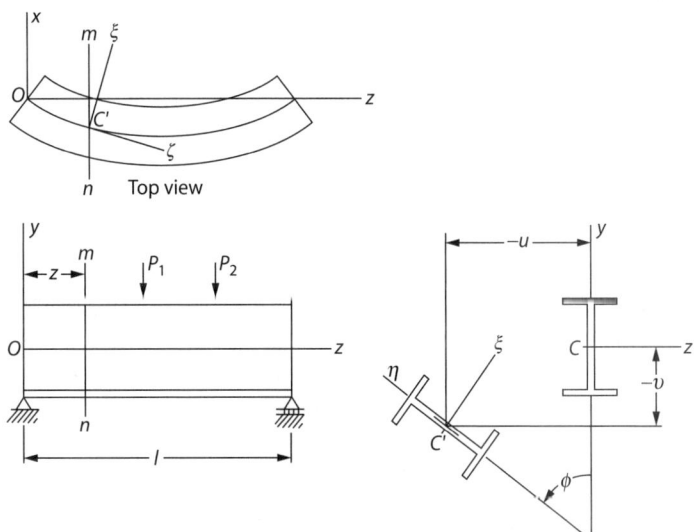

**Figure 12-9** Buckling mechanism of steel I-girder. *(From Timoshenko and Gere, Theory of Elastic Stability, McGraw-Hill, 1961)*

stressing forces, initial crookedness or imperfections in cross-sectional geometry, and interaction between local and global buckling tendencies.

Members which carry transverse loading by flexure are classified in terms of compactness, which dictates the resistance to buckling forces and is related to a member's plastic capacity. Compact sections are those which can carry moment beyond initial yield by a hinging mechanism with large rotation without fracture. Noncompact sections cannot develop this strength nor sustain large hinge rotation prior to local buckling of either a web or flange. Slender sections fail by local buckling of a shape's elements long before reaching the plastic moment.

## 12.5.2 Stability of Columns and Plates

Members or elements under compression fail either by yielding (predominantly members with low slenderness ratios) or buckling in a flexural, torsional, or flexural-torsional mode. Buckling occurs as a result of elastic or inelastic strain that causes a kink, wrinkle, bulge, or other form of loss in original shape. The *slenderness ratio*, defined as member length multiplied by a stability (effective length) factor divided by the radius of gyration ($KL/r$), is just as important in determining a member's capacity as is the cross-sectional areas of individual elements that make up a particular shape, such as webs and flanges.

The unbraced length ($L_u$) of a member's compression side against the tendency to buckle out-of-plane under transverse loading (which only occurs when a member is bent about its strong axis) works together with compactness in defining a member's limit states:

1. For large $L_u$, the limit state is elastic lateral torsional buckling.
2. For smaller $L_u$, the limit state is inelastic lateral torsional buckling.
3. For compact sections with small $L_u$, the limit state is full yielding (plastic hinge formation).
4. For noncompact sections with small $L_u$, the limit state is local buckling of the web or flange.

Though stability of tension members is of secondary concern, holes warrant special review because of the way stresses are distributed. Without holes, a tension member reaches its capacity when the entire cross-sectional area yields uniformly. The reduced cross section at a hole, defined as the net area, is important because stress concentrations exist adjacent to the hole on the order of magnitude of about three times the average stress that would ordinarily be distributed across this net section (see Fig. 12-10). As load increases, all fibers of the cross section eventually reach a uniform yield strain. When multiple holes in a single connection are not set up in a line transverse to the loading direction, there will be several different failure planes of which the one resulting in smallest net area will govern design.

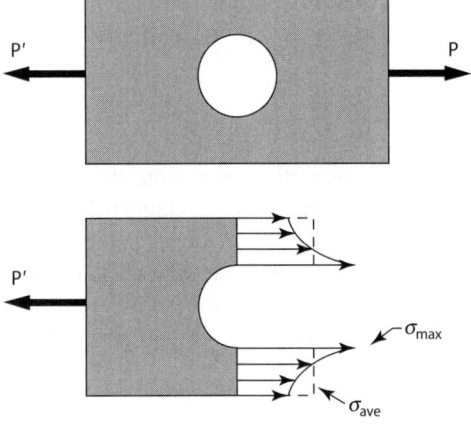

**Figure 12-10** Distribution of axial stress around hole in steel plate tension element.

The width-to-thickness ratio of a long plate segment, in addition to restraint conditions along longitudinal boundaries and elastic material properties, determines critical elastic stress expressed in Equation 12-1 (Galambos 1988, p. 91). The term $k$ is a buckling coefficient based on type of load, edge support condition, and plate geometry, $E$ is the modulus of elasticity, $\nu$ is Poisson's ratio, and $b/t$ is the width-to-thickness ratio.

$$\sigma_c = k \frac{\pi^2 E}{12(1 - \nu^2)(b/t)^2} \quad (12\text{-}1)$$

When a plate is somewhat short and stubby in the direction of applied compressive force, critical stress is measured by assuming that a unit width of plate acts as an independent column. These relationships apply equally well to plate-type elements making up larger geometric sections. Generally, plates are categorized as either stiffened (supported along two edges parallel to the compressive stress) or unstiffened (supported along one edge and free on the other edge parallel to the direction of compressive stress).

Most of the axial compressive load applied to plates is actually carried by narrow strips in the vicinity of edges, whereas the central region remains essentially unstressed. This phenomenon of useable effective width seems to have had its origin in observations of ship plating deflection, where it was discovered that longitudinal bending moments caused greater deflections than would have been calculated using gross section properties, and more accurate results were obtained by assuming an effective strip of plate with a width of about 50 times the thickness acting with stiffeners to resist applied loading (Galambos 1988, p. 93–95). Simplified equations and coefficients for use in the analysis and design of plates with different edge support and loading conditions have been reported by a number of authors (Timoshenko and Woinowsky-Krieger 1959, Young 2002).

## 12.5.3 Frames

Different frame types are described in Chapter 4 regarding lateral force resisting systems that have been defined for use in the model building codes and ASCE/SEI 7-05. After buckling of braces occurs in a concentrically braced frame, such as those with configurations shown in Fig. 12-11, the beam is pulled down which causes compression in the attached floor slab in composite construction, increasing the bending strength and stiffness of this beam, which likely affects the postbuckling behavior of the frame itself (Roeder 1989). Tests on eccentrically braced frames indicated that reserve capacity is available through strain hardening, composite action, and redistribution of forces by inelastic behavior (Foutch 1989). Column panel zones were also noted to be important sources of energy dissipation in these frames, occurring in end or midlinks, depending on configuration as shown in Fig. 12-12.

A simple rigid frame supports gravity load and resists lateral force through rigidity of the beam-to-column connections and will form plastic hinges when subjected to load combinations as shown in Fig. 12-13. Connections may be designed as completely rigid/fully restrained (Type I), simple (Type II), or semi-rigid/partially restrained (Type III) based on its ability to transmit the moment developed between framing elements. Design of simple frames is based on the assumption that beam-to-column connections do not provide moment restraint against gravity loads, but are created to have enough capacity to resist applied lateral loads (Chen 1997, p. 3–49). In all cases, it is important that the ultimate capacity of the members themselves do not exceed that of the connections. With Flexible Moment Connections (FMC), also termed *Type 2 with wind*, girders are designed as simply supported against gravity loads and the end connections for a plastic moment to resist applied wind loads.

An earthquake that occurred in Northridge, California on January 17, 1994 led to some important discoveries regarding the behavior of steel moment-resisting frames. The focal point of this 6.4 Richter magnitude event was about 19.8 miles (32 km) west-northwest of Los Angeles in the San Fernando Valley and was followed by more than 9000 aftershocks (Hall 1995, p. 4). Some areas produced

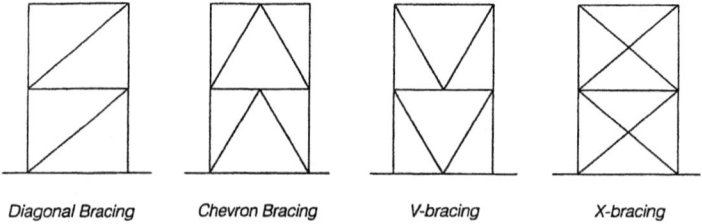

**Figure 12-11** Concentrically braced steel frame configurations. (*Courtesy U.S. Army Corps of Engineers*)

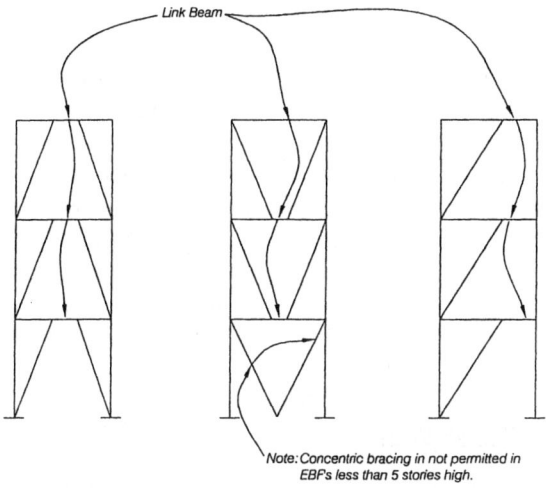

**Figure 12-12** Eccentrically braced steel frame configurations. (*Courtesy U.S. Army Corps of Engineers*)

higher accelerations than were required for design by the adopted building code, including strong vertical motion. One of the most surprising findings was that many steel moment resisting frames, originally considered quite ductile under cyclic loading, exhibited brittle fracture at column flanges, base plates, and

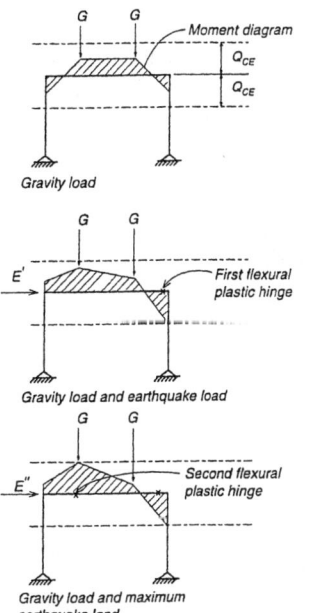

**Figure 12-13** Steel moment-resisting frame plastic hinging mechanisms. (*Courtesy Federal Emergency Management Agency*)

welded connections. In response, different agencies studied and tested the phenomena and concluded that simple measures could be implemented to prevent these failures in the future, including the incorporation of a smoothness in weld access holes, removal of backing bars, and special preparation of flanges.

According to Barsom and Pelligrino (2002), the severity of stress at the middle of a CJP weld is greatly reduced when the backing bar is removed, the root pass is back-gouged, and a reinforcing fillet is added in place of the bar. Fracture initiation then shifts to the weld access hole and improved geometry (dimension, smoothness) transfers crack initiation to imperfections within the weld metal to the toe of the reinforcing fillet or both. See Fig. 12-14 for a common diagram that describes consideration of hinge location in these beams.

### 12.5.4 Steel Panel Shear Walls

A steel-panel shear wall is an assembly where steel shapes are installed as the chord elements and shear forces are distributed through a central steel plate. Steel panels may be installed with or without stiffeners and chord members may also be filled with concrete. Different studies have shown that these assemblies have high elastic stiffness, large displacement ductility capability and display stable hysteresis behavior (Lubell et al. 2000), making them a sound choice for resistance of cyclic loads. Refer to a discussion on hysteresis in Chapter 8. Other testing performed on an unstiffened system showed pinching and tearing of the steel plate panel near the corners due to bending, but this damage did not reduce capacity or stiffness (Seilie and Hooper 2005).

System quality is influenced to a greater degree by the stiffness and capacity of the vertical boundary elements (chords) when the height/length ratio of the steel

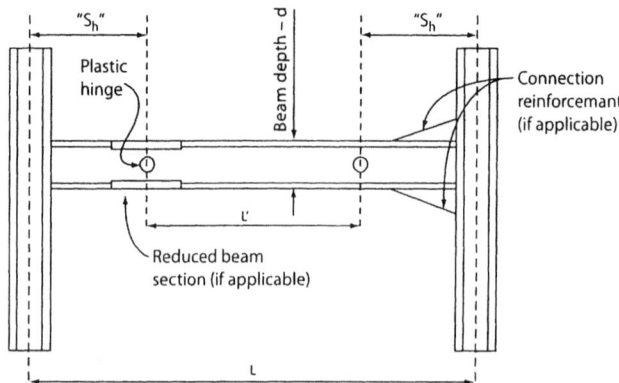

**Figure 12-14** Plastic hinge formation in steel moment-resisting frame. (*Courtesy Federal Emergency Management Agency*)

plate panel is large. For seismic behavior, multiple hinging mechanisms will occur as load is applied, preferably in the following sequence: Steel plate wall element, coupling beams, horizontal boundary elements, and lastly vertical chord elements (Seilie and Hooper 2005).

## 12.6 FABRICATION AND ERECTION

A detailer is either an employee of the fabricator or is part of another company that subcontracts the work. Design information shown on construction drawings and described within a project's set of specifications are redrawn or copied in the form of detailed shop drawings, material quantity lists, and schedules according to the needs of the fabricator. The structural engineer-of-record reviews and either approves or provides supplementary comments on the shop drawings, which should accurately reflect all design intentions and indicate compliance with common standards of tolerance, fastener spacing, and construction means for safety or erectability.

A steel fabricator may be faced with three different types of contract issues related to selection, detailing, or designing connections (Ratterman 2003). In the traditional approach, the engineer-of-record either fully designs the connections himself or specifies the type of connections needed so that the fabricator can fully develop construction details on shop drawings that are subject to the engineer's review. Another approach is where the engineer-of-record specifies connection types and related performance criteria and the fabricator's engineering staff develops and designs these connections, affixing a professional seal and signature to the documents. A different approach that may be specified in a contract is that the fabricator will essentially become the *engineer of record of connections*. Refer to Fig. 12-15.

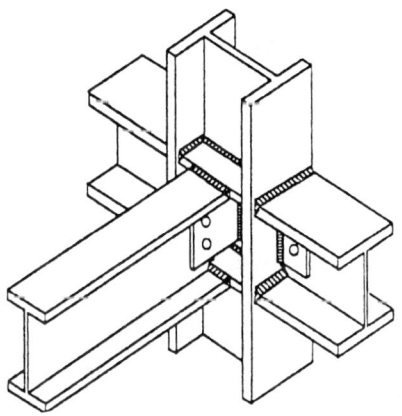

**Figure 12-15** Moment-resisting steel frame connections in two principle axes. (*Courtesy Federal Emergency Management Agency*)

Many fabricators experience common deficiencies in structural project drawings that could have been corrected with a little foresight and careful control by the project engineer. Complete dimensions that have been checked and cross-checked by the responsible design firm should be provided to the contractor. When beam reactions are needed to delegate connection design, they must be identified as service or factored and should be clearly tabulated and referenced to the right members. Repetition of common AISC connections throughout the structure, using only what is necessary to carry load (i.e., fillet welds instead of complete joint penetration welds, snug-tightened high-strength bolts instead of pretensioned) will commonly save in frustration and reduce confusion. Sometimes it is advantageous to seek advice and ask questions of the steel fabricator prior to completing structural design drawings.

It is imperative for cost containment and smooth operation of a structural steel project that as many fabrication dimensions, conditions, and responsibilities are determined as early as possible in the assembly of construction drawings. Lead times on obtaining raw materials often require consultation of early structural plan versions, even though the work of other trades may not yet be coordinated thereon, and communication will be critical. Some items that require clarification for purposes of fabrication include elevator hoistway framing and guiderail support, edge of concrete slab forming that also serves a structural function, mechanical or other deck penetrations requiring specific dimensions for clearance or anchorage, strengthening of framing members or connections for equipment housekeeping pads, and tieback devices for the work of other trades or permanently installed maintenance devices or equipment (such as window washing davits or platforms). In fact, Hazleton (2006) notes that incomplete contract documents that result in requests for information and drawing revisions create more cost and scheduling impacts than all other drawing related problems combined.

Lightweight members may require special fabrication and erection safety measures, often adding cost to a project that might not have been necessary with a heavier piece. Columns, for example, that are longer than two stories require special handling and the added cost of a splice may be less than what would be expected for erection of a longer piece (Zimmerman 1995). Light long-span beams can be dangerous, having a tendency to roll on themselves and buckle when hoisted or set in place without lateral support. Additionally, heavier sections will be a more economical choice for construction if they can be designed with fewer stiffener plates (less labor) as an alternative selection than one of lighter weight.

Bending steel is a special operation that is not typically done by the main project steel fabricator, and it is important to understand something of the industry in order to write appropriate specifications. Five methods of bending to produce simple or complex shapes (see Fig. 12-16) are commonly in use today (Alwood 2006), including cold bending (rolling), where steel is placed in a machine and

| **Figure 12-16**  Rolled steel framing elements for roof and skylight system.

curved between three rolls with differing positions between rolls; incremental bending is commonly used for cambering to very large radii by the application of a point load with a hydraulic ram or press; hot bending either by the application of a direct flame or furnace; rotary-draw bending, where a member is bent through rotation around a die and produces tight radii, mainly used in the machine and parts industry; and induction bending through heat, produced by an electric coil, applied to a short section that is cooled by water directly afterward. On the project drawings, it is important to correctly identify the member's direction of bend, overall length, and whether the steel is to be architecturally exposed with tighter tolerances in fabrication.

Architecturally exposed structural steel requires a tighter reign on fabrication tolerances than is common industry practice, such as dictated within AISC's *Code of Standard Practice for Steel Buildings and Bridges*, and these restrictions must be specified within the contract documents. An architect may not have a clear idea of what is desired and it is difficult to find acceptable dimensions that are both aesthetically pleasing and truly obtainable, especially when it comes to warpage of curved steel shapes. Suggestions can be obtained from the historical records of fabrication shops that have produced pieces under similar conditions and arrangements, but there will always be a certain amount of uncertainty in this area.

Contractors and fabricators are especially appreciative of structural engineers who consider the effects of their design on the work of those who have to build the intended assembly or structure. Shop welding is preferred to field welding due to inherent issues of quality, direct availability of materials, control of environmental conditions (contributing both to good work and comfort of workers), and economy. Overhead welding is very difficult, and joints should be designed to avoid the necessity of this position. Special requirements for preparation and finishing of a welded joint (a smooth weld access hole, removal of back-up bars, backgouging) should only be prescribed where they are absolutely essential, such as for resistance of seismic forces or architectural appearance. Additionally, in order to control distortion and add a level of economy, short segments of weld should be placed in lieu of a continuous line to the extent of what is required by the design force or code spacing requirements.

### 12.6.1 Risks in Design and During Service

A fire-resistive building may require the addition of fire proofing to structural steel members and the additional weight should be accounted for during design. The building code, enhanced by local regulations, will dictate the need for protection, but the architect might not know what type of system to specify in early stages of a project's development. Methods of fire proofing that are currently used include encasement of main framing elements in at least 2 in (50.8 mm) of concrete; adding a spray-on coating such as mineral fiber, perlite, vermiculite, or gypsum; coating with special insulating paints; or adding a fire-resistive ceiling assembly and proper sealing of joints to protect steel. Adequate measures in design and construction are necessary because of the change in behavior of steel when subjected to elevated temperatures. At about 200°F (93°C), the stress-strain curve becomes nonlinear and the yield point unstable. The modulus of elasticity, yield strength, and tensile strength reduce, especially beyond about 1000°F (540°C), and brittleness increases due to metallurgical changes.

The condition of steel exposure to corrosive elements should be evaluated and the protection called out should be commensurate with the level of exposure. Paint, galvanizing, and special alloyed steel having different degrees of corrosion protection are among available options. Galvanizing is a hot-dipping process whereby a piece of steel is immersed in a zinc ammonium chloride solution which has been heated to about 500°C (850°F). A metallurgical bond is created between the zinc coating and the steel, effectively sealing the member against corrosion even in regions where the coating may not have bonded uniformly or the coating gets damaged.

Fracture in metals is either classified as ductile or brittle with mechanisms that are termed shear fracture (ductile), cleavage fracture (brittle), and intergranular fracture (brittle). For ductile fractures, a noticeable deformation signals onset of a problem and will continue to propagate as long as external load is applied to impart energy enough to exceed plastic flow capacity. Fracture is identified as shear-type because the resulting surface of breakage is inclined at about 45° to the direction of principle stress. Brittle fractures are generally evidenced by little or no plastic deformation. They often occur quickly and may not require the continual presence of external force to cause complete breakage once damage has begun, but rather may be driven by internal elastic stress alone. A brittle failure known as lamellar tearing occurs within the base metal of a highly restrained welded joint in response to through-thickness strain induced by shrinkage of weld metal. Strain that is set up from shrinkage in a direction through the thickness of a rolled member has a far greater effect than when it exists parallel or transverse to the rolling direction, and may lead to tearing due to restraint and lack of redistribution.

## 12.7 QUALITY CONTROL

Inspection and testing of steel elements and connections can range from very minimal involvement to highly detailed operations. At the very least, an engineer is concerned that the quality of material on the market has been assured through creation according to one or more of the standards mentioned earlier. Any unidentified material must be subjected to tests according to accepted protocols, otherwise the chemical composition may lead to poor weld fusion and yield strength may be unreliable to predict or assume. Visual observation of physical qualities is not a costly process, but it must be done by trained individuals who have an eye for what appears to be out of place. Distortion or other geometrical variations can be indicative of poor storage, handling, or fabrication and should be assessed in terms of the needs of a particular structural condition and expected strength or ductility.

Testing methods for steel elements and connections are carried out through non-destructive or minimally destructive (coupon sampling) methods, each one having obvious implications for a particular assembly. Quality control measures are described briefly in Section M5 of ANSI/AISC 360-05, *Specification for Structural Steel Buildings*, and include provision for inspection of welding per AWS D1.1, inspection of high-strength slip-critical bolted connections, and steel identification. The inspection requirements for high-strength bolted connections, for example, usually involve a simple check of material data obtained from the supplier and some form of physical check on the installed tension that can be done through the use of special tension-indicating washers or other equipment.

Special inspection of welded joints is required for most structures and is to be performed in accordance with Section 6 of AWS D1.1-06 (2006 IBC, Section 1704.3.1). Welding defects that are either observed visually or by nondestructive testing (NDT) include arc blow (deflection of the arc from its normal path due to magnetic forces), brittleness, corrosion, cracking, distortion, incomplete penetration, poor fusion, porosity (cavities formed by gas entrapment), residual stresses, spatter (metal particles expelled during fusion welding that do not form part of the weld), surface irregularities, undercutting (groove melted into base metal that is unfilled by weld metal), warping of thin plates, and weld quality irregularities. Those who perform any method of NDT must have the proper training, education, experience, and have passed necessary examinations to become certified, otherwise results will be questionable. Ultrasonic testing (UT) will usually provide the most information about a weld as opposed to radiographic testing (RT), however, each method has its own set of limitations and may be dependant on a particular application.

## 12.8 CODES AND STANDARDS

Chapter 22 of the 2006 IBC and Chapter 44 of NFPA 5000 contain minimal requirements for steel design, as the lion's share of work is to be completed according to referenced standards. ANSI/AISC 360-05 and AISC's *Seismic Provisions for Structural Steel Buildings* (ANSI/AISC 341-05) are specifically referenced within these model codes and are therefore given the same legal status. These specifications have been developed and in practical use for many years, both being consensus documents to assure a uniformity of design practice in the industry. Chapter 14.1 of ASCE/SEI 7-05 contains modified requirements for seismic design and detailing of structural steel buildings, but they are not extensive. In terms of hierarchy, the model codes take precedence over applicable sections of ASCE/SEI 7-05, which takes a position in front of the AISC standards, followed by other standards or design manuals that may be available for different aspects of construction.

AISC's *Steel Construction Manual* (SCM) has been the main staple in a steel designer's diet for many years. It is a collection of the latest data related to structural steel production and design procedures to promote the use of steel and to define some level of consistency regarding engineering practice. It also includes adopted specifications and standards for design of members and connections that are of the most frequent use by engineers. The thirteenth edition has successfully combined the uses of LRFD and ASD design methodologies and addresses member dimensions and properties; design of flexural, compression, and tension members; design of members under combined loading; design with bolts, welds, and connecting elements; design of simple shear, flexible moment, fully restrained moment, bracing, and truss connections; design of bearing, base, bracket, and splice plates; and crane-rail and hanger connections.

AWS standards are intended to carry full legal authority when incorporated in or made part of any federal or state law or regulation. As far as the structural welding code for steel that is produced by AWS (D1.1) is concerned, the 2004 edition has been adopted with the same legal status of a model building code by virtue of reference (2006 IBC, Section 1704.3.1), but this is usually only valid insofar as a particular application is stipulated.

# 13 Understanding the Behavior of Wood Framing

Wood-frame systems can be somewhat of a challenge, as the material itself is anisotropic (different mechanical properties in different directions), assemblies often include the contribution of other materials (steel for strapping to define load path, as shown in Fig. 13-1), and a variety of species and grades are available for design purposes. It can be a challenge, but a little knowledge certainly goes a long way toward removing some of the mystery involved.

It is not the intention of this chapter to provide complete guidance on the design of wood-frame elements or systems. Rather, background and general information is summarized from a variety of sources to give the reader an overall picture of how wood as a construction material may be expected to behave when subjected to load. Knowledge of how wood structural elements make their way to the market adds to an understanding of their behavior, leading to more effective engineering solutions for those new to the profession as well as long-term leaders.

## 13.1 COMMON TERMS AND DEFINITIONS

*Beams and Stringers (B&S)*: Lumber that measures at least 5 in (127 mm) nominal along one side and the width is more than 2 in (51 mm) larger than the thickness. The principal use is for bending members.

*Dressed Lumber*: Formerly rough lumber that has been finished on any number of surfaces. A "4 × 12" dressed lumber piece that has been surfaced on all four longitudinal sides measures 3.5 in × 11.25 in (88.9 mm × 285.8 mm). A nominal "6 × 12" member will measure 5.5 in × 11.5 in (139.7 mm × 292.1 mm).

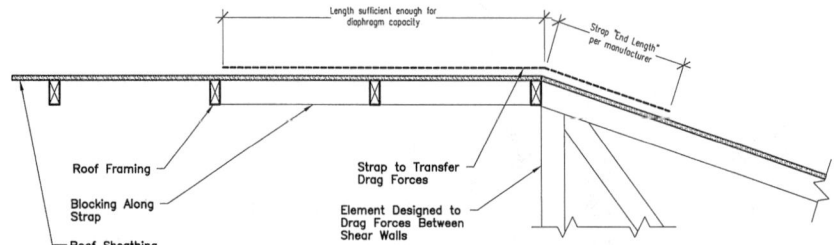

**Figure 13-1** Drag strut detail in wood-frame construction to develop force into wood structural panel diaphragm.

*Joists and Planks*: Wood elements that measure 2 to 4 in (51 to 102 mm) in nominal thickness and 5 in (127 mm) or more in nominal width.

*Kerf*: Process of cutting grooves across a wood board to allow for bending, typically cut down to about two-thirds of the member's thickness.

*Post and Timber (P&T)*: Lumber that measures at least 5 in (127 mm) nominal along one side and the width is not more than 2 in (51 mm) larger than the thickness. The principle use is for axial-load bearing elements.

*Rough Lumber*: That which has been sawn to true nominal dimensions and has not yet been surfaced. Therefore, a "4 × 12" rough-sawn member is cut to measure 4 in × 12 in (102 mm × 305 mm).

*Subfloor*: Layer of floor sheathing intended to provide structural support.

*Thickness*: Smaller dimension of a piece of lumber.

*Underlayment*: Commonly installed over a subfloor to create a smooth, durable surface to receive the finished surfacing.

*Width*: Larger dimension of a piece of lumber.

## 13.2 WHERE DOES SAWN LUMBER COME FROM?

Nature's building material can be found in many different shapes, sizes, forms, and qualities. Trees provide a basic supply, but the ingenuity of scientists and engineers has greatly extended the amount of lumber a single tree delivers and has introduced energy and environmentally conscious solutions to former problems in harvesting, production, and supply.

### 13.2.1 Lumber Supply and Harvest

Trees used to produce lumber are categorized as either *hardwood* or *softwood*, depending on the species and characteristics of living cells within. Softwood species, including pines, firs, and cedars (for example), are cone-bearing evergreens that have needlelike leaves. Softwoods are primarily used in building frame construction, paneling, doors, siding, trim, sashes, and decking. All hardwood trees, including oak, ash, maple, gums, and aspens (for example), are broad leaved and most species shed their foliage during the fall season. Hardwoods are primarily used in the manufacture of flooring, furniture, cabinetry, and railroad ties.

Lumber for use in structures is cut from the inner portions of a log, known as sapwood and heartwood tissue (xylem), either tangential to annual growth rings or radially from the center in a direction parallel to wood rays (horizontally oriented tissue through the radial plane that transfer food). New and very young trees do not contain heartwood, but it is formed after years of growth as inner sapwood cells die, possibly due to water content reduction and synthesis of metabolites that do not participate directly in tree growth (Bowyer, Shmulsky, and Haygreen 2003, pp. 34–35). Though there are virtually no strength differences between heartwood and sapwood, the former may have distinguishing characteristics such as color, special insect or decay resistance, difficulty in drying or absorbing liquids, and a slightly higher density.

### 13.2.2 Milling and Finishing

Common steps of processing harvested logs through a sawmill include debarking (which can be used as fuel for energy generation), primary breakdown of the log by a carriage rig or headrig with single or multiple passes, secondary breakdown for edging and trimming, preliminary sorting of pieces into storage bundles, drying by air and kiln, final surfacing operations, grading, and further sorting. After manufacturing and grading procedures have been completed, lumber products are marketed through wholesalers to plants, specialty outlets, retail contractors, and manufacturers of engineered lumber or assemblies such as modular housing and trusses. Sometimes the lumber is initially moved into a main distribution yard that supplies inventory for retail outlets in that region, though not all grades are readily available. The assortment of species for purchasing depends a great deal on what is grown locally, on transportation costs, and on tradition.

Most structural lumber is surfaced on four sides and is identified as S4S. It is the most common means of finishing and will most likely be supplied by default unless the engineer specifies a different combination for nominal size lumber. Other combinations distinguish between a side (S) and an edge (E), such as S2S (two sides only), S2S1E (two sides and one edge only) and may be done while the wood is still green or after it has dried (seasoned). Allowable stresses noted for some species, such as Southern Pine, differ depending on whether it was surfaced while dry at a moisture content (MC) of 19% or less (stamped with the identification "S-DRY") or in a green state. "Green" lumber is defined as freshly sawn wood in which the cell walls are saturated with water. Lumber used in appearance-quality products typically require a maximum MC of 15% to be labeled as "DRY."

Lumber that is to be used in construction is air- or kiln-dried to a specified MC, depending on its final use. Hardwood flooring, for example, is dried to between 5 and 8% in dry southwestern states, 9 and 12% in damp southern coastal states, or 6 and 9% over the remaining continental United States (Stalnaker and Harris 1989, p. 37). Softwood lumber used for structural purposes is generally targeted at 19% MC in service throughout the United States, though a multiplier is applied to allowable stress when lumber is used under wet conditions of service, defined as a condition where MC will exceed 19% for an extended period of time (NDS 2005 Supplement, p. 30). Air drying is the most economical method, but it must be carried out in a well-designed and protected yard or shed and subjected to appropriate measures of quality control, inspection and, of course, favorable environmental conditions. Whether under ordinary conditions or by the addition of fans and a small amount of controlled heat, air drying is commonly used to bring the MC of green lumber down to 20 or 25%. Wood is then transferred to a kiln for completion of drying to the desired MC. Kiln drying is controlled by high temperatures and fast air circulation in such a way that water from the interior of wood migrates to the surface and evaporates.

Seasoning is practical for members that are not greater than 4 in (102 mm) in nominal width, but can be time-consuming and expensive for larger sizes. As lumber loses moisture, it shrinks most in the tangential direction (width), almost half as much in the radial direction (thickness), and only to a slight extent in the longitudinal direction. Defects that occur as a result of drying are categorized as fracture, caused by differential shrinkage rates between core and surface fibers or rapid movement in one of the principal directions; distortion, which occurs when cell walls are crushed; warp, resulting from differential rates of shrinkage in the radial, tangential, and longitudinal directions; and discoloration, either by chemical reactions or fungal growth in the sapwood. Final dimensions after shrinkage are those used for design purposes and are reported with fair consistency in material standards.

### 13.2.3 Species

There are more than 100 different species of lumber in the United States market, but not all will be readily available in any specific region. Old-growth forests provided the raw material for many years, but their existence has diminished and it is difficult for a purchaser to find large pieces of a high quality sawn timber. In response to demand, and thanks to innovation by lumber production companies, oversized pieces of timber are engineered by combining smaller elements and strong adhesives, introduced into supply and promoted as beneficial to the environment as well as economic longevity of the entire lumber market. It is also common for most countries to import a certain supply of timber.

Density of a wood member for design purposes must be valid at the moisture content expected in service, since moisture makes up part of the reported weight even though it can vary over a range of members within the same species. Mechanical properties used for design purposes are relatively constant within a single species, therefore most of the charts, graphs, and tables that define strength are divided according to species, being distinguished primarily in terms of specific gravity. A member's *specific gravity* is defined as the oven-dry density divided by the density of ordinary water at 40°F (4.4°C). Because of moisture and physical variability, specific gravity figures are considered to be approximations, though accurate enough for purposes of design.

Douglas Fir is in wide use from the Rocky Mountains to the Pacific Coast and has a specific gravity between 0.49 and 0.52. *Douglas Fir-Larch* is the term given to a grouping of Douglas Fir and Western Larch lumber since they have nearly identical properties of strength. Douglas Fir is actually not a true fir, spruce, nor pine, but was named after one of its discoverers: It picks up the label *south* when harvested from Arizona, Colorado, Nevada, New Mexico, or Utah and is identified as *north* when obtained from Canadian forests. *Hem Fir* is considered a western wood product and exists as a grouping of West Coast hemlock and five true fir species (California Red, Grand, Noble, Pacific Silver, and White). It is grown primarily along the Pacific Coast and into the mountains of Washington, northern Idaho, and Montana. It has a specific gravity in the range of 0.43 and 0.52.

*Southern Pine* lumber includes varieties of longleaf (North Carolina to Florida and Texas), shortleaf (New York and New Jersey to Florida, Texas, and Oklahoma), loblolly (Maryland through the Atlantic Coastal Plain to Florida and Texas), and slash (Florida and South Carolina, Georgia, Alabama, Mississippi, and Louisiana) and has a specific gravity between 0.54 and 0.62. *Spruce-Pine-Fir* (SPF) is a grouping of species that includes mainly White Spruce, Lodgepole Pine, and Alpine Fir. It has a specific gravity within the range of 0.36 and 0.50 and is most commonly used in the northeastern regions of the United States and widely across Canada.

### 13.2.4 Grading Rules and Practices

Engineering properties of wood are determined by sampling and testing small pieces that are somewhat clear of defects, like knots and checks, and are straight grained. These pieces are referred to as *clear wood* due to the lack of visible surface flaws. Though many properties are still derived from testing of small clear pieces in accordance with ASTM D143, in-grade testing is also performed on a representative sample of full-size wood members using the procedures of ASTM D198 or D4761 which, in fact, has been done for all main commercial softwood dimension lumber species with a nominal thickness of less than 4 in (102 mm).

Visual grading of lumber pieces is based on the idea that mechanical properties of clear wood specimens will be greater than those in actual practice, which always contain visible defects that can be judged by the naked eye. Characteristics that are typically reviewed for identification and sorting include knots, slope of grain, checks, splits, shakes, density, decay, heartwood and sapwood, and wane. Machine-grading is also used for sorting sawn lumber, where nondestructive testing is carried out to measure strength properties and characteristics of growth are observed visually to further define the appropriate grade. Three types of machine-graded lumber are available, including machine-stress-rated, machine-evaluated-lumber, and E-rated. The main differences between lumber of this type are grade labelling, quality control measures, and coefficient of variation on the modulus of elasticity.

There are a variety of agencies in the United States who publish grading rules, following the requirements of ASTM D245 and D2555, in order to help inspectors determine allowable stresses and to make rational distinctions of wood quality within a species. All of these agencies are regulated by the U.S. Department of Commerce through the American Lumber Standard Committee (ALSC) and include the National Lumber Grades Authority (NLGA), West Coast Lumber Inspection Bureau (WCLIB), Western Wood Products Association (WWPA), Northern Softwood Lumber Bureau (NSLB), Northeastern Lumber Manufacturers Association (NELMA), Southern Pine Inspection Bureau (SPIB) and Redwood Inspection Service (RIS). Lumber producers, distributors, wholesalers, retailers, consumers, and other governmental agencies also play a role in the continual development of grading rules.

Once engineering properties are determined through grading, sorting, and labeling, reported values are categorized according to design philosophy. Reference strength for use in Load and Resistance Factor Design (LRFD) is based on the fifth percentile 5-minute bending stress, whereas that for use with Allowable Stress Design (ASD) is based on a design stress in the lower fifth percentile 10-year bending stress (USDA 1999, p. 6–1).

## 13.3 GENERAL CHARACTERISTICS OF WOOD

Wood is a highly variable building material simply due to the nature of where it is obtained. Being harvested from trees, an engineer cannot expect the physical properties of each and every piece to be identical since there is a certain amount of variability along a tree's height and even between neighboring trees of the same species. It can be controlled to a certain degree between trees on a farm, but not completely eliminated. Wood is quite durable for most applications and dissipates vibration well throughout a structure. Framing members have a high strength-to-weight ratio, which leads to lighter buildings that are sometimes advantageous in regions of strong seismic activity.

### 13.3.1 Structure

The basic structure of a tree can be found within the cell which is hollow, tubular in shape, and composed of cellulose and lignin. Cellulose is the material that gives wood its strength, whereas lignin is the glue that holds the cell fibers together. Cells are longitudinal in position, parallel to the trunk of the tree, and are quite efficient for resisting compressive forces applied along this axis. Because of the hollow, narrow shape of cells in the transverse direction, it is easy to see that compression forces applied perpendicular to the trunk of a tree will crush a cell's wall much quicker than along its longitudinal axis. Wood, therefore, is known as an orthotropic material because its physical and mechanical properties are different about three perpendicular axes—longitudinal, transverse or tangential, and radial—though strength values are typically listed in material standards according to two main directions, *parallel-to-grain* (longitudinal axis) and *perpendicular-to-grain* (a combination of the effect of transverse and radial directions).

## 13.3.2 Mechanics

Though care is taken to assure a reasonably clear picture of how a species will behave under external load, it is impossible to completely eliminate variability of properties due to natural changing influences that affect a tree's growth. Properties related to strength of wood that are of the most interest to a designer include the following:

1. Compression strength parallel to grain commonly relates to the capacity of columns or truss chord members.
2. Compression strength perpendicular to grain is used to determine the required bearing length of flexural members.
3. Creep is defined as time-dependent deformation under load that can lead to material failure if very heavy loads are carried continuously for a long time. This phenomenon occurs at the molecular level by slippage of long-chain molecules within cell walls and can be exacerbated by continual changes in humidity.
4. Fatigue is defined as progression of damage that occurs when a member is subjected to cyclic load, based on maximum experienced stress as well as frequency and number of cycles. The best example of fatigue loading may be that of a truss bridge chord member supporting railroad traffic, where continual passage of engine or rail car wheels cycle element deformation.
5. Fracture toughness describes the ability of wood to withstand failure initiated by defects.
6. Hardness is generally defined as resistance to an indentation that might be produced by the impact of a small, weighted ball.
7. Modulus of rupture identifies the maximum bending capacity of a member to the elastic limit.
8. Rolling shear strength is a member's resistance to shear acting perpendicular to grain and is of approximately equal value in the longitudinal-radial plane as in the longitudinal-tangential plane. It is of particular importance in engineered wood fabrications like plywood.

9. Shear strength parallel to grain describes an ability to resist internal slippage of one part longitudinally with respect to another.
10. Tensile strength parallel to grain is critical for truss bottom chord members subjected to ordinary service loads.
11. Tensile strength perpendicular to grain measures resistance to a tendency for splitting across the grain.
12. Torsional strength measures resistance to twisting about the longitudinal axis, which is often taken as the shear strength parallel to grain in sawn lumber.

The coefficient of friction between wood surfaces is affected by the degree of surface roughness, orientation of the fibers, age, temperature, and moisture content. It varies little between different species, except for those which contain an oily extractive. Generally, the coefficient of friction increases as moisture content increases from oven-dry to fiber saturation, but will reduce as surface water increases. Wood connections are not typically designed to rely on friction between pieces for structural value.

### 13.3.3 Moisture Content, Temperature, and Chemical Treatment

Most water that is present in green wood is removed by drying and what remains works toward equilibrium with the relative humidity in the surrounding environment. As sawn lumber dries, it has a tendency to warp, buckle, or otherwise distort, and careful attention to drying procedures, handling, and storage can help minimize unwanted effects while in service. Though ordinary fluctuations in humidity do not cause significant dimensional change in a member, seasonal extremes may be an issue where wood is not specially protected by a controlled environment or some form of treatment. The fiber saturation point of wood is commonly defined at 30% MC, which is the level at which serious decay occurs in species commonly used for timber bridges. This is defined as the point in time when the cell-wall material is completely saturated, whereas cavities are empty and any further moisture will be directed to these voids.

Conditions of moisture, temperature, and oxygen that promote decay in wood members can also initiate corrosion of embedded fasteners. Wood preservative agents, however, have also been shown to promote corrosion of fasteners, leading to the requirement of special corrosion resistance either by the steel material itself or galvanic coatings. Different oil- and water-borne preservative materials have been used over the years, including solutions containing copper, chromium, and arsenic.

Mechanical properties of wood specimens have generally been shown to decrease when heated and increase during cooling (USDA 1999, p. 4–35), exhibiting a somewhat linear relationship with temperature at a steady moisture content below about 300°F (150°C). If temperature change is rapid, mechanical properties will essentially return to their original values when cooled from 212°F (100°C).

Chemicals are driven into wood primarily for the purpose of fire or decay resistance and they will reduce mechanical properties to a certain extent. Oil-type preservatives do not chemically react with the elements within cell walls and therefore do not have a significant effect on strength, though the treating process used can have a negative impact. Waterborne preservatives affect lumber differently depending on species, chemistry and retention of material chosen, drying temperature, size and quality of material, incising, and service environmental conditions. Incising is a process by which slits are punched into the surface of a wood member to improve penetration of preservative chemicals. Heartwood of some species is more resistant to decay than sapwood, though the latter absorbs preservative treatment better.

### 13.3.4 Engineered Lumber

In addition to structural panels created of pressed wood veneers, chips, particles, or other potential waste material, glued structural members have been in wide use for many years and are assigned properties of strength based on laboratory testing and controlled manufacture. These elements consist of structural composite lumber (SCL), glued-laminated timber (glulam), or other specially fabricated assemblies subject to validation by testing. Structural composite lumber products were engineered in response to demand for high-quality timber at a time when it was becoming difficult to obtain such material from available forests. In general, SCL consists of small pieces of wood or parallel strips of veneer that are glued together to create common solid-sawn sizes, used by themselves or in the manufacture of other engineered lumber assemblies, such as I-shaped joists. Finished elements of SCL have been sold in the industry as laminated veneer lumber (LVL), parallel strand lumber (PSL), laminated strand lumber (LSL), and oriented strand lumber (OSL).

One of the main benefits of SCL is that strength-reducing defects found within sawn lumber are broken down and spread out along the length of veneer or strands of wood in such a way that the resulting product is far less affected, giving rise to high design values. Veneers used in LVL are carefully selected, often with the aid of ultrasonic testing, and end joints between individual veneers are staggered throughout to minimize the effect on design properties. Wood used in the manufacture of PSL begins with short, high-quality veneer pieces left over from plywood or LVL operations, which are then cut into strands with a common minimum length of 24 in (610 mm). These are then coated with waterproof adhesive, oriented along the length of a special press, then densified and cured into billets that are cut into shapes larger than common LVL pieces. LSL and OSL members typically have lower values of strength than other engineered wood products, but share a common stability over a wide range of environmental moisture changes that serves as a benefit to framing contractors. Quality assurance requirements in the manufacture of SCL products are dictated by ASTM D5456.

Glued-laminated timber elements are produced by gluing two or more pieces of lumber together along the length of their grain and are created using a single species or a combination thereof, such as the provision of Hem-Fir core laminations surrounded by Douglas-Fir. Production, testing, and certification of timbers is done in accordance with ANSI/AITC A190.1 and engineering design values such as modulus of elasticity, compression perpendicular-to-grain, tension and compression parallel-to-grain, bending in the strong and weak axes, shear parallel-to-grain, and radial tension are determined according to ASTM D3737. End-joining of individual laminations, subject to qualification by strength and durability testing, is used to produce longer structural members. A 1.1 in (28 mm) long fingerjoint is the most common. Waterproof adhesive used to hold the joint firmly together is cured under a combination of heat and pressure. In order to avoid the creation of a weakened plane, joints are staggered over the length of a manufactured element. Effects of drying shrinkage are greatly reduced for glulam members since the lumber used in laminating must already be seasoned prior to assembly.

## 13.4 BEHAVIOR OF WOOD ELEMENTS

Wood is described as brittle under static tension, but it has been shown to be highly nonlinear and apparently ductile under static compression or shear (Smith, Landis, and Gong 2003, p. 48). The ability of a column to carry compressive forces will be governed by its slenderness (geometrical property) and compression strength parallel-to-grain (mechanical property). If this column is carried by a beam, the beam will resist the tendency of that column to make an indentation by that beam's compression strength perpendicular-to-grain (mechanical property) and the contact area of the column (geometrical property).

Shear strength is often difficult to picture because the effect occurs horizontally along the axis of the member even though the load itself is applied perpendicular to that member. An often-used illustration is that of a beam built out of individual horizontal layers, stacked atop one another. As load is applied, these layers want to slip past each other and a resisting force, the horizontal shear force, is necessary to keep these fibers from separating.

A structural element continues to deflect, or creep, if applied loading remains in position and steady magnitude for an extended period of time. Allowable stress magnitudes or load resistance factors normally used in the design of wood-frame structures have been specified such that the rate of creep deflection, for the most part, becomes less as time passes until no further deflection would be detected and resulting failure should not be expected. However, bending members that carry a high initial flexural stress (slightly below the ultimate stress) may enter a tertiary stage of creep deflection and fail shortly thereafter.

## 13.4.1 Panels or Sheathing

Wood-based panels available for structural use are created with sheets of veneer, fibers, or particles molded together with an adhesive under high pressure. The most commonly available structural panel size is 4 × 8 ft (1.2 × 2.4 m), but other dimensions may be produced according to demand. A *wood structural panel* is technically a wood-based product manufactured from veneers, wood strands, wafers, or a combination thereof that are bonded together with a waterproof adhesive (NDS 2005, Section 9.1.3). Although other types of wood-based panels are assigned some structural value, the term wood structural panel generally only applies to plywood, oriented strand board (OSB), and composite panels. This distinction is important for determining construction requirements found within building codes, where wood structural panels are allowed greater flexibility in design than those which do not specifically carry such a title. For example, the allowable capacities for wood structural panel shear walls that are given in Table 2306.4.1 of the 2006 IBC may be increased 40% for wind design and are allowed in all seismic regions, whereas capacities that are tabulated for particleboard or fiberboard shear walls are not allowed such an increase for wind design and are restricted from use in Seismic Design Categories (SDC) D, E, or F (2006 IBC, Sections 2306.4.3 and 2306.4.4).

The strength of plywood is primarily found in the cross-laminating of veneers, where successive layers are placed with their grain directions perpendicular to one another. Each individual layer is constructed of one or more plies depending on intended use or quality and the finished panel will always contain an odd number of layers. Inner plies may actually be provided as particleboard, veneer, or other species of lumber than outer plies. Plywood grading for structural use depends on the quality of layers and order of placement, distribution of defects, type of adhesive and its resistance to moisture, and control of manufacturing conditions. Strength and behavior of plywood diaphragms have been subjected to numerous tests, such as is shown in Fig. 13-2 performed by APA – The Engineered Wood Association (APA), to establish design values and to justify certain applications.

Oriented strand board (OSB) panels are fabricated in multiple layers of cross-laminated wood flakes and are commonly assigned the same strength values. These boards are manufactured by slicing debarked and soaked logs into thin elements that are dried, blended with resin or wax, heated, and pressurized together. Waferboard panels are similar to OSB in terms of material and handling, but individual fibers are randomly oriented instead of placed according to any particular pattern. The 2006 IBC and NDS-2005 no longer recognize structural value for this precursor to OSB.

**Figure 13-2** Plywood structural diaphragm testing. (*Courtesy APA—The Engineered Wood Association*)

Particleboard was originally created from a need to dispose of sawdust, shavings, and other somewhat homogeneous mill waste. Some have even incorporated non-wood, agricultural residues and blends. It consists of discrete particles instead of fibers that are covered in adhesive, heated, and pressurized together in a fashion meeting the requirements of ANSI A208.1 and most applications are for interior use. Mechanical properties of particleboard are derived from the type of raw material used in production, additives, and manufacturing itself. When used as underlayment, it is sometimes advisable to provide glue to the subfloor in order to control some of the instability in the longitudinal direction.

The fiberboard family of sheathing includes hardboards, medium-density fiberboard, and insulation boards, manufactured in accordance with ASTM C208, and is to be identified by an approved agency if intended for structural use. Logs or other materials are reduced to chips, mixed with wax, then pressed together by a similar process as that for particleboard.

Hardboard is a product commonly used for siding or floor underlayment applications and can vary significantly in density from one manufactured element to another depending on the species, intended use, and other factors. It is manufactured primarily from interfelted fibers formed by combining lignin and cellulose, consolidated under heat and pressure, resulting in a sheet that is typically stronger in every direction than the perpendicular-to-grain strength of natural wood.

An exposure rating is assigned to manufactured panels as Exterior (designed for applications subject to permanent exposure to the weather or moisture), Exposure 1 (applications where long construction delays may be expected prior to providing protection against moisture or weather extremes) or Exposure 2 (moisture-protected construction applications). Special adhesives for laminated-type wood members used in exterior applications are manufactured according to ASTM D2559.

Orientation of plywood panels is important to the overall stiffness of a framing system, as the moment of inertia when grain direction is placed perpendicular to framing members can be 10 times larger than that when grain direction is placed parallel to supports. Values reported by manufacturers are typically calculated using a transformed section to account for the fact that the modulus of elasticity, $E$, perpendicular to grain is only about 1/35th of the modulus parallel to grain (Stalnaker and Harris 1989, p. 247).

Working together as a basic system in wood-frame construction, a structural wood panel deck serves to increase the load-carrying capacity of the joists or trusses below by providing lateral stability along the compression edge, acting in a partially composite manner due to fastening and continuity and by distributing load across multiple members. In fact, testing has demonstrated that sheathing and proper connections cause this load-sharing effect to occur not only for bending resistance, justifying the repetitive member allowable stress increase factor in design, but also for tension or compression in chord members of wood trusses (Cramer, Drozdek, and Wolfe 2000).

## 13.4.2 Connections

Fasteners for wood construction have been tested in a variety of ways over many years and ASTM D1761, originally approved in 1960, provides a unified approach for testing and reporting engineering values that can be used by the design community. Standard methods are described for testing lateral and withdrawal resistance of nails, staples and screws for evaluating strength and rigidity of bolted timber joints, with or without metal connectors and tension testing of plate-type connected joints.

Driven fasteners such as nails, spikes, and staples are to be manufactured according to ASTM F1667. Their holding capacity is based on several different factors, such as diameter, shape, and surface coating of the shank; specific gravity or moisture content of joined pieces; depth to which the fastener is driven, whether parallel or perpendicular to grain; and direction of applied load in relation to the fastener's axis. Testing and analysis have shown that nails loaded perpendicular to grain have higher ultimate strengths than those loaded parallel to grain, likely due to high initial bearing stiffness (Hunt and Bryant 1990).

The hole into which a bolt is placed has a dramatic effect on the ability of that bolt to resist lateral loads. If the hole is too large, the bolt will not bear uniformly against the wood surface whereas if the hole is too small, wood will tend to split if the bolt has to be driven in. Bolt holes should be smooth and a maximum of about 1/16 in (1.6 mm) greater in diameter than the bolt itself except that for sill plates, holes are often allowed to be oversized by 1/8 in (3.2 mm) or slightly more, though the effects of shrinkage on hole size should also be considered. ASTM D1761 explains that the behavior of a bolted joint is typically affected by member thickness, member width and margins towards ends or edges, fastener quantity and spacing, type of fastener and quantity of fastened pieces, moisture content of the wood, presence of preservative or fire-retardant treatment, presence or absence of metal side plates, and species of wood.

Behavior of a bolted joint is also related to drilled hole quality, where a roughened surface (slow drill rotation and feed rate through member) leads to greater deformation in addition to having a somewhat negative impact on wood bearing values. Cyclic testing of joints secured with ASTM A307 bolts showed that the distribution of bearing stress along the bolt length is affected by inelastic deterioration of wood fibers at the ends of the bolts (Abendroth and Wipf 1989). When these edge fibers are crushed, bearing resistance shifts toward the interior of the wood member causing significant changes in the distribution of stress along the bolt's length. This motion under many cycles led to joint failure by fatigue fracture of the bolt within the first thread root adjacent to the shank.

Ordinary wood- and lag-screws require pilot holes for proper installation and soap is sometimes used as a lubricant to help prevent binding and allow for smooth turning of the fastener. The ultimate tensile strength of a screw is obtained at a penetration depth of about 7 times the shank diameter for species with a specific gravity greater than 0.61 and 11 times for softer wood species (a specific gravity of 0.42 and less). This also corresponds with experimental driving distances necessary to achieve full tabulated lateral load resistance values.

Although withdrawal resistance design values have been based on the same testing and analysis method for many years, lateral load resistance values are currently based on a yield model derived from testing in the 1980s prior to which were derived from an empirical approach. Yield model equations for dowel-type fasteners were developed in recognition of four primary yielding modes. Modes Im and Is represent bearing-dominated yielding of wood fibers in contact with the fastener. Mode II is based on fastener pivot at the shear plane of a single-shear connection with localized crushing of exterior fibers. Modes IIIm and IIIs represent fastener yield in bending at a single plastic hinge point per shear plane and bearing-dominated yield of fibers in contact with the fastener. Mode IV identifies fastener yield in bending at two hinge points per shear plane, with limited local destruction of wood fibers. Tables summarizing results of these equations for bolts, lag screws, and drift pins are based on a steel bending yield strength of

45 ksi (310 MPa), which is approximately equivalent to the average of steel material yield strength and tensile strength (NDS-2005, Section I.4).

The use of steel side plates increases a fastener's resistance to lateral load, though the improvement is difficult to quantify. In a nailed connection, the plate acts as a somewhat impenetrable bearing surface for the head, which has a tendency to rotate when subjected to lateral loading, thereby increasing capacity. There are numerous wood-to-wood connections that call for use of light-gage metal straps, such as is shown in Fig. 13-3, where this phenomenon also holds true. In bolted joints, measured deformation is smaller when steel side plates are used instead of wood side members, therefore capacity is increased for load directed parallel to the main member's grain (USDA 1999, p. 7–15).

Older timber trusses used a birdsmouth-type connection between web and chord members where a portion of the chord was notched out to allow for the placement of the web. Testing has indicated that the basic configuration exhibits good ductility as well as semirigid behavior through rotational capacity that occurs as a function of the width and level of compression of the web, the friction and skew angle of the connection (Parisi and Piazza 2000). Modern light-frame truss assemblies use metal side plates with many small teeth that are embedded into the wood by a hydraulic press, whose design values are typically based on load-slip curves generated in a testing laboratory.

In wood construction, it is generally best to design connections that resist a simple distribution of forces in a predictable fashion and to place fasteners in a

**Figure 13-3**  Steel straps installed at wood header to jack stud and horizontal blocking in narrow shear wall.

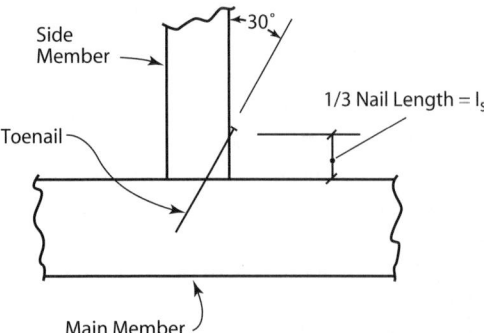

**Figure 13-4** Proper toe-nail installation according to NDS-2005 standards.

logical structural position, such as providing support for a hanging load from above the neutral axis along a bending member. Simple connections under any loading condition are preferred as there is a better assurance of proper framing in the field. Toe-nails, for example, add needed simplicity but proper installation (see Fig. 13-4) is necessary to get the most out of such a connection. These are fasteners driven through the edge or end of an attached member, at a slope of about 30° with that member, into the carrying (main) piece, though they are more effective when driven through the edge.

### 13.4.3   Influence of Defects

A knot is formed as a change in physical properties along the trunk where a branch has become ingrained within the structure of the stem, interrupting arrangement and continuity of cells (see Fig. 13-5). Wood fibers attempt to work their way around the knot, creating a pattern of distortion that includes several points of concentrated stress. Knots affect a member's tensile, bending, shear, and compressive strengths as they essentially create large voids within a cross-section. Their effect on reducing a

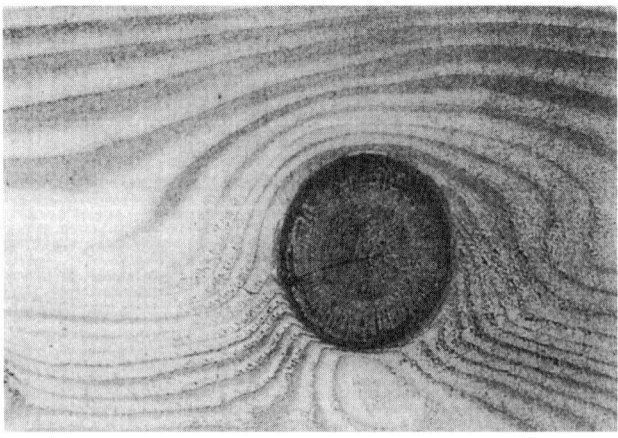

| **Figure 13-5**   Pattern of wood fibers in the vicinity of a knot.

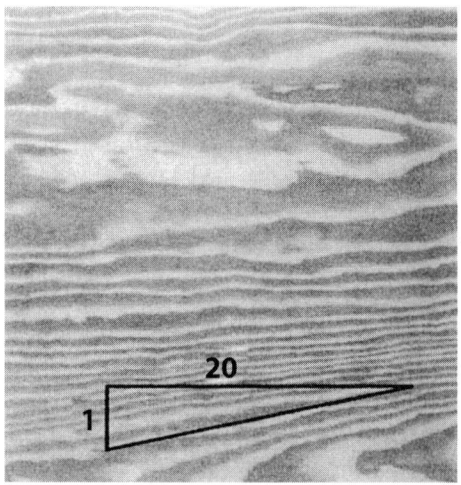

**Figure 13-6** Definition for slope-of-grain, identifying a 1 in 20 distortion.

member's capacity depends on size and location, shape and soundness, local slope of grain, and character of loading. After a branch has died and the trunk continues to grow, the resulting knot may become encased within living wood fibers and its influence over mechanical properties will be lessened. In bending members, a knot located within the region of tension fibers will have an impact on flexural capacity. The lateral stiffness of a member under axial compression will be somewhat reduced in the presence of a large knot.

Slope of grain exists when fibers of a member along the longitudinal axis are not parallel with its edges, and all lumber of reasonable lengths contain this feature to a certain extent. Wood member strength is affected when slope of grain exceeds about 1-in-20, as identified in Fig. 13-6. Grain may form a spiral pattern by winding growth of wood fibers around the core of a tree and is difficult to recognize by standard means of visual inspection. Diagonal patterns of grain result when lumber is cut at an angle to the central axis of a log or when boards are taken from initially crooked members.

Both shakes and checks are defined as separations of wood fiber and cause a reduction of shear capacity. Shakes are formed within the tree itself from stress of growth, wind loading, or frost-related deformation of the cells. Checks, on the other hand, occur during the seasoning/drying process after a tree has been harvested, usually when drying occurs rapidly or unevenly.

## 13.5 BEHAVIOR OF WOOD-FRAME SYSTEMS

The CUREE-Caltech Woodframe Project, organized by the Consortium of Universities for Research in Earthquake Engineering (CUREE), was a 4-year study on the performance of wood-frame buildings during a seismic event. The project involved testing and analysis of full scale structures, careful study of field reconnaissance information obtained after the 1994 Northridge Earthquake (California), recommendation of modifications to codes and engineering standards based on findings from research and testing, consideration of the economic aspects related to earthquake damage or building strengthening with a goal toward better performance, and dissemination of findings and conclusions through education and outreach events. One of the general findings that continued to be apparent during testing is that nonstructural finish material and interior nonbearing/nonshear walls contribute greatly to a building's seismic-resistant behavior, resulting in low deformation and relatively good performance (Cobeen, Russell, and Dolan 2004, W-30a, p. 18). A wood-frame system, therefore, does not exhibit a pure response to cyclic-type loading, but rather is influenced by other materials and assemblies. This unintentional redundancy contributes to a robustness that assures generally acceptable life-safety performance for most wood-frame structures, except for those with significant irregular configurations.

### 13.5.1 Horizontal Diaphragms

Although buckling of the thin sheathing "web" must be given consideration in wood diaphragm design, it has been shown that framing members to which the sheathing is attached act sufficiently to resist buckling forces (APA 2004, p. 4). Diaphragms that are nailed to blocking around all edges of individual sheathing pieces, such as is shown in Fig. 13-7, will resist the tendency for buckling much better than an unblocked diaphragm, thus allowing for much higher applied forces.

Tests on wood diaphragms, both horizontal and vertical (another term for shear wall), have been performed in one manner or another as far back as the 1930s. Most early tests were performed simply to determine relative stiffness and strengths of different materials and fasteners. While proving that gypsum wallboard diaphragms are less stiff than similar diaphragms constructed of plywood regardless of the existence of openings, tests performed by Falk and Itani (1988) also indicated that the stiffness of diaphragms, wood-type ones in particular, is dominated by the distribution and stiffness of fasteners used to secure the sheathing to the framing. Many innovations and design theories continue to be tested in modern times, such as the qualification of wood panel diaphragms to resist lateral forces in a flexible manner (see Fig. 13-8).

Discussion and debate continue on the issue of wood diaphragm flexibility. A diaphragm is classified as either *flexible* or *rigid* by model building codes based on the amount of horizontal deflection. If a diaphragm falls into the category of being rigid, applied forces must be distributed according to the rigidities of the vertical elements and torsional effects must be considered. It is interesting to note that conclusions offered by the CUREE-Caltech Woodframe Project include a comment that for the great majority of wood-frame structures, tributary methods of seismic force distribution (flexible analysis) appear to lead to better building performance than to redistribute applied lateral forces to different shear wall lines (Cobeen, Russell, and

Understanding the Behavior of Wood Framing | 355

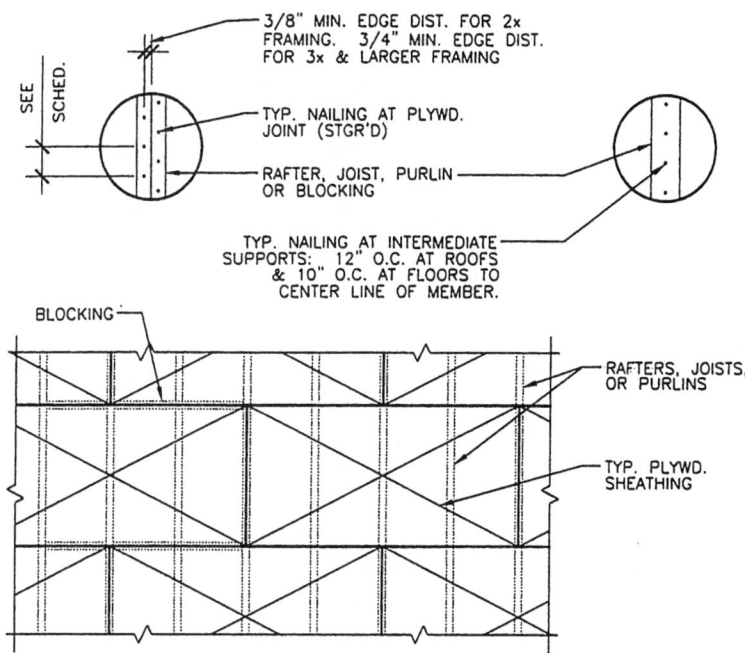

**Figure 13-7** Nailing of plywood sheathing to complete a wood-frame structural diaphragm.

**Figure 13-8** Horizontal plywood diaphragm testing. (*Courtesy APA—The Engineered Wood Association*)

Dolan 2004, W-30a, p. 45). An equation to define wood horizontal diaphragm deflection appears in Section 2305.2.2 of the 2006 IBC and typically follows several assumptions: The diaphragm under review is simply supported at the ends, the nailing pattern is uniformly distributed throughout, the sheathing panel edges are completely nailed and blocked, external forces are uniformly applied and the diaphragm is regularly shaped. In many cases, however, diaphragms that engineers must design aren't so clean and well-defined. They typically have holes, block-outs, cantilevered portions, uneven loading patterns, and uneven support positions to where caution in applying the deflection equations is warranted. Figure 13-9 provides guidance as to the classification of diaphragm rigidity.

A wood sheathed diaphragm fails either by nail heads pulling through the panel face or edge, splitting of framing members or panels, or buckling of the panels. It has been shown, and is recognized by model building codes, that a diaphragm with applied lateral forces in a direction perpendicular to continuous panel joints in an unblocked condition is stronger than a diaphragm which resists load applied parallel to these joints. At high levels of force, nailed joints between sheathing and framing exhibit nonlinear behavior, except when adhesive is used along with the nails (Kamiya 1987), and are seen as the major factor affecting lateral deformation under load.

For design of buildings using flexible diaphragms to distribute seismic forces, the overstrength factor $\Omega_0$ required under some conditions of irregular building configuration may be reduced by one-half of the original value specified in Table 12-2-1

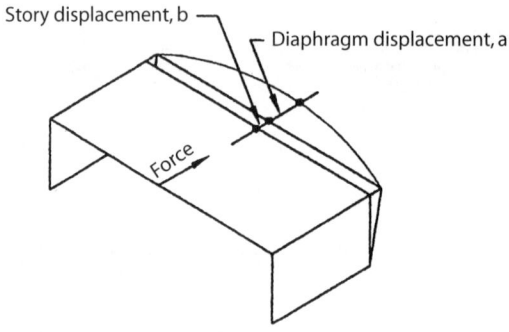

| Deflection Ratio | Flexibility |
| --- | --- |
| a > 2b | Flexible |
| 2b > a > 1/2b | Stiff |
| a < 1/2b | Rigid |

**Figure 13-9** Flexible, stiff, and rigid diaphragm definitions based on relative displacement. (*Courtesy U.S. Army Corps of Engineers*)

of ASCE/SEI 7-05, provided it is not taken as less than 2.0. Other portions of this standard allow wood-frame systems to avoid use of the overstrength factor, such as for collector elements (Section 12.14.7.3), which is a welcome allowance since calculated overstrength forces can be quite high.

### 13.5.2 Laminated Decks

Timber decks are formed by laminating together lumber strips, oriented to bend vertically about the strong axis, until a solid deck of the desired width is created. The finished product is covered by a wearing surface, such as asphalt paving over a bridge, which may also be placed atop longitudinally running planks that further protect the laminated deck below. A special type of deck is one where the laminations are prestressed in the transverse direction, compressing the lumber joints in such a way that lateral transfer of vertical shear occurs through friction, by which the stressed deck then behaves as an orthotropic plate. This has proven to be an efficient deck form for several reasons: Shorter laminations may be adhered together due to the benefit of frictional force redistribution; they can be built into relatively thin finished sections and can be easily erected. Testing conducted by Oliva and Dimakis (1988) on stressed wood bridge systems carrying AASHTO HS-20 truck loads, both single- and double-lane widths for spans up to 46 ft (14 m), indicated linear elastic behavior over the full range of loading conditions. However, some effects of prestress loss due to creep and moisture changes or deterioration are not fully understood and may play an important factor in choosing this type of deck for a project.

During the summer of 1983, eighteen timber bridges across the northern United States were inspected to determine their performance over the years since being built in the late 1960s and early 1970s (Gutowski and McCutcheon 1987). Glue-laminated decks generally appeared to be drier and in better physical condition than nail-laminated decks which may be attributed to greater passage of moisture between the nailed-lams. Moisture content and deterioration appeared to be greatest near the abutments, partially due to wear produced by entering vehicles.

### 13.5.3 Frames

Moment-resisting connections are difficult to create with wood framing and steel fasteners, mainly due to fastener slip upon loading cycles, shrinkage of wood away from fasteners, and drift limitations required by building codes. A fairly new creation, however, is the wood portal frame introduced to add rigidity along a wall with a large opening, such as a garage, as shown in Fig. 13-10. A continuous header works together with sheathing and a specific arrangement of metal straps to create a semirigid joint at the top of one or two narrow piers. Monotonic and cyclic testing has been conducted using a rational procedure that considers strength and stiffness in such a way that design values are appropriately tabulated according to the ductility of a light-frame wall with wood structural panel sheathing per Table 12.2-1 of ASCE/SEI 7-05 (APA 2005, p. 3).

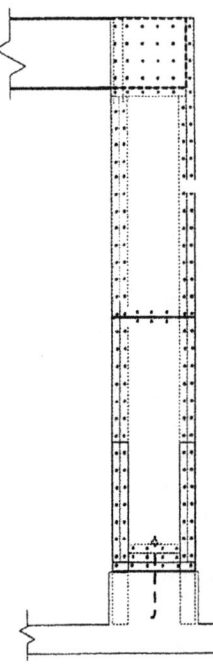

**Figure 13-10** Common wood portal frame construction detail.

## 13.5.4 Trusses

Premanufactured, light-frame engineered trusses are common roof framing (sometimes floor) elements, where the roof and ceiling framing (top and bottom chords) are held together by a system of diagonal members (webs) and light-gage metal plates mechanically pressed into both sides of each piece. Figure 13-11 shows five truss profiles with a variety of web arrangements (others are certainly

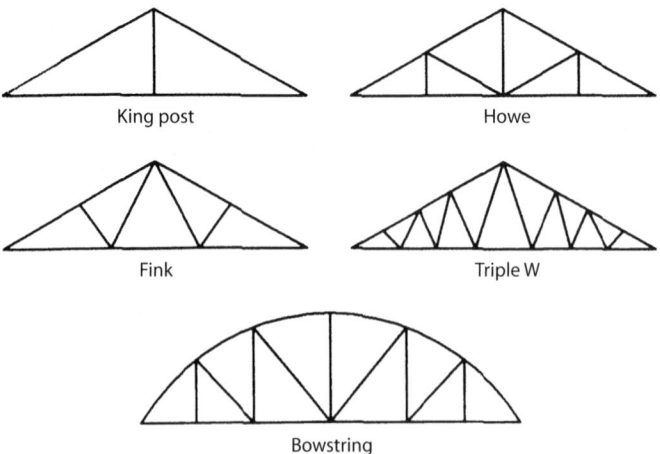

| **Figure 13-11** Miscellaneous wood-frame truss configurations.

possible), including one special chord configuration, that have different span capabilities. Structural characteristics of each type are noted below:

1. *King Post*: Economical up to about 26 ft (8 m) with wide application for residential, industrial, and commercial
2. *Howe*: Economical spans range between 16 and 48 ft (4.9 and 14.6 m), particularly suited for carrying concentrated loads applied to the bottom chord
3. *Fink*: Same range of spans as the Howe Truss, though the alternate web configuration proves to be more economical under some conditions
4. *Triple W*: Economical spans can be achieved up to about 80 ft (24.4 m) at a pitch of not less than 3 in 12
5. *Bowstring*: Commonly were used for industrial-type buildings (not used with much frequency for new construction, but existing trusses require inspection or repair from time to time), achieving spans up to about 80 ft (24.4 m), characterized by tremendous bottom chord tensile forces to resist straightening of the bowed top chord

Webs in truss assemblies are typically detailed and designed to have pinned ends, though this assumption is rarely completely true because of frictional resistance and geometric imperfections. Metal plated truss joints are tension-tested according to ASTM D1761 to further define their behavior under load, though an engineer must be careful not to extend results obtained from a given size and type to a different type of plate, unless data suggests that it can be done. After a certain load is reached, plate cross-sectional area will be the defining factor for capacity, regardless of how many teeth are present. Blocking panels and smaller engineered braced segments (see Fig. 13-12) that are placed between

**Figure 13-12** Anchored truss blocking panels to complete load path from diaphragm to shear wall.

trusses with tall heels to deliver lateral force down to a wall top plate, other collector, or shear wall.

There are two different types of lateral bracing for trusses and their systems: Temporary and permanent. Temporary lateral bracing is required for safe installation before sheathing is applied and nailed to the top and bottom chords. Permanent lateral bracing remains in place for the life of the structure and stabilizes entire trusses in order that they can function as the truss designer intends. The truss designer provides engineering for the truss system, which should include complete design of individual trusses, girder trusses, truss-to-truss connections, as well as common temporary and permanent lateral bracing. Truss design is a specialty and the engineer who runs the computer program is thoroughly familiar with member forces, member connection capabilities and behavior, and what is necessary to cause the entire truss system to function in a structurally sound fashion, being completely independent of the building below in most cases. Different conditions require coordination of effort between the truss and building designers, such as special anchorage (high uplift forces, three-point bearing truss reactions) special bracing configurations and drag truss specification and design, though the building designer typically assumes responsibility for anchorages to building elements at and below the top plate.

Jim Vogt, technical director of the Wood Truss Council of America (WTCA) at the time of his study, stated that a properly attached roof diaphragm consisting of rated wood sheathing and a properly attached gypsum board ceiling usually provides sufficient permanent lateral restraint to the top and bottom chords to prevent buckling (Vogt 1999). Publication BCSI 1-03 *Guide to Good Practice for Handling, Installing and Bracing of Metal Plate Connected Wood Trusses*, published by the WTCA and the Truss Plate Institute (TPI), gives an excellent description of different methods that may be used to provide permanent lateral bracing for web members, including continuous lateral bracing, T-reinforcement, L-reinforcement, scab reinforcement, stacked web reinforcement, and proprietary metal reinforcement products.

### 13.5.5 Structural Wood Panel Shearwalls

Historic design of light-frame shear walls include the designation of individual piers within a line of lateral force resistance that meet the aspect ratio (height-to-length) restrictions defined in model codes (2006 IBC, Table 2305.3.4), while ignoring the contribution of any other portion within that same wall line. Full-scale testing has shown, however, that wall sheathing placed above and below openings wherever they occur have a beneficial effect on the amount of drift and force distribution along the length of a wall (Cobeen, Russell, and Dolan 2004, W-30a, p. 29), causing a significant portion of applied load to be shared by all wall portions through continuous shear transfer (NAHBRC 1998, p. 15). This phenomenon had long been recognized, but seldom included in standard engineering design because of a lack of accurate or reliable scientific knowledge.

Different approaches to shear wall design are shown in Fig. 13-13. The perforated shear wall method of design introduced needed recognition and some semblance of reliability, mainly pioneered through the efforts of Hideo Sugiyama in the early 1980s. It is an empirical approach consisting of simple equations that account for reduction in stiffness and strength of a wall line that is fully sheathed with wood structural panels with a main benefit of reducing the number of overturning restraints (hold-downs). A variety of other tests have confirmed the adequacy of this method, even determining that empirical equations may actually underestimate the available strength of such walls (Dolan and Johnson 1997, TE-1996-001 and TE-1996-002). A variation of this method includes a rigorous mechanical approach to transferring force around openings and is applied to systems designated as *shear walls with openings* (2006 IBC, Section 2305.3.8.1), being subject to a different set of restrictions than the *perforated shear wall* approach (2006 IBC, Section 2305.3.8.2).

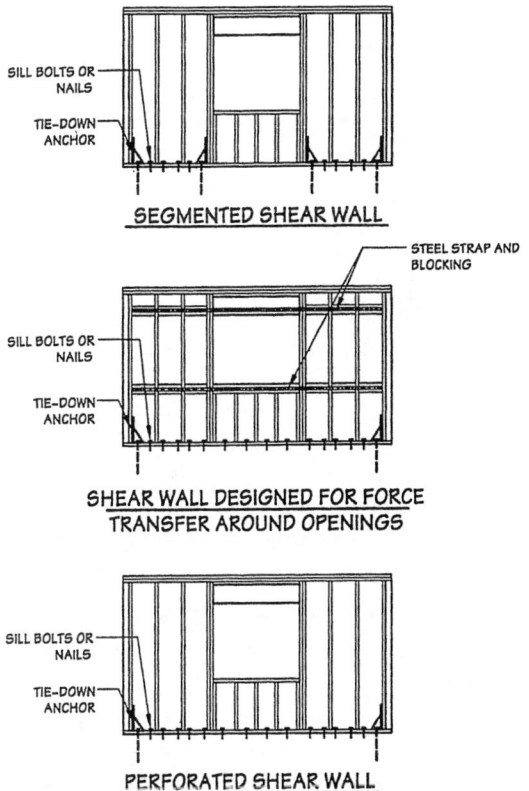

**Figure 13-13** Shear wall definitions based on design concept. (*From Breyer et al. Design of Wood Structures, McGraw-Hill, 2007*)

In SDC D, E, and F, heavily loaded shear walls defined as those with an allowable shear load of 350 plf or greater (490 plf using load and resistance factor design) require a 3x nominal stud at adjoining plywood panel edges and a 3x nominal sill plate when fastened to concrete or masonry (2006 IBC, Section 2305.3.11). The purpose of this is to reduce the risk of splitting of the framing members during a strong earthquake and is a requirement that came about in response to the 1994 earthquake that occurred in Northridge, California. This has not been a completely new requirement, but its application is now required more frequently—contractors are still getting used to it. In fact, many times these 3x studs are not installed and a retrofit is required. The technical services division of APA took notice of this problem and performed testing of a common retrofit solution for when 3x studs were omitted: Provide 2-2x studs, stitch nailed together. In their report, Martin and Skaggs concluded that ". . . the results from cyclic testing show that shear walls with double 2x studs stitch-nailed together perform about the same as those with a single 3x stud by all measures, except that those with double 2x framing had increased displacement capacity and ductility" (Martin and Skaggs 2003, p. 6).

### 13.5.6 Nonwood Panel Shearwalls

Gypsum board products are manufactured for a variety of different applications according to ASTM C1396 and tested per ASTM C473 for flexural strength, hardness, nail pull resistance, humidified deflection, end squareness, nominal thickness, geometrical properties, and water resistance. Panels are to be hung, fastened and finished per ASTM C840, which also gives the engineer a directive for quality. All gypsum board products consist of a noncombustible core, sometimes water-resistant, which is surfaced with a heavy paper. "Type X" members provide fire resistance ratings of 3/4-hour [1/2 in (12.7 mm) material] to 1 hour [5/8 in (15.9 mm) material] in assemblies that have been tested by methods described in ASTM E119.

A number of agencies have tested shear wall configurations with gypsum wallboard, cement plaster, and other sheathing materials to study behavior under wind pressure (monotonic-type loading) and seismic motion (cyclic-type loading) and determine the adequacy of current code requirements (APA 1996; Schmid 2002; Arnold, Uang, and Filiatraut 2003). Some observations that researchers believed were important to the building industry include:

1. Wall finish materials reduced lateral displacement of a building assembly subjected to earthquake motion in a laboratory by almost a factor of 3, exhibiting behavior as some form of elastic shell.
2. Allowable unit shear values for gypsum wallboard and cement plaster walls that are published in building codes (2006 IBC, Table 2306.4.5; NFPA 5000, Table 47.2.1.4.2.1) are so low that drift limitations are not likely to be exceeded during a seismic event.
3. Gypsum wallboard and cement plaster typically behave in a brittle fashion and failure of the former was often precipitated by pull-through of the nailing. When

used in combination with plywood, these materials add to the assembly's initial stiffness, but do not increase shear strength when subjected to cyclic loading.
4. Metal drip screeds, which are installed on wood-frame walls at or below the foundation plate line for the purpose of allowing trapped water to drain to the exterior, were found to pose problems in fastening metal lath of cement plaster assemblies to the wood bottom plate, resulting in poor performance during cyclic loading. Problems can be mitigated with a careful, attentive installation.

### 13.5.7 Wood Systems Combined with Other Materials

The laminating of steel plates in some fashion with adjacent wood members results in a section that can achieve greater spans than if made of wood alone. This is also a somewhat common way of reinforcing an existing wood beam to carry greater load. The transformed section method is used for design of such a member, by which the area of the steel elements are replaced by an elastically equivalent area of wood, calculated as the area of steel multiplied by the ratio of modulus of elasticity of steel to that of wood. This method is based on the idea that the deflected shapes of each material are compatible and interconnections are designed to carry the flow of shear forces along the length. When a steel plate is bolted between two wood members for bending strength purposes, the resulting assembly is known as a *flitch beam*.

## 13.6 CONSTRUCTION

Platform, or Western, framing is the easiest to construct because it allows one complete level to be framed and stabilized prior to raising the walls in the upper level. One disadvantage that must be investigated for wood-frame buildings of multiple stories, however, is that the total vertical shrinkage of framing elements can cause problems by cracking the finishing material or other difficulties related to rigid plumbing or mechanical systems running from floor to floor. A theoretical loadpath for transferring out-of-plane forces applied to a stud wall to the roof diaphragm and foundation is shown in Fig. 13-14, where a designer may choose to translate pressure into components and provide connectors accordingly.

For the past few years, hold-down anchor manufacturing companies have worked on producing devices that will transfer the uplift forces to the foundation in a concentric manner, likely in response to field observations of reduced anchor capacities due to the inherent eccentricity of loading from the center of a stud assembly or post. Nelson and Hamburger (1999) reported observations of brittle failure in wood posts through combined tension and flexure failure at the top bolt in a bolt-type hold-down device. These failures occurred at approximately two times the manufacturer's allowable hold-down loads, whereas concentric applications achieved factors of four or five times published loads.

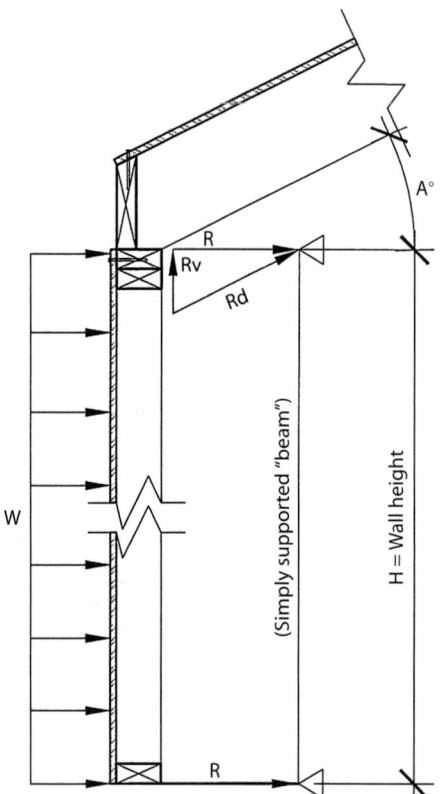

**Figure 13-14** Theoretical load path and assumptions for out-of-plane forces applied to wood-frame wall system.

Section 1203.2 of the 2006 IBC requires enclosed attic and rafter spaces to be ventilated in order to prevent damage to wood framing due to environmental effects. The basic area of ventilation required is 1/150 of the total area of space enclosed, but may be reduced to 1/300 of the total area if certain conditions are met. Fifty percent of the ventilation provided must occur 3 ft (0.91 m) above eave or cornice openings and the balance taken through the eave or cornice. Some truss manufacturers provide blocks that include holes drilled for ventilation, otherwise a contractor will need to create some and cover the openings with a corrosion-resistant screen in buildings used for human occupancy (2006 IBC, Section 1203.2.1). Ridge venting is always something of a mystery when a roof diaphragm is expected to be continuously blocked around all edges. However, it is typically sufficient to provide blocked wood panel edges between every other roof framing member and simply double the nailing requirement at this blocking to make up for what was left out at the spaces.

## 13.6.1 Risk in Design and During Service

Some of the uncertainties in designing wood assemblies are due to complicated patterns of behavior. For example, testing of narrow plywood panel shear walls indicated that the stiffness of hold-down anchor bolts against upward deflection was much less than that of the sill plate resting on a concrete foundation during downward displacement (Schmid, Nielsen, and Linderman 1994). Other uncertainties are due to conditions once thought to be acceptable and are determined to be problematic at a later time. An example of this is the use of preservative treatment used to protect sill members against decay or insect damage, once thought to accept fasteners with little concern and currently acknowledged to promote aggressive corrosion of unprotected fasteners that are embedded directly into the member. Codes and manufacturers now require a special zinc treatment on fasteners susceptible to corrosion, otherwise stainless steel or other metal that is naturally resistant to the effects may be used.

Connections can be tricky to design and it is usually best to stick with models that have already been used successfully. All connection designs begin with an applied force and are then completed through proper mechanics of resisting that load depending on fasteners chosen. Figure 13-15 shows a bottom chord brittle fracture near the location of a knot as a result of high tensile force and a poor connection design.

**Figure 13-15** Truss bottom chord tension failure in vicinity of knot. (*Courtesy Lane Engineers, Inc.*)

## 13.7 QUALITY CONTROL

Much of the quality of a completed wood-frame structure is due to measures exercised during construction. In order to minimize dimensional change due to seasoning and fluctuations in ambient humidity, for example, wood should be installed at moisture content levels similar to what will be expected in service. This will be dependent on outdoor or indoor relative humidities, exposure to sun and rain, rate of ventilation or conditioning of inside air, and other sources of moisture. Minimum fasteners are to be provided per commonly used tables from model codes, but this does not guarantee they will be driven correctly. An engineer cannot train a construction crew on proper techniques, but it is advantageous to keep fabrication details simple and predictable with what may be commonly expected in the field.

Field inspection of wood-frame structures is not commonly required, except in the case of high-load plywood diaphragms, special site-built or shop-built members (trusses, glued arches), or if the structure is of high importance due to occupancy or state requirements. Educational or institutional facilities commonly require special attention during construction and the regulating agency can dictate a need for observation by the structural engineer of record. These needs must be properly discovered and coordinated.

## 13.8 CODES AND STANDARDS

Early forms of wood construction were based on traditional methods developed over the years by carpenters and journeymen which worked well for somewhat regular-shaped buildings with short to intermediate spans. Unification of some direction in design and allowable wood strength values entered the construction industry through the U.S. Department of Agriculture's publication in 1935, Wood Handbook, which was followed by other helpful manuals.

Basic requirements for construction, design, and quality of materials for wood-frame construction are found in Chapter 23 of the 2006 IBC or Chapter 45 of NFPA 5000 with different levels of description, often helpful for understanding referenced design standards. Both model codes make reference to AF&PA *National Design Specification* (NDS-2005) and relevant supplements for both Allowable Stress Design (ASD) and Load and Resistance Factor Design (LRFD) for detailed design, construction, and quality assurance guidelines. It is a standard that defines methods for structural design of visually and mechanically graded lumber, glued laminated timber, timber poles and piles, prefabricated wood I-joists, structural composite lumber, wood structural panels, and single or multiple fastener connections. Chapter 14.5 of ASCE/SEI 7-05 contains some additional information regarding seismic design and detailing of wood-frame structures that govern over the original text. In terms of hierarchy, the model codes take precedence over applicable sections of ASCE/SEI 7-05, which takes a position in front of the NDS-2005 and associated supplement material, followed by other standards or design manuals that may be available for different aspects of construction. Design of these structures has predominantly been done by the ASD method for many years, but with the strength-approach of model codes, the LRFD approach is being taught and promoted on a fairly aggressive scale. Most engineers are familiar with common load duration factors that are allowed by the standard for ASD (see Fig. 13-16) to such an extent that they are somewhat emblazoned into memory, making it difficult to change design methods in the future.

One of the most tricky aspects of designing wood-frame structures is the consideration of irregular features as described in Tables 12.3-1 and 12.3-2 of ASCE/SEI 7-05. Many of these features involve an offset of elements within the lateral force resisting system, or those that are supported over a softer, more flexible layout of framing, such as the shear wall installed atop a beam shown in Fig. 13-17. In this particular photo, premanufactured roof trusses also require

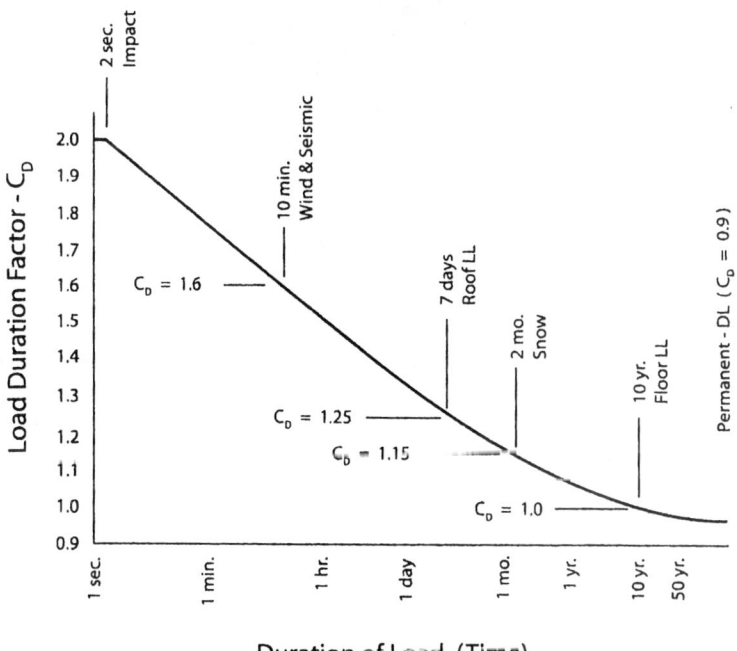

**Figure 13-16** Load duration factors for allowable stress design according to NDS-2005. (*Breyer et al, Design of Wood Structures, McGraw-Hill, 2007*)

**Figure 13-17** Wood-frame trusses attached to structural beam, requiring hangers and consideration of shear wall placement for diaphragm shear transfer.

positive fastening to the beam and special blocking or strapping is necessary to tie the lower roof diaphragm and the floor or upper roof diaphragm together in such a way that the building functions as a whole. Caution is warranted in designing wood elements and ties with the overstrength factor identified in Table 12.2-1 of ASCE/SEI 7-05, as forces can become quite high and difficult to manage properly.

# References

Abendroth, Robert and Terry J. Wipf. "Cyclic Load Behavior of Bolted Timber Joints." *Journal of Structural Engineering* 115, no. 10 (October 1989): 2496–510.

Aboulella, Fakhry. "Analysis of Cable-Stayed Bridges Supported by Flexible Towers." *Journal of Structural Engineering* 114, no. 12 (December 1988): 2741–54.

Adams, Dave K. "Performance of Structures Before and During Failure." *Practice Periodical on Structural Design and Construction* 11, no. 1 (February 2006): 20–4.

Addis, William. *Structural Engineering—The Nature and Theory of Design*. Oxford: Elsevier Books, 1990.

Aghayere, Abieyuwa and James G. MacGregor. "Tests of Reinforced Concrete Plates Under Combined In-Plane and Transverse Loads." *ACI Structural Journal* 87, no. 6 (November–December 1990): 615–22.

Ahmed, Tarig, Eldon Burley, and Stephen Rigden. "Effect of Alkali-Silica Reaction on Bearing Capacity of Plain and Reinforced Concrete." *ACI Structural Journal* 96, no. 4 (July–August 1999): 557–70.

Ali, Mir M. "Design of Foundations in Seismic Zones." *Concrete International* 19, no. 1 (January 1997): 44–8.

Akin, Ethan. *In Defense of Mindless Rote*. New York: The City College, March 30, 2001.

Alwood, Todd A. "What Engineers Should Know About Bending Steel." *Modern Steel Construction* 46, no. 5 (May 2006): 24–6.

American Association of School Administrators (AASA). *Learning Styles: Putting Research and Common Sense into Practice*. Arlington, VA: AASA Publications, 1991.

American Concrete Institute (ACI) Committee 211. *Standard Practice for Selecting Proportions for Normal, Heavyweight, and Mass Concrete, ACI 211.1-91*. Farmington Hills, MI: American Concrete Institute, 1997.

American Concrete Institute (ACI) Committee 212. *Chemical Admixtures for Concrete, ACI 212.3R-91*. Farmington Hills, MI: American Concrete Institute, 1991.

American Concrete Institute (ACI) Committee 318. *Building Code Requirements for Structural Concrete, ACI 318-05*. Farmington Hills, MI: American Concrete Institute, 2005.

American Concrete Institute (ACI) Committee 334. *Concrete Shell Structures Practice and Commentary, ACI 334.1R-92*. Farmington Hills, MI: American Concrete Institute, 1997.

American Concrete Institute (ACI) Committee 336. *Suggested Analysis and Design Procedures for Combined Footings and Mats, ACI 336.2R-88*. Farmington Hills, MI: American Concrete Institute, 1993.

American Concrete Institute (ACI) Committee 336. *Design and Construction of Drilled Piers, ACI 336.3R-98*. Farmington Hills, MI: American Concrete Institute, 1998.

American Concrete Institute (ACI) Committee 341. *Seismic Analysis and Design of Concrete Bridge Systems, ACI 341.2R-97*. Farmington Hills, MI: American Concrete Institute, 1997.

American Concrete Institute (ACI) Committee 343. *Analysis and Design of Reinforced Concrete Bridge Structures, ACI 343R-95*. Farmington Hills, MI: American Concrete Institute, 1995.

American Institute of Steel Construction (AISC). *Code of Standard Practice for Steel Buildings and Bridges (AISC 303-05)*. Chicago, IL: American Institute of Steel Construction, March 18, 2005.

American Institute of Steel Construction (AISC). *Manual of Steel Construction*. 13th ed. Chicago, IL: American Institute of Steel Construction, 2005.

American Institute of Steel Construction (AISC). *Seismic Provisions for Structural Steel Buildings (ANSI/AISC 341-05)*. Chicago, IL: American Institute of Steel Construction, 2005.

American Institute of Steel Construction (AISC). *Specification for Structural Steel Buildings (ANSI/AISC 360-05)*. Chicago, IL: American Institute of Steel Construction, March 9, 2005.

American Society of Civil Engineers (ASCE). *Minimum Design Loads for Buildings and Other Structures (ASCE/SEI 7-05)*. Reston, VA: American Society of Civil Engineers, 2005.

American Welding Society (AWS) Committee on Structural Welding. *Structural Welding Code—Steel (AWS D1.1-06)*. Miami, FL: American Welding Society, 2006.

American Wood Council. *National Design Specification for Wood Construction (NDS-2005)*. Washington, DC: American Forest and Paper Association, 2005.

Amrhein, James and Dimitry Vergun. *Reinforced Masonry Engineering Handbook*. 5th ed. New York: John Wiley & Sons, 1994.

APA—The Engineered Wood Association (APA). *Diaphragms and Shear Walls—Design/Construction Guide (Form L350)*. Tacoma, WA: APA—The Engineered Wood Association, November 2004.

APA—The Engineered Wood Association (APA). *A Portal Frame with Hold-Downs for Wall Bracing or Engineered Applications (Form TT-100)*. Tacoma, WA: APA—The Engineered Wood Association, January 2005.

APA—The Engineered Wood Association (APA). *Structural Panel Shear Walls with Gypsum Wallboard and Window/Door Openings (Report 157)*. Tacoma, WA: APA—The Engineered Wood Association, 1996.

Arnold, Andrew E., Chia-Ming Uang, and Andre Filiatraut. *Cyclic Behavior and Repair of Stucco and Gypsum Sheathed Wood-Framed Walls: Phase 1*. La Jolla, CA: Consortium of Universities for Research in Earthquake Engineering (CUREE), May 2003.

Arora, Madan L. "Writing Effective Specifications." *Civil Engineering* 64, no. 3 (March 1994): 69–71.

Ashby, Michael F. and David R.H. Jones. *Engineering Materials I—International Series on Materials Science and Technology*, Vol. 34. Oxford: Pergamon Press, 1980.

Astaneh, Abolhassan. *Steel Tips: Seismic Behavior and Design of Steel Shear Walls*. Moraga, CA: Structural Steel Educational Council, 2001.

Astaneh, Abolhassan and Marwan N. Nader. "Cyclic Behavior of Double Angle Connections." *Journal of Structural Engineering* 115, no. 5 (May 1989): 1101–18.

Ayers, Chesley. *Specifications for Architecture, Engineering, and Construction*. 2nd ed. New York: McGraw-Hill, 1984.

Azizinamini, Atorod and James B. Radziminski. "Static and Cyclic Performance of Semi-Rigid Steel Beam to Column Connections." *Journal of Structural Engineering* 115, no. 12 (December 1989): 2979–99.

Baker, Michael Jr. Inc. *LRFD Design Example for Steel Girder Superstructure Bridge (FHWA NHI-04-041)*. Washington, DC: FHWA/National Highway Institute, December 2003.

Bakht, Baidar and Leslie G. Jaeger. "Bearing Restraint in Slab-on-Girder Bridges." *Journal of Structural Engineering* 114, no. 12 (December 1988): 2724–40.

Bakht, Baidar and Leslie G. Jaeger. *Bridge Analysis Simplified*. New York: McGraw-Hill, 1985.
Banks, James H. *Introduction to Transportation Engineering*. 2nd ed. New York: McGraw-Hill, 2002.
Barsom, John M., and J. V. Pelligrino Jr. "Failure Analysis of Welded Steel Moment-Resisting Frame Connections." *Journal of Materials in Civil Engineering* 14, no. 1 (January/February 2002): 24–34.
Beaudoin, James J. "Why Engineers Need Materials Science." *Concrete International* 21, no. 8 (August 1999): 86–9.
Beer, Ferdinand P. and E. Russell Johnson, Jr. *Mechanics of Materials*. New York: McGraw-Hill, 1981.
Beer, Ferdinand P. and E. Russell Johnson, Jr. *Vector Mechanics for Engineers—Statics & Dynamics*. 4th ed. New York: McGraw-Hill, 1984.
Biehl, Bobb. *Mentoring—Confidence in Finding a Mentor and Becoming One*. Nashville, TN: Broadman & Holman Publishers, 1996.
Billington, David P. "Bridges as Art." *Civil Engineer* 60, no. 3 (March 1990): 50–3.
Bimel, Carl. "Yes, But After All, You Are the Expert." *Concrete International* 19, no. 1 (January 1997): 42–3.
Blodgett, Omer. *Design of Welded Structures*. Cleveland, OH: The James F. Lincoln Arc Welding Foundation, 1966.
Bonelli, Patricio, Rene Tobar, and Gilberto Leiva. "Experimental Study on Failure of a Reinforced Concrete Building." *ACI Structural Journal* 96, no.1 (January–February 1999):3–8.
Bortman, Jacob. "Analysis of Fastened Structural Connections." *American Institute of Aeronautics & Astronautics Journal* 30, no. 11 (November 1992):
Bowles, Joseph E. *Foundation Analysis and Design*. 2nd ed. New York: McGraw-Hill, 1982.
Bowyer, Jim L., Rubin Shmulsky, and John G. Haygreen. *Forest Products and Wood Science—An Introduction*. 4th ed. Ames, IA: Iowa State Press, 2003.
Breyer, Donald et al. *Design of Wood Structures*. 6th ed. New York: McGraw-Hill, 2007.
Brown, Robert. *Foundation Behavior and Repair*. 3rd ed. New York: McGraw-Hill, 1997.
Buchheit, John A. "Old Enough to Drive." *Roads & Bridges* 43, no. 9 (September 2005): 24–7.
Budinski, Kenneth G. and Michael K. Budinski. *Engineering Materials—Properties and Selection*. 6th ed. Upper Saddle River, NJ: Prentice Hall, 1999.
Building Seismic Safety Council (BSSC). *NEHRP Recommended Provisions for Seismic Regulations for New Buildings and Other Structures (FEMA 450)*. 2003 ed. Washington, DC: Federal Emergency Management Agency, 2004.
California Society of Professional Engineers. *An Introduction to Professional Engineering Licensing in California and Other States*. Arlington, VA: National Society of Professional Engineers, November 5, 2001.
Carter, Charles J. and Robert O. Disque. "Flexible Moment Connections." *Modern Steel Construction* 45, no. 9 (September 2005): 33–4.
Case, John, Lord Chiller and Carl T.F. Ross. *Strength of Materials and Structures*. 4th ed. New York: John Wiley & Sons, 1999.
Chai, Y. H. and D.T. Elayer. "Lateral Stability of Reinforced Concrete Columns Under Axial Reversed Cyclic Tension and Compression." *ACI Structural Journal* 96, no. 5 (September–October 1999): 780–9.
Chaffee, John. *Thinking Critically*. 3rd ed. Boston, MA: Houghton Mifflin Company, 1991.
Chen, Wai Fa, ed. *Handbook of Structural Engineering*. Boca Raton, FL: CRC Press, 1997.

Chen, Wai-Fah and Lian Duan, eds. *Bridge Engineering Handbook*. Boca Raton, FL: CRC Press, 1999.

Chen, Wai-Fa and Seung-Eock Kim. *LRFD Steel Design Using Advanced Analysis*. New York: CRC Press, 1997.

Cheok, Geraldine S. and William C. Stone. "Behavior of 1/6-Scale Model Bridge Columns Subjected to Inelastic Cyclic Loading." *ACI Structural Journal* 87, no. 6 (November–December 1990): 630–8.

Cheung, Sai On, Tak Wing Yiu Yiu, and Sau Fung Yeung. "A Study of Styles and Outcomes in Construction Dispute Negotiation." *Journal of Construction Engineering and Management* 132, no. 8 (August 2006): 805–14.

Cheung, Victor W.T. and W.K. Tso. "Lateral Load Analysis for Buildings with Setback." *Journal of Structural Engineering* 113, no. 2 (February 1987): 209–27.

Cobeen, Kelly, James Russell, and J. Daniel Dolan. *Recommendations for Earthquake Resistance in the Design and Construction of Woodframe Buildings (CUREE Pub. No. W30a & W30b)*. Richmond, CA: Consortium of Universities for Research in Earthquake Engineering (CUREE), 2004.

Collins, Michael P. and Daniel Kuchma. "How Safe Are Our Large, Lightly Reinforced Concrete Beams, Slabs, and Footings?" *ACI Structural Journal* 96, no. 4 (July–August 1999): 482–90.

Cooper, Norm. "Causes of Lawsuits In Masonry Construction." *Structural Engineer* 4, no. 1 (February 2003): 26–9.

Coren, Stanley. "Sleep Deprivation, Psychosis, and Mental Efficiency". *Psychiatric Times* 15, no. 3 (March 1998).

Cramer, Steven M., John M. Drozdek, and Ronald W. Wolfe. "Load Sharing Effects in Light-Frame Wood-Truss Assemblies." *Journal of Structural Engineering* 126, no. 12 (December 2000): 1388–94.

Darwin, David. "Concrete in Compression." *Concrete International* 21, no. 8 (August 1999): 82–5.

Deng, Kam, Gilbert Y. Grondin, and Robert G. Driver. "Effect of Loading Angle on the Behavior of Fillet Welds." *Engineering Journal* 43, no. 1 (First Quarter 2006): 9–23.

Dodd, Lawrence L. and Nigel Cooke. "Capacity of Circular Bridge Columns Subjected to Base Excitation." *ACI Structural Journal* 97, no. 2 (March–April 2000): 297–307.

Dolan, J.D., S.T. Gutshall, and T.E. McLain. "*Monotonic and Cyclic Tests to Determine Short-Term Load Duration Performance of Nail & Bolt Connections, Timber Engineering Report No. TE-1994-001.*" Virginia Polytechnic Institute and State University, October 1995.

Dolan, J.D. and A.C. Johnson. "*Cyclic Tests of Long Shear Walls with Openings, Timber Engineering Report TE-1996-002.*" Virginia Polytechnic Institute and State University, June 16, 1997.

Dolan, J.D. and A.C. Johnson. "*Monotonic Tests of Long Shear Walls with Openings, Timber Engineering Report TE-1996-001.*" Virginia Polytechnic Institute and State University, June 16, 1997.

Drysdale, Robert G., Ahmad A. Hamid, and Lawrie R. Baker. *Masonry Structures—Behavior and Design*. Boulder, CO: The Masonry Society, 1999.

El-Attar, Adel G., Richard N. White, and Peter Gergely. "Behavior of Gravity Load Designed Reinforced Concrete Buildings Subjected to Earthquakes." *ACI Structural Journal* 94, no. 2 (March–April 1997): 133–45.

Ellingwood, Bruce. "Design and Construction Error Effects on Structural Reliability." *Journal of Structural Engineering* 113, no. 2 (February 1987): 409–22.

Englekirk, Robert E. *Seismic Design of Reinforced & Precast Concrete Buildings*. New York: John Wiley & Sons, 2003.

Esmaeily, Asad and Yan Xiao. "Behavior of Reinforced Concrete Columns Under Variable Axial Loads: Analysis." *ACI Structural Journal* 102, no. 5 (September 2005): 736–44.

Falk, Robert H. and Rafik Y. Itani. "Dynamic Characteristics of Wood and Gypsum Diaphragms." *Journal of Structural Engineering* 113, no. 6 (June 1988): 1357–72.

Felder, Richard M. *"Does Engineering Education Have Anything to Do with Either One? Towards a Systems Approach to Training Engineers."* Lecture delivered at North Carolina State University at Raleigh, October 12, 1982.

Federal Emergency Management Agency (FEMA). *Building Performance: Hurricane Andrew in Florida (FIA-22)*. Washington, DC: Federal Emergency Management Agency, December 21, 1992.

Fisher, James M. "Expansion Joints: Where, When and How." *Modern Steel Construction* 45, no. 4 (April 2005): 29–34.

Fischer, David et al. *"Shake Table Tests of a Two-Story Woodframe House (CUREE Pub. No. W-06)."* Richmond, CA: Consortium of Universities for Research in Earthquake Engineering (CUREE), 2001.

Foutch, Douglas A. "Seismic Behavior of Eccentrically Braced Steel Buildings." *Journal of Structural Engineering* 115, no. 8 (August 1989): 1857–76.

Frangopol, Dan M. and James P. Curley. "Effects of Damage and Redundancy on Structural Reliability." *Journal of Structural Engineering* 113, no. 7 (July 1987): 1533–49.

Franklin, James. *The Science of Conjecture: Evidence and Probability Before Pascal*. Baltimore, MD: Johns Hopkins University Press, 2002.

French, Thomas E. et al. *Mechanical Drawing*. New York: McGraw-Hill Book Company, 1980.

Galambos, Theodore V. *Guide to Stability Design Criteria for Metal Structures*. New York: John Wiley & Sons, 1988.

Gang-Nail Systems, Inc. *A Handbook of Prefabricated Wood Trusses*. Miami, FL: Gang-Nail Systems, Inc., 1981.

Gaylord, Edwin, Jr. and Charles N. Gaylord, eds. *Structural Engineering Handbook*. 3rd ed. New York: McGraw-Hill, 1990.

Ghoneim, Mashhour G. and James G. MacGregor. "Behavior of Reinforced Concrete Plates Under Combined In-plane and Lateral Loads." *ACI Structural Journal* 91, no. 2 (March–April 1994): 188–97.

Ginger, John D. "Internal Pressures and Cladding Net Wind Loads on a Full-Scale Low-Rise Building." *Journal of Structural Engineering* 126, no. 4 (April 2000): 538–43.

Goldbloom, Joseph. "Improving Specifications." *Civil Engineering* 62, no. 9 (September 1992): 68–70.

Gonis, A., P.E.A. Turchi, and Josef Kudrnovsky, eds. *Stability of Materials*. New York: Plenum Press, 1996.

Goschy, Bela. *Design of Buildings to Withstand Abnormal Loading*. London: Butterworth & Co., 1990.

Government of South Australia, Southern Adelaide Health Service. *"Caffeine"*. Parkside: Drug & Alcohol Services Council, June 6, 2002.

Green, Eric C. "Excessive Movement." *Structural Engineer* 6, no. 9 (October 2005): 34–40.

Gregory, S. A. *Creativity and Innovation in Engineering*. London: C. Tinling & Co., 1972.

Griemann, L.F. et al. *Design of Piles for Integral Abutment Bridges, Final Report*. Ames, Iowa: Iowa State University, 1984.

Gulkan, Polat and Mete A. Sozen. "Procedure for Determining Seismic Vulnerability of Building Structures." *ACI Structural Journal* 96, no. 3 (May–June 1999): 336–42.

Gulde, Jim. "The Most Common Misunderstandings in Masonry." *Structural Engineer* 1, no. 3 (April 2000): 38–41.

Guralnick, Sidney A. and Abbes Yala. "Plastic Collapse, Incremental Collapse and Shakedown of Reinforced Concrete Structures." *ACI Structural Journal* 95, no. 2 (March–April 1998): 163–74.

Gutowski, Richard M. and William J. McCutchen. "Comparative Performance of Timber Bridges." *Journal of Structural Engineering* 113, no. 7 (July 1987): 1468–86.

Hall, John F., ed. "Northridge Earthquake of January 17, 1994 Reconnaissance Report, Vol. 1." *Earthquake Spectra* 11, nos. S2 and S3 (April 1995).

Halvorson, Robert A. "Structural Details: Is Anything Missing?" *Civil Engineering* 60, no. 11 (November 1990): 70–1.

Hamburger, Ronald and Andrew Whittaker. "Design of Steel Structures for Blast-Related Progressive Collapse Resistance." *Modern Steel Construction* 44, no. 3 (March 2004): 45–51.

Hanagan, Linda M. "Floor Vibration Serviceability: Tips and Tools for Negotiating a Successful Design." *Modern Steel Construction* 43, no. 4 (April 2003): 39–48.

Hanson, Robert D. "Supplemental Damping for Improved Seismic Performance." *Earthquake Spectra* 9, no. 3 (August 1993): 319–34.

Harrington, John R. "Design-Build and the Structural Engineer." *Structural Engineer* 1, no. 3 (April 2000): 26–37.

Harrison, Patrick J. "For the Ideal Slab-on-Ground Mixture." *Concrete International* 26, no. 3 (March 2004): 49–55.

Hazleton, Robert. "What's Wrong with Steel Drawings?" *Structure* 13, no. 2 (February 2006): 10–12.

Hendry, Arnold W., ed. *Reinforced and Prestressed Masonry*. Essex, Great Britain: Longman Scientific & Technical, 1991.

Hertzberg, Richard W. *Deformation and Fracture Mechanics of Engineering Materials*. New York: John Wiley & Sons, 1996.

Heyer, Paul. *Architects on Architecture: New Directions in America*. New York: Van Nostrand Reinhold, 1993.

Higbee, Kenneth L. *Your Memory—How it Works and How to Improve It*. 2nd ed. New York: Marlowe & Company, 1996.

Highet, Gilbert. *The Art of Teaching*. New York: Vintage Books, 1950.

Hoke, John, ed. *Architectural Graphic Standards*. 8th ed. New York: John Wiley & Sons, 1988.

Holmsteat, Laurie. "Winning the Word War." *Civil Engineering* 60, no. 3 (March 1990): 70–2.

Holtz, Robert D. and William D. Kovacs. *An Introduction to Geotechnical Engineering*. Englewood Cliffs, NJ: Prentice-Hall, 1981.

Hosford, William F. and Robert M. Caddell. *Metal Forming—Mechanics and Metallurgy*. Englewood Cliffs, NJ: Prentice-Hall, 1983.

Howell, Edward B. and Richard P. Howell. *Lessons in Professional Liability: A Notebook for Design Professionals*. Monterey, CA: Design Professionals Insurance Company, 1988.

Hritonenko, Natali and Yuri Yatsenko. *Applied Mathematical Modeling of Engineering Problems*. Norwell, MA: Kluwer Academic Publishers, 2003.

Hunaidi, Osama. "Traffic Vibrations in Buildings." *NRC Construction Technology Update No. 39*, Ottawa, Canada: Institute for Research in Construction, June 2000.

Hunt, Richard D. and Anthony H. Bryant. "Laterally Loaded Nail Joints in Wood." *Journal of Structural Engineering* 116, no. 1 (January 1990): 111–24.

Ibell, Tim J., Chris T. Morley, and Campbell R. Middleton. "Strength and Behavior in Shear of Concrete Beam-and-Slab Bridges." *ACI Structural Journal* 96, no. 3 (May–June 1999): 386–91.

Inbanathan, Mahil J. and Martin Wieland. "Bridge Vibrations Due to Vehicle Moving Over Rough Surface." *Journal of Structural Engineering* 113, no. 9 (September 1987): 1994–2008.

Inter-Jurisdictional Regulatory Collaboration Committee (IRCC). *Guidelines for the Introduction of Performance-Based Building Regulations (Discussion Paper)*. May 1998. http://www.ircc.gov.au

International Code Council (ICC). *International Building Code.* 2006 ed. Whittier, CA: International Code Council, 2006.

Ioannides, Socrates A. and Sandeep Mehta. "Restrained vs. Unrestrained Fire Ratings: A Practical Approach." *Modern Steel Construction* 37, no. 5 (May 1997): 52–64.

Jaradat, Omar A., David I. McLean, and M. Lee Marsh. "Performance of Existing Bridge Columns Under Cyclic Loading—Part 1: Experimental Results and Observed Behavior." *ACI Structural Journal* 95, no.6 (November–December 1998): 695–704.

Jeffers, Andrew T., Angela G. Safferman, and Steven I. Safferman. "Understanding K-12 Engineering Outreach Programs." *Journal of Professional Issues in Engineering Education and Practice* 130, no. 2 (April 2004): 95–108.

Jones, Russell H. ed. *Environmental Effects on Engineered Materials.* New York: Marcel Dekker, 2001.

Kaminetzky, Dov. *Design and Construction Failures.* New York: McGraw-Hill, 1991.

Kamiya, Fumio. "Buckling Theory of Sheathed Walls: Linear Analysis." *Journal of Structural Engineering* 113, no. 9 (September 1987): 2009–22.

Kardon, Joshua B. *The Structural Engineer's Standard of Care.* Online Ethics Center International Conference on Ethics in Engineering and Computer Science, March 1999. http://www.onlineethics.dnsalias.com

Kelley, David. *The Art of Reasoning.* 2nd ed. New York: W. W. Norton and Company, 1994.

Kelly, Dominic J. and Joseph J. Zona. "Design Tips for Steel in Low or Moderate Seismicity Regions." *Modern Steel Construction* 46, no. 2 (February 2006): 50–6.

Kemper, John D. *Introduction to the Engineering Profession.* New York: Holt, Rinehart & Winston, 1985.

Kim, Daejoong, Woo Kim, and Richard N. White. "Arch Action in Reinforced Concrete Beams—A Rational Prediction of Shear Strength." *ACI Structural Journal* 96, no. 4 (July–August 1999): 586–93.

King, Bruce. *Buildings of Earth and Straw.* Sausalito, CA: Ecological Design Press, 1996.

Kiureghian, Armen Der. "Measures of Structural Safety Under Imperfect States of Knowledge." *Journal of Structural Engineering* 115, no. 5 (May 1989): 1119–40.

Kline, Eric S. and James D. Machen. "Shop Painting: Myth Versus Reality." *Modern Steel Construction* 33, no. 12 (December 1993): 32–8.

Koglin, Terry L. *Movable Bridge Engineering.* New York: John Wiley & Sons, 2003.

Kowalsky, Mervyn J., M. J. Nigel Priestley, and Frieder Seible. "Shear and Flexural Behavior of Lightweight Concrete Bridge Columns in Seismic Regions." *ACI Structural Journal* 96, no.1 (January–February 1999): 136–48.

Kurama, Yahya, Richard Sause, et al. "Lateral Load Behavior and Seismic Design of Unbonded Post-Tensioned Precast Concrete Walls." *ACI Structural Journal* 96, no. 4 (July–August 1999): 622–32.

Kuwamura, Hitoshi and Theodore V. Galambos. "Earthquake Load for Structural Reliability." *Journal of Structural Engineering* 115, no. 6 (June 1989): 1446–62.

Lassen, T., Ph. Darcis, and N. Recho. "Fatigue Behavior of Welded Joints Part 1—Statistical Methods for Fatigue Life Prediction." *Welding Journal* 84, no. 12 (December 2005): s183–7.

Levins, Alexander and William Chalk. *Graphics in Engineering Design*. 3rd ed. New York: John Wiley & Sons, 1980.

Li, W.Y., L.G. Tham, and Y.K. Cheung. "Curved Box Girder Bridges." *Journal of Structural Engineering* 114, no. 6 (June 1988): 1324–38.

Linnert, George E. *Welding Metallurgy*. 4th ed. Miami, FL: American Welding Society, 1994.

Linzell, D., R.T. Leon, and A.H. Zureick. "Experimental and Analytical Studies of a Horizontally Curved Steel I-Girder Bridge During Erection." *Journal of Bridge Engineering* 9, no. 6 (November/December 2004): 521–30.

Liu, Judy. "Cyclic Testing of Simple Connections Including Effects of Slab." *Journal of Structural Engineering* 126, no. 1 (January 2000): 32–9.

Locke, Dennis. *Project Management*. 6th ed. New York: John Wiley & Sons, 1996.

Lubell, Adam S., Helmut G. L. Prion, Carlos E. Ventura et al. "Unstiffened Steel Plate Shear Wall Performance Under Cyclic Loading." *Journal of Structural Engineering* 126, no. 4 (April 2000): 453–60.

Maaddawy, Tamer El, Khaled Soudki, and Timothy Topper. "Long-Term Performance of Corrosion-Damaged Reinforced Concrete Beams." *ACI Structural Journal* 102, no. 5 (September 2005): 649–56.

Magnusson, Jon. "Learning From Structures Subjected to Loads Extremely Beyond Design." *Modern Steel Construction* 44, no. 3 (March 2004): 31–4.

Mangonon, Pat L. *The Principles of Materials Selection for Engineering Design*. Upper Saddle River, NJ: Prentice Hall, 1999.

Mansur, M. A. "Design of Reinforced Concrete Beams with Small Openings Under Combined Loading." *ACI Structural Journal* 96, no. 5 (September–October 1999): 675–82.

Martin, Mike W. and Roland Schinzinger. *Ethics in Engineering*. 4th ed. Boston, MA: McGraw-Hill, 2005.

Martin, Zeno A. and Thomas D. Skaggs. *Shear Wall Lumber Framing: Double 2x's vs. Single 3x's at Adjoining Panel Edges, APA Report T2003-22*. Tacoma, WA: American Plywood Association, May 8, 2003.

Marxhausen, Peter D. and Aaron Bagley. "Foundation Design—Understanding Geotechnical Factors of Safety in the Design of Foundations." *Structure* 13, no. 6 (June 2006): 47–50.

Masonry Standards Joint Committee. *Building Code Requirements for Masonry Structures (ACI 530/ASCE 5/TMS 402)*. Farmington Hills, MI: American Concrete Institute; Reston, VA: Structural Engineering Institute of the American Society of Engineers; Boulder, CO: The Masonry Society 2005.

Merritt, Frederick S. *Building Design & Construction Handbook*. New York: McGraw-Hill, 1994.

Miller, Duane K. "What Every Engineer Should Know About Welding." *Modern Steel Construction* 37, no. 5 (May 1997): 38–48.

Miller, Duane K. "What Every Engineer Should Know About Welding Procedures." *AWS Welding Journal* 78, no. 8 (August 1999): 37–44.

Miller, Duane K. "Welding Considerations for Designers and Fabricators." *Modern Steel Construction* 46, no. 2 (February 2006): 37–42.

Mistry, Vasant. "Economical Steel Bridge Design." *Modern Steel Construction* 34, no. 3 (March 1994): 42–7.

Monasa, Frank F. *Approximate Methods & Verification Procedures of Structural Analysis & Design*. New York : American Society of Civil Engineers, 1991.

Mueller, David S. and Chad R. Wagner. "Field Observations and Evaluations of Streambed Scour at Bridges." McLean, VA: U.S. Department of Transportation Federal Highway Administration, May 2005.

Murdough Center for Engineering Professionalism. *Conduct and Ethics in Engineering Practice Related to the North American Free Trade Agreement*. Lubbock, TX: Texas Tech University, 1995. http://www.niee.org

Murphy, Cathy. "The Struggle with Ethics." *Structural Engineer* 1, no. 4 (May 2000): 26–31.

Murphy, Cathy. "The Architect and the Bridge (Part Two)." *Structural Engineer* 1, no. 10 (November 2000): 32–5.

Naeim, Farzad. *Seismic Design Handbook*. 2nd ed. Boston, MA: Kluwer Academic Publishers, 2001.

Naeim, Farzad. *Steel Tips: Design Practice to Prevent Floor Vibrations*. Moraga, CA: Structural Steel Educational Council, September 1991.

Narayanan, R., ed. *Plated Structures—Stability and Strength*. London: Applied Science Publishers, 1983.

National Association of Home Builders Research Center, Inc (NAHBRC). *Performance of Perforated Shear Walls with Narrow Wall Segments, Reduced Base Restraint, and Alternate Framing Methods*. Upper Marlboro, MD: National Association of Home Builders Research Center, Inc., May 1998.

National Council of Examiners for Engineering and Surveying (NCEES). *Continuing Professional Competency Guidelines*. Clemson, NC: National Council of Examiners for Engineering and Surveying, November 2004.

National Fire Protection Association (NFPA). *Building Construction and Safety Code (NFPA 5000)*. Quincy, MA: National Fire Protection Association, 2005.

National Institute of Standards and Technology (NIST). *Performance of Physical Structures in Hurricane Katrina and Hurricane Rita: A Reconnaissance Report (NIST TN 1476)*. Gaithersburg, MD: U.S. Department of Commerce, June 2006.

National Science Board Committee on Education and Human Resources (NSB). *The Science and Engineering Workforce: Realizing America's Potential*. Arlington, VA: National Science Foundation, August 14, 2003.

Nawy, Edward G. *Prestressed Concrete— A Fundamental Approach*. Upper Saddle River, NJ: Prentice Hall, 2003.

Nelson, Robert H. *Zoning and Property Rights*. Cambridge, MA: The MIT Press, 1977.

Nelson, Ronald F. and Ronald O. Hamburger. "Hold-down Eccentricity and the Capacity of the Vertical Wood Member." *Building Standards* LXVIII, no. 6 (November/December 1999): 27–8.

Nevling, D., D. Linzell, and J. Laman. "Examination of Level of Analysis Accuracy for Curved I-Girder Bridges Through Comparisons to Field Data." *Journal of Bridge Engineering* 11, no. 2 (March/April 2006): 160–8.

Nilson, Arthur H. *Design of Prestressed Concrete*. 2nd ed. New York: John Wiley & Sons, 1987.

Noeth, Richard J., Ty Cruce, and Matt T. Harmston. *Maintaining a Strong Engineering Workforce—ACT Policy Report*. Iowa City, IA: American College Testing, Inc. (ACT), 2003.

Nowak, Andrzej S. and Kevin R. Collins. *Reliability of Structures*. Boston, MA: McGraw-Hill, 2000.

O'Brien, Annette. *AWS Welding Handbook*. 4th vol. Miami, FL: American Welding Society, 2004.

O'Brien, Robert, ed. *Jefferson's Welding Encyclopedia*. 18th ed. Miami, FL: American Welding Society, 1997.

Oh, Byung Hwan, Kwang Soo Kim, and Young Lew. "Ultimate Load Behavior of Post-Tensioned Prestressed Concrete Girder Bridge through In-Place Failure Test." *ACI Structural Journal* 99, no. 2 (March–April 2002): 172–80.

Oh, Byung Hwan and Sung Tae Chae. "Structural Behavior of Tendon Coupling Joints in Prestressed Concrete Bridge Girders." *ACI Structural Journal* 98, no.1 (January–February 2001): 87–95.

O'Leary, John E. and Sergio Zoruba. "High-Strength Bolting Made Easy." *Modern Steel Construction* 46, no. 5 (May 2006): 41–4.

Oliva, M.G. and A. Dimakis. "Behavior of Stress-Laminated Timber Highway Bridge." *Journal of Structural Engineering* 114, no. 8 (August 1988): 1850–69.

Olshansky, Robert B. *Promoting the Adoption and Enforcement of Seismic Building Codes: A Guidebook for State Earthquake and Mitigation Managers (FEMA 313)*. Washington DC: Federal Emergency Management Agency, January 1998.

Parisi, Maria A. and Maurizio Piazza. "Mechanics of Plain and Retrofitted Traditional Timber Connections." *Journal of Structural Engineering* 126, no. 12 (December 2000): 1395–1403.

Paz, Mario. *Structural Dynamics—Theory and Computation*. 3rd ed. New York: Van Nostrand Reinhold, 1991.

Petroski, Henry. *To Engineer is Human—The Role of Failure in Successful Design*. New York: Vintage Books, 1992.

Phillips, William R. and David A. Sheppard. *Plant-Cast Precast and Prestressed Concrete—A Design Guide*. Chicago, IL: The Prestressed Concrete Manufacturers Association of California, 1980.

Pilakoutas K. and A. S. Elnashai. "Cyclic Behavior of Reinforced Concrete Cantilever Walls, Part II: Discussions and Theoretical Comparisons." *ACI Structural Journal* 92, no. 4 (July–August 1995): 425–34.

Pincheira, Jose, Michael G. Oliva, and Wei Zheng. "Behavior of Double-Tee Flange Connectors Subjected to In-Plane Monotonic and Reversed Cyclic Loads." *PCI Journal* 50, no. 6 (November–December 2005): 32–54.

Podolny Jr., Walter. "Cable-Stayed Bridges." *Engineering Journal* (1st quarter 1974): 1–11.

Podolny Jr., Walter. "Making the Most of HPC." *Civil Engineering* 71, no. 2 (February 2001): 54–7.

Popper, Karl. *Conjectures and Refutations*. London: Taylor and Francis Group, 2002.

Prescott, Samuel T. *Federal Land Management—Current Issues and Background*. New York: Nova Science Publishers, 2003.

Priestley, M. J. Nigel. "Myths and Fallacies in Earthquake Engineering." *Concrete International* 19, no. 2 (February 1997):54–63.

Protter, Murray H. and Charles B. Morrey, Jr. *Analytic Geometry*. 2nd ed. Reading, MA: Addison-Wesley Publishing Company, 1974.

Pujol, Santiago et al. "Behavior of Low-Rise Reinforced Concrete Buildings." *Concrete International* 22, no. 1 (January 2000): 40–44.

Rahal, Khaldoun N. "Longitudinal Steel Stresses in Beams Due to Shear and Torsion in AASHTO-LRFD Specifications." *ACI Structural Journal* 102, no. 5 (September 2005): 689–98.

Ratay, Robert T. *Forensic Structural Engineering Handbook*. New York: McGraw-Hill, 2000.

Rahal, Khaldoun N. and Michael P. Collins. "Analysis of Sections Subjected to Combined Shear and Torsion—A Theoretical Model." *ACI Structural Journal* 92, no. 4 (July–August 1995): 459–69.

Ratterman, David B. "Managing Risk: Insurance and Indemnity Clauses in Construction Contracts." *Modern Steel Construction* 43, no. 4 (April 2003): 74–80.

Research Council on Structural Connections (RCSC) Committee A.1. *Specification for Structural Joints Using ASTM A325 or A490 Bolts*. Chicago, IL: Research Council on Structural Connections, c/o American Institute of Steel Construction, June 30, 2004.

Richardson, Jesse J., Jr., Meghan Zimmerman Gough, and Robert Puentes. *Is Home Rule the Answer? Clarifying the Influence of Dillon's Rule on Growth Management*. Washington, DC. The Brookings Institute, January 2003.

Roeder, Charles W. "Seismic Behavior of Concentrically Braced Frames." *Journal of Structural Engineering* 115, no. 8 (August 1989): 1837–56.

Rosignoli, Marco. "Presizing of Prestressed Concrete Launched Bridges." *ACI Structural Journal* 96, no. 5 (September–October 1999): 705–10.

Rosignoli, Marco. "Incremental Bridge Launching." *Concrete International* 19, no. 2 (February 1997): 36–40.

Rossow, Edwin C. *Analysis and Behavior of Structures*. Upper Saddle River, NJ: Prentice Hall, 1996.

Salmon, Charles G. and John E. Johnson. *Steel Structures—Design and Behavior*. 4th ed. New York: HarperCollins, 1996.

Salvadori, Mario. *Why Buildings Stand Up—The Strength of Architecture*. New York: W.W. Norton & Company, 1990.

Sande, Ken. *The Peacemaker—A Biblical Guide to Resolving Personal Conflict*. Grand Rapids, MI: Baker Book House, 1991.

Sargent, Dennis D., Carlos E. Ventura, Aftab A. Mufti, and Baidar Bakht. "Testing of Steel-Free Bridge Decks." *Concrete International* 21, no. 8 (August 1999): 55–61.

Sarkisian, Mark. "Learning from Kobe's Tragedy." *Structural Engineering Forum* 2, no. 1 (May–June 1996): 18–23.

Saul, Reiner, Holger S. Svensson, and Karl Humpf. "Cable-Stayed Bridges: The German Connection." *Concrete International* 20, no. 2 (February 1998): 71–5.

Scott, Stuart. *Communication and Conflict Resolution—A Biblical Perspective*. Bemidji, MN: Focus Publishing, 2005.

Schijve, Jaap. *Fatigue of Structures and Materials*. London: Kluwer Academic Publishers, 2001.

Schmid, Ben L. "Results and Ramifications of Cyclic Tests of Typical Sheathing Material Used on One- and Two-Story Wood Framed Residences." *Proceedings of the 71st Annual SEAOC Convention*, Santa Barbara, CA, 2002.

Schmid, Ben L., Richard J. Nielsen, and Robert R. Linderman. "Narrow Plywood Shear Panels." *Earthquake Spectra* 10, no. 3 (August 1994): 569–88.

Schuller, Michael P. "Masonry Tips for Structural Engineers." *Structure* 13, no. 5 (May 2006): 17–9.

Seilie, Ignasius F. and John D. Hooper. "Steel Plate Shear Walls: Practical Design and Construction." *Modern Steel Construction* 45, no. 4 (April 2005): 37–43.
Segui, William T. *LRFD Steel Design*. 2nd ed. Pacific Grove, CA: Brooks/Cole Publishing Company, 1999.
Shneur, Victor. "24 Tips for Simplifying Braced Frame Connections." *Modern Steel Construction* 46, no. 5 (May 2006):33–6.
Simiu, Emil and Robert H. Scanlan. *Wind Effects on Structures*. New York: Wiley-Interscience, 1978.
Simons, Bryce. "Spotting Specs Offenders." *Civil Engineering* 63, no. 2 (February 1993): 68–9.
Smith, Ian, Eric Landis, and Meng Gong. *Fracture and Fatigue in Wood*. West Sussex: John Wiley & Sons, 2003.
Spence, Gerry. *How to Argue & Win Every Time*. New York: St. Martin's Press, 1995.
Stalnaker, Judith J. and Ernest C. Harris. *Structural Design in Wood*. New York: Van Nostrand Reinhold, 1989.
Stanley, Paul D. and J. Robert Clinton. *Connecting—The Mentoring Relationships You Need to Succeed in Life*. Colorado Springs, CO: Navpress, 1992.
Stiller, William B., Janos Gergely, and Rodger Rochelle. "Testing, Analysis, and Evaluation of a GFRP Deck on Steel Girders." *Journal of Bridge Engineering* 11, no. 4 (July/August 2006): 394–400.
Stine, Tabitha S. "Specifying HSS." *Modern Steel Construction* 45, no. 7 (July 2005): 39–40.
Structural Engineers Association of California (SEAOC) Seismology Committee. *1999 Recommended Lateral Force Requirements and Commentary*. 7th ed. Sacramento, CA: Structural Engineers Association of California, 1999.
Structural Engineers Association of California (SEAOC). *Recommended Guidelines for the Practice of Structural Engineering in California*. Sacramento, CA: Structural Engineers Association of California, 1999.
Strunk, William Jr. and E.B. White. *The Elements of Style*. 3rd ed. New York: Macmillan Publishing Company, 1979.
Taly, Narendra. *Loads and Load Paths in Buildings—Principles of Structural Design*. Country Club Hills, IL: International Code Council, 2003.
Tang, Margaret et al. "Designing for Progressive Collapse." *Structure* 13, no. 4 (April 2006): 13–7.
Taylor, John R. *An Introduction to Error Analysis—The Study of Uncertainties in Physical Measurements*. Mill Valley, CA: University Science Books, 1982.
Timoshenko, Stephen P. *Strength of Materials*. New York: McGraw-Hill, 1940.
Timoshenko, Stephen P. and James M. Gere. *Theory of Elastic Stability*. 2nd ed. New York: McGraw-Hill, 1961.
Timoshenko, S. and S. Woinowsky-Krieger. *Theory of Plates and Shells*. 2nd ed. New York: McGraw-Hill, 1959.
Todd, Diana et al. *1994 Northridge Earthquake: Performance of Structures, Lifelines, and Fire Protection Systems (NIST 862)*. Gaithersburg, MD: National Institute of Standards and Technology, May 1994.
Torroja, Eduardo. *Philosophy of Structures*. Berkeley, CA: University of California Press, 1967.
Tremblay, Robert and Denis Mitchell. "Collapse During Construction of a Precast Girder Bridge." *Journal of Performance of Constructed Facilities* 20, no. 2 (May 2006): 113–25.

Troitsky, M.S. *Orthotropic Bridges—Theory and Design*. 2nd ed. Cleveland, OH: The James F. Lincoln Arc Welding Foundation, 1987.

Troup, Emile W. J. and George Metzger. "Design in Half the Time." *Structure* 8, no. 3 (April 2001): 44–7.

US Army Corps of Engineers, Engineering Division. *Seismic Design for Buildings (TI809-04)*. Washington, D.C.: US Army Corps of Engineers, December 1998.

US Army Corps of Engineers, Engineering Division. *Welding—Design Procedures and Inspections (TI809-26)*. Washington, D.C.: US Army Corps of Engineers, March 2000.

US Department of the Navy. *Soil Mechanics, Foundations & Earth Structures (NAVFAC DM-7)*. Washington, D.C.: US Department of the Navy, 1971.

USDA Forest Service (USDA), Forest Products Laboratory. *Wood Handbook—Wood as an Engineering Material*. Madison, WI: Forest Products Society, 1999.

Valentine, Lawrence. "Improving Concrete Durability." *Structural Engineer* 7, no. 3 (March 2006): 24–6.

Van Vlack, Lawrence H. *Elements of Materials Science and Engineering*. 6th ed. Reading, MA: Addison-Wesley Publishing Company, 1989.

Victor O. Schinnerer and Company, Inc. *Understanding and Managing Risk*. Chevy Chase, MD: Victor O. Schinnerer & Company, Inc., 1998.

Villaggio, Pierro. *Mathematical Models for Elastic Structures*. Cambridge: Cambridge University Press, 1997.

Vogt, Jim. "Permanent Bracing Concepts for Metal Plate Connected Wood Trusses." *Structure* (Fall), 1999: 24–5.

Wang, Chu-Kia and Charles G. Salmon. *Reinforced Concrete Design*. 6th ed. New York: Harper & Row Publishers, 2002.

Wang, John X. *What Every Engineer Should Know—Decision Making Under Uncertainty*. New York: Marcel Dekker, 2002.

Watson, Joseph P. "Design/Build and the Structural Engineer." *Modern Steel Construction* 40, no. 6 (June 2000): 54–7.

West, Harry. *Analysis of Structures—An Integration of Classical and Modern Methods*. 2nd ed. New York: John Wiley & Sons, 1989.

Westcott, Scott. "The Flip Side of Failure." *MyBusiness* (April–May 2006): 36–8.

Wentworth, George and David Eugene Smith. *Plane and Spherical Trigonometry with Tables*. Boston, MA: Ginn & Company, 1915.

West, Michael A. "Deflection Criteria for Spandrel Construction." *Modern Steel Construction* 34, no. 6 (June 1994): 28–9.

Wilson, Edward L. *Three Dimensional Static and Dynamic Analysis of Structures—A Physical Approach with Emphasis on Earthquake Engineering*. Berkeley, CA. Computers & Structures, 1998.

Winterkorn, Hans F. and Hsai-Ying Fang, eds. *Foundation Engineering Handbook*. New York: Van Nostrand Reinhold, 1975.

Wulpi, Donald J. *Understanding How Components Fail*. Metals Park, OH: American Society for Metals, 1985.

Yee, Chung Yan and Charles Y.J. Cheah. "Interactions between Business and Financial Strategies of Large Engineering and Construction Firms." *Journal of Management in Engineering* 22, no. 3 (July 2006): 148–55.

Young, Warren. *Roarke's Formulas for Stress & Strain*. 7th ed. New York: McGraw-Hill, 2002.

Zalka, Karoly A. *Global Structural Analysis of Buildings*. New York: Taylor & Francis Group, 2000.

Zalka, Karoly A. and G.S.T. Armer. *Stability of Large Structures*. Oxford: Butterworth-Heinemann, 1992.

Zeitz, Paul. *The Art and Craft of Problem Solving*. New York: John Wiley & Sons, 1999.

Zimmerman, William G. II. "Steel Erection Awareness: An Erector's View." *Modern Steel Construction* 35, no. 5 (May 1995): 30–3.

# Index

10th Amendment of U.S. Constitution, ratification of, 36–37
2006 IBC, significance of, 167

AAC masonry, definition of, 279–280
AAHSHTO (American Association of State Highway and Transportation Officials), role in highway-bridge construction, 107–108
ABET (Accreditation Board for Engineering and Technology), significance of, 26
abutments, constructing for bridges, 117
academic tools, using, 8–10
accessibility, requirements for, 85
acronyms, use of, 162
ACT (American College Testing, Inc.), significance of, 47
adjectives and adverbs, use of, 162
admixtures, characteristics of, 250–251
advanced educational degrees, obtaining, 141–142
advocacy, significance of, 49–50
aggregate, using, 247–249, 290
agreements, negotiating, 62
aider and abetter, definitions of, 181
air conditioners, features of, 81
air entraining chemicals, adding to concrete batches, 250
alkali-silica reaction, occurrence of, 275
Allowable Stress Design (ASD) relationship to loads, 202
ALTA survey, purpose of, 96
American Association of State Highway and Transportation Officials (AAHSHTO), role in highway-bridge construction, 107–108
American College Testing, Inc. (ACT), significance of, 47

American Railway Engineering and Maintenance of Way Association (AREMA) formation of, 110
American Railway Engineering Association (AREA), organization of, 110
American Society for Testing and Materials International (ASTM), significance of, 42
American Society of Civil Engineers (ASCE), significance of, 141
American Welding Society (AWS) standards, purpose of, 336
analysis and design, performing, 167–169
analytic learners, qualities of, 17
ANSI (American National Standards Institute), significance of, 42
application and theory, making transition between, 2–6
approval phase of projects for bridges, 128–130
  building department review, 99
  environment restrictions for buildings, 99–100
  planning department review for buildings, 98–99
arbitration, resorting to, 73
architects, role on building team, 91–93
architectural features, anchorage of, 65–66
arch-type bridges, construction of, 113
AREA (American Railway Engineering Association), organization of, 110
AREMA (American Railway Engineering and Maintenance of Way Association) formation of, 110
as-built drawings, purpose of, 106
ASCE (American Society of Civil Engineers), significance of, 141
ASD (Allowable Stress Design) relationship to loads, 202

asphaltic roof systems, using, 88
associations, examples of, 27–28
ASTM (American Society for Testing and Materials International), significance of, 42
ASTM steel reinforcement, providing for reinforced concrete, 254–257, 263–264
attics, ventilation of, 364
attitude, role in training philosophy, 12
autoclaving, using with masonry, 284
AWS (American Welding Society) standards, purpose of, 336
axial loading
 applying to plates, 326
 behavior of reinforced concrete under, 260–261

B&S (beams and stringers), definition of, 337
backfill material, choosing, 238–239
bascule bridges, construction of, 123–124
basements, risk to flood damage, 241
battledeck floor, use with bridge decks, 116
beam behavior, occurrence in reinforced concrete, 258
beams
 behavior of, 209–210
 beams and columns in masonry assemblies, 294–296
 function of, 295
 plate girders as, 316–317
 shear stress in, 296
 stability in steel-framed systems, 325–326
 and stringers (B&S), 337
bearing wall, definition of, 78
bearings, using with bridges, 120
bed joint, definition of, 280
beliefs, relationship to problem solving, 149
Bernoulli's Assumption, relationship to solid body mechanics, 205
bidding documents, contents of, 173–174
bidding phase of projects, explanation of, 100–101

billet steel, definition of, 246
BIM (Building Information Modeling), explanation of, 181
birdsmouth-type connections, use of, 351
bolted joints, behavior of, 350
bolts, using in steel connections, 319–320
bond beam, definition of, 280
bonder/header, definition of, 280
boundary survey, completing, 95
bowstring truss, description of, 359
brain, exercising, 154
breaks, taking, 154
bridge elements
 abutments and retaining structures, 117
 deck support components, 118–121
 decks, 115–117
 piers and foundations, 117–118
bridge piers, design of, 267–268
bridge systems
 behavior of, 216–217
 combining vehicle live loading and seismic forces in, 187
 construction phase of, 132–137
 costs associated with, 133
 design phase of, 130–132
 drainage considerations, 126–127
 economics and aesthetics of, 125
 environmental impact of, 130
 field work related to, 136–137
 including notes from fabrication facilities for, 135–136
 incremental launching of, 133–134
 maintaining, 136–137
 planning, financing, and politics of, 128–130
 scour considerations for, 114
 shop drawings and submittals for, 135–136
 traffic/use review of, 129–130
 using expansion joints with, 127–128
bridge types
 arch-type bridges, 113
 floating bridges, 112
 highway bridges, 107–109
 pedestrian bridges, 111
 railway bridges, 110

bridges
    balanced cantilevered segmental construction of, 134–135
    codes for, 41
    deflection of, 115
    dynamic load condition of, 115
    performance of, 114
    spans of, 112–113
    substructure of, 114
    superstructure of, 112, 114
    suspension systems of, 216
bridge-system components
    movable bridge decks, 123–124
    prestressed concrete, 122–123
    slab spans, 121
    steel, 121–122
    timber, 123
Brooklyn Bridge, image of, 18
Brooks Act, ratification of, 61
building codes. *See* codes
building columns, design of, 267–268
*Building Construction and Safety Code*, 41
building department review, explanation of, 99
building design, relationship to architecture, 92
building envelope, definition of, 85
building form, definition by architects, 92
building frame, definition of, 78
Building Information Modeling (BIM), explanation of, 181
building materials
    availability of, 203–204
    relationship to architecture, 92
building methods and planning, relationship to architecture, 92
building plans, laying out, 177
building systems
    behavior of, 215–216
    codes for, 40–41
    regulation of, 75
building team members
    architects, 91–93
    contractors and subcontractors, 94
    engineers, 93
    owners, 90–91

buried structures, effects of earth movement on, 243
business hierarchy, outline of, 52–54. *See also* employment
business knowledge, getting training in, 11
buttress elements, using with retaining walls, 240

cable-stayed bridges, construction of, 118, 216–217
caffeine absorption, relationship to sleep requirements, 152–154
caisson, using with bridges, 118
calcium silicate masonry, characteristics of, 284–286
Canadian ring ceremony, significance of, 47–48
cantilevered columns
    definition of, 78
    versus rigid frames, 266–267
CaO (quicklime), using in masonry, 288
carbon steel bolts, using, 319–320
carbon steel, classification of, 313–314
career growth. *See* technical growth opportunities
cement, characteristics of, 249. *See also* concrete mix
cementitious materials, examples of, 250
ceramics, using as construction materials, 203
certification, obtaining, 27
CEU (Continuing education unit), value of, 140
CFR (Code of Federal Regulations)
    pertaining to highway bridges, 107–108
    significance of, 36
change order, using, 94
chemistry, significance of, 9–10
China skyscraper, image of, 26
circuit, purpose in electrical systems, 82
Citicorp building in New York, image of, 28
City of Mounds View v. Walijarvi, 31
civil engineers, responsibilities of, 96
civil liability. *See also* liability and ethics
    and managing risk, 32–33
    and standard of care, 30–32
cladding systems, components of, 86–87

clarity
  consistency of, 162
  developing, 155–156
Class A-C ratings, applying to fire protection, 83
clay and silt soil types, characteristics of, 222–223
clay or shale masonry, characteristics of, 281–283
cleanout, definition of, 280
clients
  and consultants, 61–62
  managing risks associated with, 32–33
  reviewing work of, 180–181
clinker, definition of, 246
Code of Federal Regulations (CFR)
  pertaining to highway bridges, 107–108
  significance of, 36
codes. *See also* standards; regulations
  for bridges, 41
  for buildings, 40–41, 75
  conceptualizing, 75
  design standards referenced by, 43
  geotechnical considerations in, 243–244
  justifying compliance with, 169–170
  and load combinations, 190
cofferdam, using with bridges, 118
collapse avoidance, determining for seismic load, 196
collar joint, definition of, 280
colleges and universities, employment in, 56–57
Colombia, earthquakes in, 216
columns, considering as vertical systems, 210, 212–213
columns and beams, using in masonry assemblies, 294–296
columns and plates, stability in steel-framed systems, 326–327
combustible construction, definition of, 77
common-sense learners, qualities of, 17
communicating and delivering products, getting training in, 11
communication skills
  applying, 159–160
  importance of, 59–60, 104
  philosophy of, 160–161
  verbal skills, 161
  writing skills, 161–163

compaction of soil layers, purpose of, 228–230
compatibility, considering as physical law, 206
compensation for work, determining, 68
complexity, relationship to simplicity, 5
computers, avoiding dependence on, 156
conclusions, relationship to logic, 10
concrete. *See also* reinforced concrete
  behavior of, 263
  codes and standards for, 277–278
  cracks in, 277
  finishing, 273–274
  formation of, 245
  plain concrete, 252–253
  precast and prestressed concrete, 261–265
  quality control of, 276–277
  terminology related to, 246–247
  using in bridge systems, 122–123
concrete batches, adding air entraining chemicals to, 250
concrete box girder-type deck, construction of, 122
concrete frames, behavior of, 266. *See also* frames
concrete masonry, characteristics of, 283–284
concrete mix. *See also* cement
  characteristics of, 251–252
  compressive strength of, 275
  consolidation of, 272–273
  use of water in, 249–250
concrete slab-on-grade, performance of, 275
concrete surfaces, floating, 274
concrete types
  admixtures, 250–251
  aggregate, 247–249
  hydraulic cement, 249
concrete-system behaviors
  horizontal diaphragms, 270–271
  rigid frames or cantilevered columns, 266–267
  risks in design and during service, 275–276
  shear walls, 268–270
  shell-type structures, 271–272
conferences, attending, 142–143

confidence
  building, 156–159
  significance of, 7–8
conflicts, resolving, 69–71, 174
consideration, including in contracts, 32
consistency
  developing, 155–156
  in project drawings, 179–180
Constitution, 10th Amendment of, 36–37
construction costs, determining, 98
construction documents
  quality control of, 181
  reviewing, 98
  and services, 92
construction manager, responsibilities of, 94
construction materials, availability of, 203–204
construction phase of projects
  for bridge systems, 132–137
  field observation and inspection, 104–105
  occupancy and continued use, 105–106
  postcontract, 105–106
  revisions, 103
  shop drawings and RFIs, 101–102
construction standards, overview of, 42–44
construction-site preparation
  clearing and excavation, 226–227
  compaction, 228–230
  grading, 227–228
  producing geotechnical reports, 224–226
consultants
  reviewing work of, 180–181
  working with, 61–62
consulting firms, employment in, 54–55
contacts, establishing for projects, 58–59
Continuing education unit (CEU), value of, 140
continuous (strip) footings, behavior of, 232–233
contractors
  responsibilities of, 101
  and subcontractors on teams, 94
contracts
  executing, 32
  negotiating, 63
contractual conditions, examples of, 173–174

control survey, requirement of, 95
converging type, relationship to cable-stayed bridges, 118, 120
Cooley, Thomas, 39
cooling rate, determining for equipment, 81
corporation, definition of, 55
correcting, relationship to training method, 13
cost of work, estimating, 68
cost-plus-fee agreement, explanation of, 68
courage, significance of, 8
creativity, significance of, 7
credibility, importance of, 157
creep, relationship to wood, 343
crisis management
  conflict resolution, 69–71
  litigation, 74
  working with difficult people, 71–72
critical thinking, applying to problem solving, 147–149
CUREE-Caltech Woodframe Project, significance of, 353–354
curiosity, significance of, 7
curtain walls, supporting, 87

damping
  relationship to dynamic-type loads, 190–191
  of vibrations, 208
darby, definition of, 246
dead lines, establishing, 150
dead loads, applying to structural members, 186
deck material, maintaining for bridge systems, 137
deck support components, constructing for bridges, 118–121
decks of bridges, construction of, 115–117, 118
deep foundations, characteristics of, 235–238
deflection
  of concrete shear walls, 269
  of flexural members, 259–260
  impact of, 208
  of wall, 212
degree-of-freedom dynamic system, example of, 191

degrees, obtaining, 141–142
Department of Transportation (DOT), role in bridge construction, 132
design development phase of projects, components of, 97
design methods
  applying to loads and forces, 202–203
  confusion of, 243
design phase of projects
  for bridge systems, 130–132
  components of, 97–98
  pacing, 151–152
design regulations, examples of, 34. *See also* regulations
design specifications, applying to vehicle live loading, 109
design standards, overview of, 42–44
design-bid-build approach, implementing, 90
design-build approach, implementing, 90–91
designs
  and analyses, 167–169
  presenting, 169–171
  project specifications for, 171–175
  structural calculations for, 166–171
diet, paying attention to, 154
Dillon's Rule, 38–40
direct professional knowledge, significance of, 181
direct supervisory control, significance of, 181
Disabilities Act of 1990, 85
disagreements, resolving, 69–71, 174
disputes, resolving, 69–71, 174
distribution piping, determining, 80
dobies, definition of, 246
DOT (Department of Transportation), role in bridge construction, 132
Douglas Fir, use of, 341
dowel-type fasteners, yield model equations for, 350
drafters, procedures for, 177
drainage, considering for bridge systems, 126–127
drawings, types of, 106. *See also* project drawings
dressed lumber, definition of, 337

drilled piers, description of, 235
DRY label, applying to wood, 339
drying shrinkage, occurrence in concrete, 277
dual system, definition of, 78
ductility of steel, explanation of, 309
dynamic learners, qualities of, 17
dynamic-type loads. *See also* load combinations
  blast, impact, and extreme loads, 201–202
  seismic, 194–200
  and structural dynamics, 190–191
  wind, 191–194

EA/L elastic spring constant, using with mats, 235
Earthquake Hazards Reduction Act, significance of, 35–36
earthquakes
  in Colombia, 216
  defining response related to, 198
  in Kobe, Japan, 266
  measuring on Richter scale, 194–195
  in Vina del Mar, Chile, 266
Economic Loss Doctrine, explanation of, 30
educational degrees, obtaining, 141–142
egress and circulation, requirements for, 84–85
EI (expansion index) of soil, measurement of, 222
elastomeric bearing pads, using with bridges, 120
electrical systems
  components of, electrical, 81–82
  relationship to architecture, 92
elevator support, considering in scope of services, 66–67
elevator systems, elements of, 85
Empire State Building, image of, 5
employees, managing, 58
employment. *See also* business hierarchy; engineering businesses
  in colleges and universities, 56–57
  in government agencies, 54
  in industries, 55–56
  in private consulting firms, 54–55

engineering, foundational gospel of, 4
engineering businesses, survival of, 57–61.
    *See also* employment
engineering education, placing in
    perspective, 1–2
engineering failure, example of, 182–183
engineering mechanics, static loads,
    185–190
engineering reports. *See also* reports
    evaluating, 145
    organization of, 183–184
engineering services
    estimating worth of, 68
    proposal for, 62
engineers
    credibility of, 157
    guarantee offered by, 64
    role on building team, 93
engineers' abilities
    applying knowledge, 3
    comprehension, 3
    defining solutions, 3–4
    elaborating on ideas or solutions, 4
    knowledge retention and recall, 2–3
    originality of thought, 4
engineers' tools
    academic tools, 8–10
    psychological tools, 7–8
environment restrictions, considering,
    99–100
environmental impact, assessing relative to
    bridges, 130
environmentally sensitive materials, using,
    204
EO (Executive Order), purpose of, 37
equations. *See* formulas
equilibrium, determining for static
    members, 206
equipment, determining cooling rate for, 81
errors, occurrences of, 22–24, 30
estimation principles, applying to
    engineering problems, 148–149
ethical behavior, actions associated
    with, 28
ethics and liability, getting training in, 11
Euler, Leonard, 212
evaluating, relationship to training
    method, 14

evaporator, capabilities of, 81
Executive Order (EO), purpose of, 37
exercise schedule, maintaining, 154
expansion index (EI) of soil, measurement
    of, 222
expansion joints, using in bridge systems,
    127–128
expansion-type bearings, using with
    bridges, 120
experiments, performing, 143–146

fabrication facility notes, including for
    bridge systems, 135–136
facilities, determining constructed cost
    of, 68
Factory Mutual (FM), significance of,
    44–45
fan type, relationship to cable-stayed
    bridges, 120
fasteners, using in woodframe
    construction, 349–352
fatigue life, considering for structural
    elements, 201–202
fatigue of wood, definition of, 343
federal versus state government, 35–37
feeder, purpose in electrical systems, 82
FEMA (Federal Emergency Management
    Agency)
    relationship to NFIP (National Flood
        Insurance Program), 45
    significance of, 35
FHWA (Federal Highway Administration),
    role in bridge construction, 107–108
fiberboard sheathing, examples of, 348
field observation, requirement of, 104
field work, performing for bridge systems,
    136–137
Figures
    aggregate grading chart, 248
    anchoring mechanical equipment, 65
    beam loading and reaction diagrams,
        170
    bracing pipe systems, 56
    brick masonry with metal ties, 292
    bridge bearings, 120
    bridge deck launching process, 134
    bridge deck with joint, 127
    bridge piers, 117

Figures (*Cont.*):
    bridge system structural elements, 132
    Brooklyn Bridge, 18
    buckling stiffness of columns, 211
    building story, 228
    business hierarchical structure, 53
    cable-stayed pedestrian bridge, 119
    China skyscraper, 26
    Citicorp building in New York, 28
    cladding system, 86
    concrete bridge damage and failure, 267–268
    concrete masonry unit, 285
    concrete shear wall deflection, 269
    continuous footing, 233
    data points with range of error, 23
    degree-of-freedom dynamic system, 191
    earthquakes measured on Richter scale, 194
    Empire State Building, 5
    floor plan, 178
    Geisel Library (UCSD), 166
    glass masonry walls, 287
    grade-crossing bridges, 113
    groove welds, 322
    hollow-core prestressed concrete, 263
    horizontal plywood diaphragm testing, 355
    hysteresis diagram, 207
    load duration factors, 367
    load support mechanisms for piers, 237
    London Bridge, 124
    masonry wall construction, 293
    masonry wall construction elements, 282
    metal flashing systems, 89
    Millau Viaduct, 133
    movable bridge with towers, 125
    mullion and steel framing, 66
    pad footing, 231
    plate I-girder section, 316
    plywood roof diaphragm testing, 43
    plywood structural diaphragm testing, 348
    prestressing steel anchorage connections, 265
    reinforced concrete, 257

Figures (*Cont.*):
    reinforced concrete with shear stress cracking, 259
    reinforcing steel, 255
    response spectra for seismic motion, 199
    retaining wall with soil moisture draining system, 239
    rigid diaphragm theory, 200
    rolled steel framing elements, 333
    roof diaphragms, 211
    roof drain, 89
    Royal Mint Building, 5
    seismic joint, 79
    Severn Bridge between Wales and England, 116
    shear wall definitions, 361
    shear wall piers, 213
    slope-of-grain for wood, 353
    soil stress-strain curve, 230
    static flood loading on basement wall, 189
    steel beam connection, 308
    steel frame configurations, 328–329
    steel stress-strain diagram, 313
    stress-strain curve, 206
    stress-strain curve for concrete masonry units, 285
    stress-strain curve for plain concrete, 253
    superposition in horizontal systems, 209
    suspension bridge tower configurations, 119
    truck weight distribution (AASHTO), 109
    truss bottom chord tension failure, 365
    Vierendeel truss configuration, 218
    waste plumbing layout, 80
    weld zones for steel, 321
    wood fibers in vicinity of knot, 352
    wood portal frame construction detail, 358
    wood-frame structural diaphragm, 355
    wood-frame truss configurations, 358
    wood-frame trusses, 368
fillet weld, description of, 322
final design phase of projects, components of, 97–98

finish, definition of, 247
fink truss, description of, 359
fire proofing, applying to structural steel members, 334
fire protection
 implementation of, 201
 requirements for, 77, 82–84
FIRMs (Flood Insurance Rate Maps), publication of, 45–46
fixed-type bearings, using with bridges, 120
flashing, using in roofing systems, 88–89
flexural members, deflection of, 259–260
flexural tension, failure in masonry units, 303
floods, loads delivered by, 189
floor framing systems, choosing, 77
floor slab systems, using lift slab method with, 274–275
floor systems, general observation of, 105
floor-framing elements, deflection of, 208
FM (Factory Mutual), significance of, 44–45
footings
 behavior during seismic events, 232
 behavior of, 230
 behavior of continuous (strip) footings, 232–233
 behavior of spread footings, 231–232
 combined or mat-type footings, 233–235
 considering in scope of services, 67
 deformation of, 234–235
forces and loads, combining, 202–203
formulas
 active lateral soil loads for moist conditions, 240
 for ASTM A706 steel, 256
 column loading, 212
 Ideal Gas Law, 3
 USD (Ultimate Strength Design), 203
 wall deflection, 212
 for wind service-load pressure, 193
foundation design considerations
 codes and standards, 243–244
 consequences of poor soils, 241–242
 risk, 242–243
 settlement, 241–242
foundation observation, preparing for, 105

foundation plans, providing for project drawings, 176
foundation types
 combined or mat-type footings, 233–235
 continuous (strip) footings, 232–233
 deep foundations, 235–238
 pier and beam, 238
 spread footings, 231–232
foundations, constructing for bridges, 117–118
FRA (Federal Railroad Administration), creation of, 110
fracture toughness of wood, definition of, 343
frames. *See also* concrete frames
 behavior of, 213
 and masonry, 298–299
 resistance to cyclic loads, 215
 in steel-framed systems, 328–330
 in wood-frame systems, 357
framing elevations or sections, providing for project drawings, 177
framing members, supporting, 83
framing plans, providing for project drawings, 176–177
F-ratings, applying to tornadoes, 192–193
free-body diagram, determining equilibrium by means of, 206
friction, coefficient of, 259
Fujita, classification of tornadoes by, 192–193
Fuller, R. Buckminster "Bucky," 49
fundamental period of vibration (T), significance of, 190

gas piping, sizing, 79
general contractor, responsibilities of, 94
geotechnical engineering, explanation of, 219
geotechnical reports, preparing, 224–226
GFRP (glass fiber reinforced polymer) panels, using with bridges, 115
glass, using in masonry, 286
government agencies, employment in, 54
government control
 federal versus state, 35–37
 hierarchy of, 35
 state versus local, 37–38

grade separations, types of, 113
grading, requirements for, 227–228
gravel soil type, characteristics of, 221–222
gravity, impact on engineering, 4
green lumber, definition of, 339
groove weld, description of, 322
ground, purpose in electrical systems, 82
grout, using in masonry, 289–291
grout lift, definition of, 280
grout pour
   definition of, 281
   height of masonry units for, 300
grubbing, occurrence of, 226–227
gypsum board products, manufacture of, 362

H-1B visa program, significance of, 49
hardboard, use of, 348
hardness of wood, definition of, 343
hardwood, definition of, 339
harp type, relationship to cable-stayed bridges, 118, 120
head joint, definition of, 281
heating, ventilating, and air-conditioning systems (HVAC), features of, 81
high-strength bolted connections, pretensioning, 324
highway bridges, construction of, 107–109
hold-harmless clause, explanation of, 63
home rule, adoption of, 39
honeycomb/rock-pocket, definition of, 247
Hooke's Law, relationship to solid body mechanics, 205
horizontal diaphragms
   in concrete systems, 270–271
   in woodframe systems, 354–357
horizontal systems, behavior of, 209–210
hot mud, definition of, 247
hot-rolled steel, behavior of, 315–316
howe truss, description of, 359
HPC (high-performance concrete), characteristics of, 251–252
HSS (Hollow Structural Sections), availability of, 317–318
hurricanes, categorization of, 192
HVAC (heating, ventilating, and air-conditioning systems), features of, 81
hydraulic cement, characteristics of, 249
hydrology studies, conducting for bridge systems, 131–132
hysteresis, considering in solid body mechanics, 207–208

IBC (International Building Code), significance of, 40–41
ICC (International Code Conference)
   Evaluation Service (ES) administered by, 43–44
   significance of, 40–41
Ideal Gas Law, formula for, 3
ideas, elaborating on, 4
imaginative learners, qualities of, 17
Immigration Act of 1990, 49
incompetence, relationship to ethics, 29
industries, employment in, 55–56
inferences, relationship to logic, 10
information, communicating, 60
infrastructure, planning and financing, 128–129
ingots
   definition of, 307
   hot or cold rolling of, 312
   shapes of, 315–316
inspections, requirement of, 104
insurance, managing risk by means of, 33
integral-type abutment, constructing for bridges, 117, 127
International Building Code (IBC), significance of, 40–41
International Code Conference (ICC)
   Evaluation Service (ES) administered by, 43–44
   significance of, 40–41
international issues, considering, 48–49
ISO (International Organization for Standardization), significance of, 44

jargon, avoiding use of, 162
joints
   including in designs, 78–79
   using in bridge systems, 127–128
joists and planks, definition of, 338

Ke factor, relationship to vertical systems, 210, 212

kerf, definition of, 338
killed steel, definition of, 307
kiln, definition of, 247
king post truss, description of, 359
knots, formation in wood, 352
Kobe, Japan, earthquake in, 266

ladle refining, using with structural steel, 311–312
laminated decks, characteristics of, 357
laminated strand lumber (LSL) members, strength values for, 345–346
land development
  gathering information about, 95
  ownership and legal interests, 95
  surveying, 95–96
Land Ordinance Act of 1785, 95
land use
  planning, 37
  studying relative to bridge systems, 129–130
  zoning of, 37–38
lateral forces, relationship to architecture, 92
lateral load resisting systems, structural aspects of, 77–78
leadership, tiers and responsibilities of, 52–54, 58. *See also* managers
learners, types of, 16–17
LEED (Leadership in Energy and Environmental Design), significance of, 75
legal capacity, including in contracts, 32
legal treatises, on responsible care, 31
legally binding arbitration, explanation of, 73
legislative home rule, explanation of, 39
Letter of Map Revision (LOMR), issuing of, 46
lever rule, applying to vehicular live load distribution, 109
liability and ethics, getting training in, 11. *See also* civil liability
licensing
  continuing education regulations for, 140–141
  origin of, 25
  process of, 27
  requirements for, 26

lien rights, defining, 73
lift slab method, using with floor slab systems, 274–275
limestone, using in masonry, 285, 287
limitations, being honest about, 150
linguistics, significance of, 10
lintels, using with masonry, 294–295
listening, importance to mentors, 15
litigation, resorting to, 74
live load
  applying, 186–187
  specifying for railway bridges, 110
LL (liquid limit) of soil, determining, 222–223
load carrying capacity, limits of, 21
load combinations, relationship to model building codes, 190. *See also* dynamic-type loads
load path, explanation of, 22
load resisting systems, structural aspects of, 77–78
loads
  applying to wall surfaces, 297
  combining with forces, 202–203
loam soil type, characteristics of, 223
local versus state government, 37–38
logic, significance of, 10
logs, processing, 339–340
LOMR (Letter of Map Revision), issuing of, 46
London Bridge, construction of, 123–124
LRFD (Load and Resistance Factor Design), applying, 203
LRFD specifications, mandate for, 107–108
LSL (laminated strand lumber) members, strength values for, 345–346
lumber. *See also* wood
  drying, 340
  grading rules and practices, 341–342
  grading rules for, 342
  LSL (laminated strand lumber), 345
  LVL (laminated veneer lumber), 345
  milling and finishing, 339–340
  OSL (oriented strand lumber), 345
  PSL (parallel strand lumber), 345
  SCL (structural composite lumber), 345
  seasoning, 340

lumber (*Cont.*):
  species of, 340–341
  supply and harvest of, 338–339
lumber pieces, grading, 341–342
lumped-mass approach, relationship to dynamic motion, 191

main, purpose in electrical systems, 82
main wind force resisting system (MWFRS), significance of, 193–194
managers, responsibilities of, 58. *See also* leadership
marine soil type, characteristics of, 223–224
masonry assemblies
  beams and columns, 294–296
  behavior of, 293–294
  compressive strength of, 293, 304
masonry columns, regulation of, 296
masonry units
  calcium silicate (sand lime), 284–286
  clay or shale, 281–283
  codes and standards for, 305–306
  cold conditions for, 300–301
  concrete, 283–284
  construction of, 300–303
  crack appearance in, 303
  and flexural tension failure, 303
  frames, 298–299
  glass, 286
  grout, 289–291
  heights of, 300
  modulus of rupture for, 295
  mortar, 288–289
  prestressed assemblies, 299
  quality control of, 304–305
  reinforcement of, 291–292
  risk in design and during service, 303
  stone, 286–288
  terminology related to, 279–280
  tolerances during construction of, 300
  using aggregate in, 290
  using steel reinforcing bars with, 292
  walls, 297–298
masonry walls
  characteristics of, 297
  openings through, 301

Masterformat system, using with technical specifications, 174
mat foundations, use of, 235
mathematics, significance of, 8–9
MC (moisture content) of wood, significance of, 339
measurements, errors associated with, 23
mechanical code, requirements of, 81
mechanical equipment, anchoring, 65
mechanical stabilized earth (MSE) structures, formation of, 117
mechanical systems, relationship to architecture, 92
Mechanic's Liens, creation of, 73–74
mediation, resorting to, 73
members, designing quickly, 151
memory, improving, 156
mentoring, importance of, 12, 14–16
mentors, concerns of, 14–15
metallic iron, using with structural steel, 311
metals
  fracture in, 334–335
  mechanical properties of, 310
  using as construction materials, 203
methods
  imperfection of, 22
  role in training philosophy, 12–14
Millau Bridge, launched construction approach used with, 133
misconduct, claims of, 29
mistakes, occurrences of, 22–24, 30
MM (Modified Mercalli), reporting earthquake magnitudes on, 195
model building codes. *See* codes
Modes, applying to dowel-type fasteners, 350
modes of structure vibration, explanation of, 190
modulus of rupture, relationship to wood, 343
moisture content (MC) of wood, significance of, 339
molecular materials, using in construction, 203
moment-resisting frame, definition of, 78
mortar
  proportioning of, 304
  using in masonry, 288–289

movable bridge decks, using, 123–124
movable bridges, maintaining, 137
MSE (mechanical stabilized earth) structures, formation of, 117
muck soil type, characteristics of, 223
mullion and steel framing, example of, 65–66
mutual assent, including in contracts, 32
MWFRS (main wind force resisting system), significance of, 193–194

NAFTA (North American Free Trade Agreement), ethics of, 29
*National Electrical Code of 1897*, 40–41
National Fire Protection Association (NFPA), significance of, 40–41
National Flood Insurance Program (NFIP), significance of, 45–46
National Institute of Standards and Technology (NIST), significance of, 45
NCEES (National Council of Examiners for Engineering and Surveying), significance of, 140
N-cycles, applying to structural elements, 201–202
NDT (nondestructive testing), performing on structural steel, 335–336
negligence, relationship to ethics, 29
The Nevada State Board of Agriculture v. United States, 36
New York, suspension bridges in, 21
Newton's laws, relationship to solid body mechanics, 205
NFIP (National Flood Insurance Program), significance of, 45–46
NFPA (National Fire Protection Association), significance of, 40–41
NFPA 70, drafting of, 82
NFPA testing, conducting for fire protection, 77
NIST (National Institute of Standards and Technology), significance of, 45
nondestructive testing (NDT), performing on structural steel, 335–336

occupancy and continued use, requirements for, 105–106
OL (occupant load), definition of, 84

Order of the Engineer, significance of, 47
ore, selecting for structural steel, 310–311
orthotropic plate, use with bridge decks, 116
OSB (oriented strand board) panels, fabrication of, 347
OSL (oriented strand lumber) members, strength values of, 345–346
owners, role on building team, 90–91

P&T (post and timber), definition of, 338
pad footings, width and depth of, 105
panels or sheathing, exposure rating for, 349
parallel method, explanation of, 214
parallel strand lumber (PSL), wood used in, 345
particleboard, creation of, 348
partnership, definition of, 55
patent deficiency, explanation of, 30
Paxton v. County of Alameda, 31
PD (Plastic Design), applying to steel members, 203
PDH (professional development hours), relationship to CEU, 140–141
peat soil type, characteristics of, 223
pedestrian bridges, construction of, 111
performance-based system, considering in design regulation, 34
permafrost areas
  characteristics of, 224
  use of slab-on-grade systems in, 238
permissive constitutional home rule, explanation of, 39
photographs, including in engineering reports, 184
physics, significance of, 9
PI (plasticity index) of soil, determining, 222–223
pictures versus words, 162
pier and beam foundation types, characteristics of, 238
piers
  constructing for bridges, 117–118
  load support mechanisms for, 236–237
pig iron, using with structural steel, 311
pilaster, definition of, 281

pile foundations, characteristics of, 235–236
piles, supporting mats with, 235
pin bearings, using with bridges, 120–121
pipe sizes, manufacturing of, 318
pipe systems, bracing, 55–56
piping, determining, 80
planks and joists, definition of, 338
planning department review, explanation of, 98–99
plans, providing for project drawings, 176–177
Plastic Design (PD), applying to steel members, 203
plasticity index (PI) of soil, determining, 222–223
plate girders, behavior of, 316–317
plates and columns, stability in steel-framed systems, 326–327
Platform framing, constructing, 363
plumbing, designing and installing, 79–80
plywood, strength of, 347
plywood panels, orientation of, 349
Portland cement, characteristics of, 249
post and timber (P&T), definition of, 338
posttensioning, description of, 263
precast concrete, characteristics of, 261–265
pre-design, relationship to architecture, 92
premises, relationship to logic, 10
prescriptive-based design, explanation of, 34
presentation skills, teaching, 180
presentations, reading aloud, 163
prestressed assemblies, using with masonry, 299
prestressed concrete
　characteristics of, 261–265
　using in bridge systems, 122–123
prestressing, applying to masonry, 292
pretensioning
　of high-strength bolts, 324
　transfer of force in, 264
　using to apply axial force, 262
problem solving
　and critical thinking, 147–149
　process of, 3–4, 146–147
　reaching conclusions in, 149

procedures, following, 59
productivity factors
　considering, 149–151
　developing consistency and clarity, 155–156
　time management, 152–155
professional development hours (PDH), relationship to CEU, 140–141
Professional Engineers Act, adoption of, 25
professional involvement, participating in, 142
professionals, characteristics of, 27
progressive collapse, study of, 217
progressive method, using with bridges, 135
project conditions, examples of, 173–174
project drawings. *See also* drawings
　contents of, 175–177
　goals and methods of, 177–178
　presenting, 179–180
project notes, flaws in, 172
project phases
　approval of buildings, 98–100
　approval phase for bridges, 128–130
　bidding for buildings, 100–101
　construction of buildings, 101–105
　design of buildings, 97–98
project specifications. *See* specifications
projects. *See also* work
　accepting and rejecting, 150–151
　limitations at beginning of, 62
　responsibility for, 181–182
　success of, 58–59
proposals, developing, 62
PSL (parallel strand lumber), wood used in, 345
psychological tools, examples of, 7–8

qualifications, importance of, 60–61
quenching, definition of, 308
quicklime (CaO), using in masonry, 288

raceway, purpose in electrical systems, 82
rafter spaces, ventilation of, 364
railway bridges, construction of, 110
rain loading, impact on flat roofs, 189
rain water, drainage of, 126

ready-mixes concrete, preparation of, 251
rebar congestion, effect of, 302
rebuking, relationship to training method, 13
record drawings, purpose of, 106
recovered materials, definition of, 204
recycling, definition of, 204
redundancy of structural systems, explanation of, 214–215
refined analysis method, applying to vehicular live load distribution, 109
registration, relationship to licensing, 27
regulations. *See also* codes; design regulations; standards
    ISO (International Organization for Standardization), 44
    NFIP (National Flood Insurance Program), 45–46
    NIST (National Institute of Standards and Technology), 45
    UL (Underwriters Laboratories, Inc.), 44
regulatory documents, hierarchy of, 34–35
regulatory powers, limitation of, 39
reinforced concrete. *See also* concrete
    behavior under axial loading, 260–261
    flexural effects in, 257–260
    openings in, 258
    steel reinforcements for, 254–257
reinforcing bars, type and placement of, 105
reinforcing steel, using with masonry walls, 302
reliability of structural systems, explanation of, 214–215
*Report of Committee on Automatic Sprinkler Protection*, 40
reports, reading aloud, 163. *See also* engineering reports
research
    conducting, 142–143
    making proper use of, 143–146
residual stress, definition of, 308
responsibilities
    delegating, 150
    for projects, 181–182
    to society, 46–48

responsible charge of work, explanation of, 33
rest, importance of, 154
restrained versus unrestrained framing members, 83
retaining structures, constructing for bridges, 117
retaining walls
    drainage for, 239
    using buttress elements with, 240
revisions, making, 103
RFIs (requests-for-information)
    importance of, 94
    submitting, 102
Richter scale, measuring earthquakes on, 194–195
rigid diaphragm theory, applying to seismic motion, 200
rigid frames
    versus cantilevered columns, 266–267
    in steel-framed systems, 328
ring ceremony, significance of, 47–48
risk
    considering in design regulation, 34
    considering in foundation design, 242–243
    managing, 32–33, 57
road buster, definition of, 247
rock soil type, characteristics of, 221
rocker bearings, using with bridges, 120
rolling shear strength, relationship to wood, 343
roof covering materials, fire resistance classifications of, 83
roof systems
    characteristics of, 87–88
    general observation of, 105
    impact of rain loading on, 189
    requirements of, 77
rough lumber, definition of, 338
Royal Mint Building, image of, 5

S4S structural lumber, meaning of, 339
Saffir/Simpson Scale, categorizing hurricanes by, 192
sand lime masonry, characteristics of, 284–286
sandstone, using in masonry, 287–288

sawn lumber. *See* lumber
SCL (structural composite lumber), benefits of, 345
scope of services, determining, 64–67
scour, relationship to bridges, 114
seat-type abutment, constructing for bridges, 117
segmental concrete bridge construction, use of, 122–123
seismic design, basis of, 198
seismic dynamic-type load
  application of, 198–200
  characteristics of, 195–196
  considering, 194–195
  derivation of, 196–198
seismic events, behavior of isolated footings in, 232
seismic force minimum levels, complying with, 169–170
seismic forces
  providing resistance to, 216
  studying effects of, 215–216
seismic joint, example of, 79
seismic motion
  idealization of, 199
  impact of underlying soil on, 244
  response spectrum of, 199
self-executing home rule, explanation of, 39
seminars, attending, 142–143
septic tanks, using, 96
series method, explanation of, 214
settlement, considering in foundation design, 242
SFHA (Special Flood Hazard Areas), explanation of, 45–46
shale or clay masonry, characteristics of, 281–283
shall versus will, legal implications of, 172
shear strength
  and reinforced concrete, 259
  and wood elements, 346
shear stress, occurrence in beams, 296
shear walls
  behavior of, 212
  cracking in, 268–269
  design of, 297

shear walls (*Cont.*):
  nonwood panel shear walls, 362–363
  structural wood panel shear walls, 360–362
  testing, 362–363
shear-friction, relationship to reinforced concrete, 259
sheathing or panels, exposure rating for, 349
shell, definition of, 281
shell-type structures, behavior of, 271–272
shingles, using, 87–88
shop drawings
  for bridge systems, 135–136
  for building systems, 101–102
silt and clay soil types, characteristics of, 222–223
simplicity, relationship to complexity, 5
sintering, definition of, 247
site planning, relationship to architecture, 92
site visits, requirement of, 104
slab spans, using in bridge systems, 121
slab-on-girder bridges, test results on, 217
slab-on-grade systems, using in permafrost areas, 238
slate, using in masonry, 288
sleep, importance of, 154–155
slenderness ratio, explanation of, 326
sliding bearings, using with bridges, 120
slip-critical joint, definition of, 324
slope of grain, existence of, 353
slump test, definition of, 247
snow loads
  considering, 187–188
  project drawings related to, 176
society, responsibilities to, 46–48
softwood, definition of, 338–339
soil, defining mechanical properties of, 230
soil bearing pressures, relationship to footings, 232
soil layers, compaction of, 228–230
soil observation, completion of, 105
soil pressure, considering, 188

soil quality, considering in foundation design, 241–242
soil response, approximating for footings, 234
soil surveys, obtaining, 225–226
soil types
    classification of, 243–244
    directing water away from, 227–228
    gravel, 221–222
    loam, 223
    marine soil, 223–224
    muck, 223
    peat, 223
    rehabilitating, 242
    rock/granite, 221
    sand, 222
    silt and clay, 222–223
sole proprietorship, definition of, 54–55
solid body mechanics
    serviceability, 208
    hysteresis, 207–208
    principles of, 205–206
    stress and strain, 206–207
solutions
    defining, 3–4
    elaborating on, 4
    repeatability of, 8
Southern Pine lumber, use of, 341
span-by-span method, using with bridges, 135
Special Flood Hazard Areas (SFHA), explanation of, 45–46
special sections or conditions, accounting for, 175
specification books, organization of, 173
specifications. *See also* technical specifications
    for designs, 171–175
    flaws in, 172
spread footings, behavior of, 231–232
Stages of teaching, explanations of, 17–18
staircases, considering in scope of services, 66–67
standard of care, definitions of, 31–32
standards. *See also* codes; regulations
    ANSI (American National Standards Institute), 42

standards (*Cont.*):
    ASTM (American Society for Testing and Materials International), 42
    ICC-ES (Evaluation Service), 43–44
star type, relationship to cable-stayed bridges, 120
state
    versus federal government, 35–37
    versus local government, 37–38
static loads
    dead loads, 186
    live loads, 186–187
    snow loading, 187–188
    soil pressure, 188
static members, determining equilibrium of, 206
steel. *See also* structural steel
    bending, 332–333
    cutting holes in, 319
    stress-strain diagram for, 313
    types of, 314
    using in bridge systems, 121–122
steel columns, economizing, 318–319
steel connections
    bolts, 319–320
    high-strength bolted connections, 324
    welds, 320–324
steel corrosion, process of, 276
steel fabricators, concerns of, 331–332
steel members, applying PD (Plastic Design) to, 203
steel reinforcement
    providing for reinforced concrete, 254–257
    using with masonry, 292
steel shapes
    composite members, 318–319
    hot-rolled, 315–316
    plate girders, 316–317
    tubular and pipe sections, 317–318
steel side plates, using, 351
steel-framed systems
    frames, 328–330
    stability of beams in, 325–326
    stability of columns and plates in, 326–327
    steel-panel shear walls, 330–331

steel-panel shear walls, characteristics of, 330–331
stirrup, definition of, 247
stone, using in masonry, 286
stone aggregate, grading and quality of, 248
storm drains, types of, 96
strain and stress, considering in solid body mechanics, 206–207
strain hardening, definition of, 308
straw, considering as molecular material, 204
stress, managing, 57
stress and strain
 on concrete masonry element, 284–285
 on plain concrete, 252–253
 in solid body mechanics, 206–207
stringers and beams, definition of, 337
strip (continuous) footings, behavior of, 232–233
structural analysis
 and design, 170–171
 reliability of, 22
structural calculations, performing, 166–171
structural composite lumber (SCL), benefits of, 345
structural design, method of, 21
structural details, providing in drawings, 178
structural drawings, assembling, 177
structural elements
 behavior of, 205–208
 fatigue life of, 201–202
structural engineering
 analysis and design components of, 20
 definition of, 19
 egress and circulation, 84–85
 electrical, 81–82
 experience of, 24
 fire protection, 82–84
 of joints, 78–79
 of lateral force resisting systems, 77–78
 mechanical, 81
 uncertainty and error components of, 20–24

structural engineering (*Cont.*):
 of vertical load resisting systems, 77
 weatherproofing, 85–89
structural failures, causes of, 151
structural members, applying dead loads to, 186
structural notes, providing for project drawings, 176
structural steel. *See also* steel
 at atomic level, 308–309
 codes and standards for, 336
 ductility of, 309
 exposure to corrosive elements, 334
 fabrication and erection of, 331–335
 fire proofing, 334
 inspection and testing of, 335
 mills and suppliers, 312–313
 mining and refining, 310–312
 quality control of, 335–336
 regulations, 313–315
 risks in design and during service, 334–335
 terminology related to, 307–308
 yield stress of, 313
structural systems
 horizontal systems, 209–210
 redundancy and reliability of, 214–215
 vertical systems, 210, 212–213
structure design, understanding process of, 2
structure vibration, modes of, 190
structures
 avoiding types of, 156
 behaviors of, 4
 dynamic properties of, 190
 progressive collapse of, 217
 safety of, 22
subcontractors, role on building team, 94
subfloor, definition of, 338
subsurface conditions, problems associated with, 243
superposition, relationship to horizontal systems, 209–210
surety bond, relationship to bidding phase, 100–101
surveying land, 95–96

suspension bridges
    construction of, 118
    in New York, 21
swing bridges, construction of, 123–124

T (fundamental period of vibration), significance of, 190
teaching
    methods of, 12–13, 16–19
    opportunities for, 16–19, 56–57
technical growth opportunities
    advanced educational degrees, 141–142
    continuing education for licensure, 140–141
    professional involvement, 142
    seminars, conferences, and personal research, 142–143
technical knowledge, getting training in, 11
technical research, making proper use of, 143–146
technical specifications, assembling, 174–175. *See also* specifications
tempering, definition of, 308
tensile steel, using with masonry units, 295
tensile strength of wood, explanation of, 344
tension members, stability in steel-framed systems, 326
terminology, specificity of, 172–173
testing, performing, 143–146
theory and application, making transition between, 2–6
thermal expansion, coefficients of, 79
thermal insulation, using, 89
thickness of lumber, definition of, 338
through-penetrations, position and reinforcement around, 105
tiles, using in roofing systems, 88
timber, using in bridge systems, 123
timber and post, definition of, 338
timber decks, formation of, 357
time and cost of work, estimating, 68
time management factors
    caffeine absorption, 152–154
    sleep, 154–155
ton of refrigerant, explanation of, 81

tools of trade, being familiar with, 150
topographic survey, requirement of, 95
Tornado Alley, location of, 192
torsional rigidity, significance in bridge systems, 122
training, areas of, 11
training philosophy, components of, 12–14
transportation planning studies, initiating, 130
transportation projects, evaluating, 130
trees, structure of, 342–343
triple W truss, description of, 359
truck weight distribution, specifications for, 109
trusses
    characteristics of, 358–360
    lateral bracing for, 360

UL (Underwriters Laboratories, Inc.), services offered by, 44
ultimate limit, relationship to load carrying capacity, 21
underlayment, definition of, 338
underwater bridge foundations, requirements for, 118
United States v. Darby, 36
universities and colleges, employment in, 56–57
U.S. Constitution, 10th Amendment of, 36–37
USD (Ultimate Strength Design), formula for, 203
UT (ultrasonic testing), performing on structural steel, 336

vehicle live loading
    combining with seismic forces in bridge design, 187
    design specifications for, 109
ventilation, providing for attics and rafter spaces, 364
verbal skills, importance of, 161
vertical lift bridges, construction of, 123–124
vertical load resisting systems, structural aspects of, 77
vertical systems, behavior of, 212–213
vibrations, damping, 208

visualization, teaching, 180
voltage, purpose in electrical systems, 82

wall segments, rigidity of, 212
wall surfaces, applying loads to, 297
wall ties, using with masonry units, 291
walls
   behavior of shear walls, 212
   and masonry, 297–298
waste, management by civil
     engineers, 96
water
   drainage of, 126–127
   in green wood, 344
   management by civil engineers, 96
   use in concrete mix, 249–250
water main service piping,
     determining, 80
water-reducing admixtures, using,
     250–251
water-to-cement ratio (w/c), significance
     of, 249–250
w/c (water-to-cement ratio), significance
     of, 249–250
weakest link method, explanation of,
     214
weatherproofing
   cladding systems, 86–87
   roofing and flashing, 87–89
   thermal insulation, 89
webs
   definition of, 281
   in truss assemblies, 359
welded joints, inspection of, 335–336
welding procedure specifications (WPS),
     overview of, 323–324
welds, using in steel connections, 320–324
Western framing, constructing, 363
width of lumber, definition of, 338
will versus shall, legal implications of,
     172
wind
   characteristics and deviation of, 193
   considering as dynamic-type load,
     191–194
   impact on tall structures, 215
wind pressures, application of, 193–194
window framing mullions, accounting or, 65

wood. *See also* lumber
   chemical treatment of, 344–345
   engineering properties of, 341
   mechanics of, 343–344
   moisture content of, 344–345
   strength of, 343
   structure of, 342–343
   temperature of, 344–345
wood- and lag-screws, using, 350
wood diaphragms, tests on, 354
wood elements
   behavior of, 346–347
   connections in, 349–352
   influence of defects on, 352–353
   panels or sheathing, 347–349
wood fiber, separations of, 353
wood framing
   codes and standards for, 366–368
   construction of, 363–364
   designing connections for, 365
   quality control of, 366
   risk in design and during service,
     364–365
   terminology related to, 337–338
wood members, density of, 340–341
wood sheathed diaphragm, failure of,
     356
wood-frame construction, fasteners for,
     349–352
wood-frame structures, field inspection of,
     366
wood-frame systems
   behavior of, 353–354
   combining with other materials, 363
   frames, 357
   horizontal diaphragms, 354–357
   laminated decks, 357
   nonwood panel shear walls, 362–363
   structural wood panel shear walls,
     360–362
   trusses, 358–360
words versus pictures, 162
work. *See also* projects
   accepting and rejecting, 150–151
   estimating time and cost of, 68
workforce, health of, 46–47
WPS (welding procedure specifications),
     overview of, 323–324

writing skills
    applying to project specifications, 172
    importance of, 161–163
wythe, definition of, 281
yield limit, relationship to load carrying
    capacity, 21

yielding modes, using with dowel-type
    fasteners, 350
Young's Modulus, representing in column
    loading, 212

zoning ordinances, institution of, 37–38